8th Grade Alaska Math for Beginners

Standardized Testing and Home Schooling Study Guide

MATHFA

8th Grade Alaska Math for Beginners

By Mathfa, Email: Support@mathfa.com

Web: www.mathnotion.com

Copyright © 2025 by Mathfa, an imprint of Math Notion Inc. All rights reserved. No part of this publication may be reproduced, stored in a retrieval system, or transmitted in any form or by any means, electronic, mechanical, photocopying, recording, scanning, or otherwise, except as permitted under Section 107 or 108 of the 1976 United States Copyright Act, without permission of the author.

All inquiries should be addressed to Mathfa.

ISBN: 978-1-63620-325-6

www.mathnotion.com

… So Much More Online!

- ✓ FREE Mathematics Worksheets
- ✓ More Math Learning Books!
- ✓ FREE Math Lessons
- ✓ Online Math Tutors
- ✓ **PDF Version of This Book**

No Registration Required!

Visit us here!

Welcome to Math Grade 8!

A year full of exciting challenges and important skills that will prepare you for high school and beyond. Whether you're confident in math or still trying to find your footing, this book is here to guide you every step of the way.

Inside, you'll explore a full range of topics, from equations and geometry to ratios, statistics, fractions, and decimals. Each chapter is thoughtfully organized to help you build your understanding gradually. You'll find clear explanations, worked-out examples, and guided practice that makes it easier to follow along—without needing to look for extra resources.

If you're struggling with a concept, don't worry. The book includes step-by-step examples and helpful tips exactly where you need them. If you're ready to go further, special challenge sections offer deeper problems to stretch your thinking and push you to the next level.

This book isn't just about solving math problems—it's about making connections. With real-world examples, word problems, and hands-on activities, you'll see how math plays a role in everyday life and why it's worth mastering.

It's a great resource for all types of learners: students working independently, homeschoolers needing a structured guide, classroom teachers planning lessons, and parents helping at home. No matter your learning style or pace, this book was made to support you.

Our goal is simple: to make 8th grade math clear, practical, and yes—even enjoyable. Let this book be your companion through a year of growth, discovery, and confidence-building in math. Let's get started on this journey—together.

How to Get the Most Out of the book?

This book is more than just a collection of lessons and practice problems; it's your personal guide through one of the most important years of middle school math. Whether you're learning on your own, at home, or using it alongside classroom lessons, here are some smart strategies to make the most of your study experience:

1) **Get Familiar with the Book.**

   Start by flipping through the pages and taking a few minutes to explore what's inside. Notice how each chapter is laid out with examples, explanations, practice problems, and review sections. This will help you feel more comfortable as you begin.

2) **Set Your Goals.**

   Set clear intentions for what you want to achieve. Are you reviewing for a test? Getting ahead? Filling in some gaps? Knowing your goal helps you focus and stay motivated.

3) **Make a Study Plan and Use the Book at Your Own Pace.**

   Create a study schedule that fits your life, whether it's daily or a few times a week. Stick to your plan but remember this guide is flexible, you can move faster through parts you understand and slow down where you need more time. You're in control of your learning journey, so find a pace that works best for you.

4) **Learn Actively with Each Lesson.**

   Math isn't just something to read, it's something to do. Solve the examples as you follow along, underline important steps, and write notes or questions as you go. The more you interact with the material, the more it will stick.

5) **Practice as You Learn.**

   Each chapter includes guided practice and independent exercises to reinforce your understanding. Don't skip them; this is where true learning happens. Check your answers and review any mistakes to better understand the concept.

6) **Stay Persistent Through Difficult Topics.**
   When something feels confusing, don't skip it. Take your time with the explanations, go back to earlier steps, and try breaking the problem into smaller parts. Use the examples in the book as models to guide you.

7) **Make Time for Review.**
   As you, complete chapters, revisiting earlier chapters keeps your skills sharp. Reviewing past work strengthens memory and reinforces concepts you may have struggled with at first.

8) **Track Your Progress.**
   Use end-of-chapter reviews and summaries to check your understanding. These checkpoints help you recognize areas of growth and areas that may need more attention.

9) **Built on Standards You Can Trust.**
   Every topic in this book is carefully chosen and structured to reflect the key skills expected in 8th grade state learning standards. The lessons support a strong foundation that aligns with what's taught in classrooms nationwide.

10) **Be Ready for What Comes Next.**
    Because this guide is based on widely used math standards, it also helps prepare you for year-end evaluations. By working through the material, you're not only building math skills, but you're also getting ready to show what you know when it counts.

11) **Relate Math to Everyday Life.**
    Wherever you can, connect math concepts to real-world situations. Whether you're measuring ingredients, comparing prices, or analyzing data, math is part of everyday decisions.

12) **Stay Curious and Keep Going.**
    Math is a subject of discovery. Stay curious, keep asking "why," and don't be afraid to make mistakes—they're part of learning. With effort and the right approach, you'll be amazed at what you can accomplish.

This book is here to support you in your way. When you take the time to explore, practice, and stay curious, you'll be surprised how much you can grow. With every page, you're building not just math skills, but confidence for what's ahead.

Contents

Chapter 1: Integers

Topics that you'll learn in this chapter:

- ✓ Integers and Additive Inverse
- ✓ Adding and Subtracting Integers
- ✓ Multiplying and Dividing Integers
- ✓ Order of Operations
- ✓ Ordering Integers and Absolute Value
- ✓ Absolute Value Operations
- ✓ Real World Application
- ✓ Worksheets
- ✓ Answer of Worksheets

Integers and Additive Inverse

Integers: Integers are the set of whole numbers and their opposites. This includes positive numbers, negative numbers, and zero.

Examples:

- **Positive integers:** $1, 2, 3, 4, \ldots$
- **Negative integers**: $-1, -2, -3, -4, \ldots$
- **Zero:** 0 (It is neither positive nor negative)

Opposite Integers (Additive Inverse): Opposite integers, also known as additive inverses, are pairs of numbers that are the same distance from zero on the number line but in opposite directions. When you add an integer and its opposite, the result is always zero. For example, the opposite of 5 is -5 because $5 + (-5) = 0$

Visualizing on a Number Line:

Imagine a number line with zero in the center. Positive integers are to the right of zero, and negative integers are to the left.

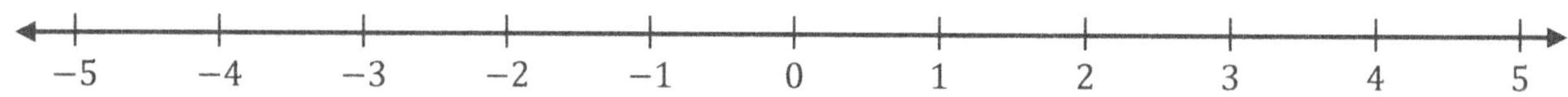

Examples:

1) Which of the following numbers is an integer?

$$-17, \quad -\sqrt{49}, \quad \frac{1}{20}, \quad 12.12, \quad 0, \quad \frac{24}{6}, \quad -(-(-15)), \quad \sqrt{200}$$

Solution:

Integer numbers do not include any fractional or decimal parts so, $-17, -\sqrt{49}$ (this number is equal to -7), 0, $\frac{24}{6}$ (after simplifying this fraction, it is equal to 4) and $-(-(-15))$ (this number is equal to -5) are integers.

2) Find the reflection of the number 4 with respect to the number -2.

Solution: We must find out how far and in what direction 4 is from -2 and then applying that same distance in the opposite direction:

- Calculate the distance between -2 and 4: Distance= $4 - (-2) = 4 + 2 = 6$
- Apply the same distance in the opposite direction: Since 4 is 6 units to the right of -2, its reflection should be 6 units to the left of -2: Reflection $= -2 - 6 = -8$.

Adding and Subtracting Integers

Adding Integers:

1. **Same Signs:** When adding two integers with the same sign, you add their absolute values and keep the sign.
   - Example: $3 + 5 = 8$ and $-4 + (-6) = -10$
2. **Different Signs:** When adding two integers with different signs, you subtract the smaller absolute value from the larger absolute value and keep the sign of the integer with the larger absolute value.
   - Example: $7 + (-3) = 4$ and $-8 + 5 = -3$

Subtracting Integers:

1. **Change to Addition:** Subtracting an integer is the same as adding its opposite (additive inverse).
   - Example: $5 - 3$ is the same as $5 + (-3) = 2$
2. **Keep the Change:** When subtracting integers, change the subtraction sign to an addition sign and change the sign of the second integer, then follow the rules for adding integers.
   - Example: $7 - (-2)$ is the same as $7 + 2 = 9$

Example: Find the result of the expressions:

a) $-5 + 6 - 8 - (-9)$

b) $48 - 19 - 17 + 29 - 75$

Solution a:

- The expression $-8 - (-9)$ can be rewritten as $-8 + 9$. Now, the expression becomes:
  $-5 + 6 - 8 + 9$
- Perform the addition and subtraction from left to right:
  $-5 + 6 - 8 + 9 = 1 - 8 + 9 = -7 + 9 = 2$

Solution b:

Since the numbers in this expression are large, we can first find the sum of the negative and positive numbers separately:

- $48 + 29 = 77$
- $-19 - 17 - 75 = -111$
- Result: $48 - 19 - 17 + 29 - 75 = 77 - 111 = -34$

Multiplying and Dividing Integers

Multiplying Integers:

1. **Same Signs**:
   - Multiply the absolute values of the integers.
   - The result is positive.

   Example: $(4) \times (7) = 28$ or $(-6) \times (-2) = 12$

2. **Different Signs**:
   - Multiply the absolute values of the integers.
   - The result is negative.

   Example: $(2) \times (-3) = -6$ or $(-8) \times (3) = -24$

Dividing Integers:

1. **Same Signs**:
   - Divide the absolute values of the integers.
   - The result is positive.

   Example: $(15) \div (3) = 5$ or $(-18) \div (-3) = 6$

2. **Different Signs**:
   - Divide the absolute values of the integers.
   - The result is negative.

   Example: $(32) \div (-2) = -16$ or $(-36) \div (3) = -12$

Example:

Find the answers.

a) $-5 \times (-2) \times (-3) \times 8$

b) $-36 \div 9 \div (-4)$

Solution a:

Perform the multiplications from left to right using the rules:

$-5 \times (-2) \times (-3) \times 8 = 10 \times (-3) \times 8 = -30 \times 8 = -240$

Solution b:

Perform the divisions from left to right using the rules:

$-36 \div 9 \div (-4) = -4 \div (-4) = 1$

Order of Operations

The order of operations is a set of rules that helps you determine the correct sequence to solve a math problem. It's often remembered by the acronym **PEMDAS**:

1. **P**arentheses: Solve any operations inside parentheses first. Example: $(3 + 2) = 5$
2. **E**xponents: next, solve any exponents (powers or roots). Example: $2^3 = 8$
3. **M**ultiplication and **D**ivision: From left to right, solve multiplication and division. Example: $8 \div 4 \times 2 = 2 \times 2 = 4$
4. **A**ddition and **S**ubtraction: From left to right, solve addition and subtraction. Example: $7 - 3 + 4 = 4 + 4 = 8$

Example:

Evaluate the following expressions:

a) $12 - 24 \div (3^2 - 2 \times 3) =$

b) $(-48 - 4 \times 2) - (-9 \times 8 \div 12 + 2^4) =$

Solution a:

1. Parentheses are first: $3^2 - 2 \times 3$
2. Exponents: $3^2 = 9$
3. Multiplication: $2 \times 3 = 6$

   The expression inside the parentheses is now: $3^2 - 2 \times 3 = 9 - 6 = 3$
4. Now the expression is: $12 - 24 \div 3$
5. Division: $24 \div 3 = 8$
6. Subtraction: $12 - 8 = 4$

So, $12 - 24 \div (3^2 - 2 \times 3)$ simplifies to 4.

Solution b:

1. Parentheses first: Inside the parentheses, we have two expressions: $-48 - 4 \times 2$ and $-9 \times 8 \div 12 + 2^4$
2. Exponents in the second expression: $2^4 = 16$. So the second expression becomes: $-9 \times 8 \div 12 + 16$
3. Multiplication and division inside each parenthesis:

   First parentheses: $-48 - 4 \times 2 = -48 - 8 = -56$

   Second parentheses: $-9 \times 8 \div 12 + 16 = -72 \div 12 + 16 = -6 + 16 = 10$
4. Subtraction: $-56 - 10 = -66$

So, $(-48 - 4 \times 2) - (-9 \times 8 \div 12 + 2^4)$ simplifies to -66.

Ordering Integers and Absolute Value

Ordering integers means arranging them from the smallest to the largest (ascending order) or from the largest to the smallest (descending order).

Tips for ordering integers:

1. **Negative numbers are always less than positive numbers.**
2. **The farther left a number is on the number line, the smaller it is.**
3. **Zero is the middle point**: Negative numbers are to the left of zero, and positive numbers are to the right.

Absolute Value is the distance of a number from zero on the number line, regardless of direction. It is always non-negative.

Notation: The absolute value of a number x is written as $|\ x\ |$.

Examples:

- $|-3| = 3$ (The distance of -3 from 0 is 3 units)
- $|4| = 4$ (The distance of 4 from 0 is 4 units)
- $|0| = 0$ (The distance of 0 from itself is 0 units)

Examples:

1) Order the following integers from smallest to greatest: $-2,\ 15,\ |-18|,\ |+10|,\ 0,\ -|-3|$

   Solution:

   - $|-18| = 18$
   - $|+10| = 10$
   - $-|-3| = -3$

   List the result: $-2 < -3 < 0 < 10 < 15 < 18$

2) Order the following integers from greatest to smallest: $5-5, |10| - |-11|, -3^2, -15 \div 5, \frac{-18}{-6}$

   Solution:

   - $5 - 5 = 0$
   - $|10| - |-11| = 10 - 11 = -1$
   - $-3^2 = -9$
   - $-15 \div 5 = -3$
   - $\frac{-18}{-6} = 3$

   List the result: $3 > 0 > -1 > -3 > -9$

Absolute Value Operations

Absolute value is the distance of a number from zero on the number line, regardless of direction. It is always non-negative.

Properties of absolute value:

1. $|a| \geq 0$
2. $|a| = a \text{ if } a \geq 0$
3. $|a| = -a \text{ if } a < 0$
4. $|0| = 0$

Addition and Subtraction:

When you add or subtract absolute values, you first evaluate the absolute values, then perform the operations.

Notation: $|a + b| \neq |a| + |b| \text{ and } |a - b| \neq |a| - |b|$

Multiplication and Division:

When you multiply or divide absolute values, you first evaluate the absolute values, then perform the operations.

Absolute value of expressions:

For expressions involving multiple operations, always simplify inside the absolute value symbols first and then apply the absolute value rules. For example, for doing the $|-3 + 2|$ we can perform following steps:

1. Simplify inside the absolute value symbols: $-3 + 2 = -1$
2. Find the absolute value: $|-1| = 1$

Example:

Do the following expressions:

a) $|-9| + |-4|$

b) $|-5 - 3|$

c) $|-2| - |14|$

d) $|-45| \div |9|$

e) $-|8| \times |-6|$

f) $-|-12| - |30 \div (-6)| \times |-3|$

Solution:

a) $|-9| + |-4| = 9 + 4 = 13$

b) $|-5 - 3| = |-8| = 8$

c) $|-2| - |14| = 2 - 14 = -12$

d) $|-45| \div |9| = 45 \div 9 = 5$

e) $-|8| \times |-6| = -8 \times 6 = -48$

f) $-|-12| - |30 \div (-6)| \times |-3| = -12 - |-5| \times 3 = -12 - 5 \times 3 = -12 - 15 = -27$

Real World Application

Integers are incredibly useful in real-world applications. Here are a few examples:

1. **Finance:** Balancing a budget, tracking income and expenses, and calculating net worth all involved integers. For instance, the difference between your salary (a positive integer) and your expenses (a negative integer) shows your financial health.
2. **Temperature:** Weather forecasts use integers to report temperatures. For example, you might see a temperature of -5°C in the winter and 25°C in the summer.
3. **Sports:** Scores in games often use integers. For instance, in soccer, the score could be $3 - 1$, or in basketball, a player might score 23 points in a game.
4. **Travel:** Altitudes above and below sea level are measured using integers. For example, the height of a mountain might be 2,000 meters above sea level ($+2000$), while the depth of a submarine could be 500 meters below sea level (-500).
5. **Technology:** Error codes in software, IP addresses, and version numbers in programming are all examples of integers in technology.
6. **Population Counts:** Census data and demographic statistics often use integers to represent the number of people living in a certain area.

Example:

John is planning a road trip across three cities. He starts from City A and drives to City B, then from City B to City C, and finally from City C back to City A. Here are the details of his trip:

- From City A to City B, he drives 120 kilometers.
- From City B to City C, he drives 80 kilometers.
- From City C back to City A, he drives 150 kilometers.

However, on his way from City A to City B, he has to make a detour due to road construction, which adds an additional 25 kilometers to his trip.

1. Calculate the total distance John drives, including the detour.
2. If his car consumes 1 liter of fuel for every 15 kilometers, how many liters of fuel does he need for the entire trip?

Solution:

1. Calculate the total distance with the detour:
   - Distance from City A to City B (including detour): 120 km + 25 km = 145 km
   - Distance from City B to City C: 80 km and Distance from City C to City A: 150 km
   - Total distance = 145 km + 80 km + 150 km = 375 km
2. Calculate the total fuel consumption:
   - Fuel consumption rate: 1 liter for every 15 kilometers
   - Total distance: 375 kilometers
   - Total fuel needed = Total distance / Fuel consumption rate
   - Total fuel needed = 375 km / 15 km per liter = 25 liters

   John needs to drive a total of 375 kilometers and will need 25 liters of fuel for the entire trip.

Worksheets

Integers and Additive Inverse

Identify if the numbers are integers or not:

1) $\frac{75}{30}$
2) $\sqrt[3]{-64}$
3) $\pi - \pi$
4) 100.01
5) $\sqrt{\frac{81}{49}}$

Find the reflection of following numbers with respect to the number -3:

6) 0
7) -1
8) -3
9) 5
10) 3

Adding and Subtracting Integers

Do the following additions and subtractions:

1) $19 - (-2)$
2) $-5 - 3 + 8$
3) $12 - (-1) - 9$
4) $-15 - (-2 + 3)$
5) $-(-10) - 5 - (-(-2))$
6) $-(4 - 9 + 10) - (-6)$
7) $[-15 + 18 - (3 - 7)] - 17$
8) $48 - 74 + 36 - 120 + 14 - (-92) - 32$
9) $-6 + 2 - 8 + 4 - 10 + 6 - 12 + 8 - \cdots - 88 + 84$
10) $5 - 2 + 8 - 5 + 11 - 8 + \cdots + 302 - 299$

Multiplying and Dividing Integers

Do the following multiplications and divisions:

1) $-8 \times (-3)$
2) $10 \times 2 \times (-4)$
3) $36 \div (-12) \times (-3)$
4) $-8 \times (-10) \div (-16)$
5) $-[15 \div (-5)] \times 6$
6) $(-9 \times 2 \div 6) \div 3 \times (-4)$
7) $[-54 \div 9 \times 4] \times [-8 \div 4 \times (-2)]$
8) $(-18 - 1) \times (-17 - 1) \times (-16 - 1) \times \ldots \times (17 - 1) \times (18 - 1)$
9) $(-2 - 1) \times (-1 - 2) \times (0 - 3) \times (1 - 4) \times \ldots \times (20 - 23)$
10) $(-30 \div 30) \times (-31 \div 31) \times (-32 \div 32) \times \ldots \times (-60 \div 60)$

Order of Operations

Solve.

1) $-9 + 2 \times 3$
2) $12 \div (-4) + (-20) \div (-5)$
3) $-243 \div (-8 - 1)^2$
4) $-4 - (16 \div (-4)) \times 2^2$
5) $(-3)^3 \div \big(-6 + (-3)\big) - (-5 + 4)$
6) $6 - \big[24 - \big(6 \div (-3)\big)\big] + 8$

7) $5 \times (-8 \div 2) \div 10 + 6 \times (-3)$
8) $-(-18 \div 3^2) + (-1 + 5)^2 \div (-8)$
9) $\frac{-45 \div 9 + 16 \div (-4)}{-(7-4)^2}$
10) $-3 \times (-8 + 7 \times 5) - (-12 \div 6) + (-1)^3 - 8$

Ordering Integers and Absolute Value

Order each set of numbers from least to greatest:

1) $-100, 0, -(-5), 16, -18, -101$
2) $|-6|, -2, |-8+9|, -23, -5, -2+9$
3) $-|3|, 175, -\sqrt{25}, -(-12), 4 \times (-9)$
4) $\frac{-25}{-5}, -750, 750, -|3-(-5)|, -4^3, 39$
5) $|17| - |-9|, |-6 \times 2|, -148, \frac{36}{-18}, \sqrt{10000}, -99$

Absolute Value Operations

Evaluate each expression:

1) $|-5| + |15|$
2) $-|-9| \times |-2|$
3) $|-28 \div 7|$
4) $|-12 + 6 \div (-2)|$
5) $-|-9| \times |-72 \div (-8-1)|$
6) $\left|\frac{-20 \div 2 - 15}{-6-1}\right|$
7) $|-(-3 \times (-8) + 4^2 - 27 \div (-9))|$
8) $|63 \div (-10 + 1)| - |3 \times (-10)|$
9) $\left|\frac{-(4 \times 10 \div 8) - 19 + 3 \times 4}{-5 - (14 \div 2)}\right|$
10) $|1 - 8 \times 3| + |-\sqrt{25-16} + 7|$

Real Word Application

Do the following word problems about integers:

1) A building has 20 floors. The ground floor is numbered 0, floors above it is numbered 1, 2, 3, etc., and floors below it is numbered $-1, -2, -3$, etc. If Ali takes the elevator from the $3rd$ floor to the $-2nd$ floor, how many floors does he travel?
2) John has an initial balance of \$600 on his credit card. He makes three purchases: \$120 for groceries, \$85 for gas, and \$200 for a new phone. What is his current balance after these transactions?
3) Emma is buying groceries. She has \$50 to spend. She buys the following items:
   - 3 packs of pasta, each costing \$4
   - 2 jars of sauce, each costing \$3
   - 4 cans of beans, each costing \$2

   After buying these items, how much money does Emma have left? If she wants to buy 2 more packs of pasta, does she have enough money left?
4) The temperature in New York at 6 AM was $-2°C$. By noon, the temperature had risen by 5°C. In the evening, it dropped by 7°C from the noon temperature. What was the temperature in the evening?
5) A mountaineer starts at the base of a mountain at 1,000 meters above sea level. He climbs 1,800 meters to reach a plateau, then descends 2,400 meters to a lower valley and finally ascends 600 meters to a campsite. What is his final elevation?

6) Maya invests $500 in a savings account that earns 5% interest per year, compounded annually. At the end of each year, she withdraws $50 for personal use. Calculate her account balance at the end of the third year.
7) Rachel took out a loan of $4,000. After making a payment of $500, her balance increased by $200 due to an interest charge. What is the remaining balance on her loan?
8) A group of hikers starts at the base of a mountain at an altitude of 200 meters above sea level. On the first day, they ascend 350 meters. On the second day, they descend 150 meters due to a steep slope. On the third day, they ascend 400 meters to reach the peak. Calculate their final altitude.
9) The temperature in Chicago fluctuates over a week as follows:
   - Monday: -5°F
   - Tuesday: 7°F higher than Monday
   - Wednesday: 3°F lower than Tuesday
   - Thursday: 4°F higher than Wednesday
   - Friday: 2°F lower than Thursday
   - Saturday: 8°F higher than Friday
   - Sunday: 6°F lower than Saturday

   Calculate the temperature on each day of the week. What is the average temperature for the week?
10) A store sells two types of products: Product A and Product B. Product A costs $30, and Product B costs $50. The store offers a 10% discount on Product A and a 20% discount on Product B. If a customer buys 3 units of Product A and 2 units of Product B, how much do they pay after the discounts are applied?

Answer of Worksheets

Integers and Opposite Integers (Additive Inverse)

1) Not integer
2) Integer
3) Integer
4) Not integer
5) Not integer
6) -6
7) -5
8) -3
9) -11
10) -9

Adding and Subtracting Integers

1) 21
2) 0
3) 4
4) -16
5) 3
6) 1
7) -10
8) -36
9) -168
10) 300

Multiplying and Dividing Integers

1) 24
2) -80
3) 9
4) -5
5) 18
6) 4
7) -96
8) 0
9) $(-3)^{23}$
10) -1

Order of Operations

1) -3
2) 1
3) -3
4) 12
5) 4
6) -12
7) -20
8) 0
9) 1
10) -88

Ordering Integers and Absolute Value

1) $-101 < -100 < -18 < 0 < 5 < 16$
2) $-23 < -5 < -2 < 1 < 6 < 7$
3) $-36 < -5 < -3 < 12 < 175$
4) $-750 < -64 < -8 < 5 < 39 < 750$
5) $-148 < -99 < -2 < 8 < 12 < 100$

Absolute Value Operations

1) 20
2) -18
3) 4
4) 15
5) -72
6) $\frac{25|}{7}$
7) 43
8) -23
9) 1
10) 27

Real Word Application

1) 5 floors.
2) \$195.
3) Yes, she has \$24 left.
4) -4°C.
5) 1,000 meters.
6) \$421.19.
7) \$3,700.
8) 800 meters.
9) Temperatures for the week:
   - Monday: -5°F
   - Tuesday: 2°F
   - Wednesday: -1°F
   - Thursday: 3°F
   - Friday: 1°F
   - Saturday: 9°F
   - Sunday: 3°F

   The average temperature: $\approx$ 1.71°F
10) \$161.

Chapter 2: Rational and Irrational Numbers

Topics that you'll learn in this chapter:

- ✓ Convert Repeating Decimal
- ✓ Identify Rational Numbers
- ✓ Identify Irrational Numbers
- ✓ Irrational Numbers on Number lines
- ✓ Approximate Irrational Numbers
- ✓ Compare and Order Rational and Irrational Numbers
- ✓ Adding and Subtracting Rational Numbers
- ✓ Multiplying and Dividing Rational Numbers
- ✓ Mixed Operations on Rational Numbers
- ✓ Adding and Subtracting Irrational Numbers
- ✓ Multiplying and Dividing Irrational Numbers
- ✓ Mixed Operations on Irrational Numbers
- ✓ Absolute Value of Rational / Irrational Numbers
- ✓ Real Word Applications
- ✓ Worksheets
- ✓ Answer of Worksheets

Convert Repeating Decimal

Converting a Repeating Decimal to a Fraction:

Convert 2.345345 ... (where 345 is repeating) to a fraction.

1. **Separate the Whole Number:**
   - The mixed number has a whole part (2) and a repeating decimal part (0.345345 ...)
   - Focus on the repeating decimal part: $0.\overline{345}$
2. **Set Up an Equation:**
   - Let $x = 0.\overline{345}$
   - Notice that $1000x = 345.\overline{345}$ (shift the decimal point three places to the right, as the repeating part is three digits long).
3. **Form and Solve the System of Equations:**
   - $1000x = 345.\overline{345}$
   - $x = 0.\overline{345}$
   - Subtract the second equation from the first:

     $1000x - x = 345.\overline{345} - 0.\overline{345}$

     $999x = 345$

     $x = \frac{345}{999} = \frac{115}{333}$
4. **Combine with the Whole Number:**
   - The original number was $2.\overline{345}$ and we found that $0.\overline{345} = \frac{115}{333}$
   - The entire mixed number is $2 + \frac{115}{333} = 2\frac{115}{333} = \frac{781}{333}$

   So, 2.345345 ... or $2.\overline{345}$ as a fraction is $\frac{781}{333}$.

Converting a Fraction to a Repeating Decimal:

Example: Convert $\frac{215}{18}$ to a repeating decimal:

Divide the Numerator by the Denominator:

- Perform long division of 215 by 18.
- $215 \div 18 = 11.9\overline{4}$

So, $\frac{215}{18}$ as a decimal is $11.9\overline{4}$.

Identify Rational Numbers

Rational Numbers

A number is rational if it can be expressed as the quotient or fraction $\frac{p}{q}$ of two integers p and q, where $q \neq 0$. This includes:

- Integers: $1, -5, 0, etc.$
- Fractions: $\frac{3}{4}, -\frac{7}{2}$ etc
- Terminating decimals: $0.75, 3.6, etc.$
- Repeating decimals: $0.333 \ldots = \frac{1}{3}, 0.142857142857 \ldots = \frac{1}{7}, etc.$

Reciprocals

The reciprocal of a number is obtained by flipping the numerator and the denominator of its fraction representation. For example:

- Reciprocal of $\frac{3}{4}$ is $\frac{4}{3}$
- Reciprocal of 5 (or $\frac{5}{1}$) is $\frac{1}{5}$
- Reciprocal of -7 (or $-\frac{7}{1}$) is $-\frac{1}{7}$

Multiplicative Inverses

The multiplicative inverse of a number a is a number b such that $a \times b = 1$. For rational numbers, the multiplicative inverse is the same as the reciprocal.

Example:

Identify whether these numbers are rational and determine their reciprocals/multiplicative inverses:

$1\frac{2}{5}, -0.9, 0, \frac{\pi}{2}$

Solution:

- $1\frac{2}{5}$ or $\frac{7}{5}$ is rational and its reciprocal is $\frac{5}{7}$
- -0.9 or $-\frac{9}{10}$ is rational and its reciprocal is $\frac{10}{9}$
- 0 or $\frac{0}{1}$ is rational and doesn't have reciprocal!
- $\frac{\pi}{2}$ is not rational (because its nominator (π) is not integer.

Identify Irrational Numbers

Irrational numbers are numbers that cannot be expressed as a simple fraction (i.e., they cannot be written as $\frac{p}{q}$, where p and q are integers and $q \neq 0$). These numbers have non-repeating, non-terminating decimal expansions. To identify irrational numbers, here are some key characteristics:

1. **Non-terminating and Non-repeating Decimals**:
   If the decimal representation of a number goes on forever without repeating, it's an irrational number. For example, $\pi \approx 3.14159 \ldots$ and it never repeats.
2. **Square Roots of Non-perfect Squares**:
   The square root of a number that is not a perfect square is always irrational. For example, $\sqrt{2} \approx 1.414 \ldots$ and it never ends or repeats.
3. **Numbers Like π *and* e:**
   - The numbers π (the ratio of the circumference of a circle to its diameter) and e (the base of the natural logarithm) are famous irrational numbers.
   - Both π and e have infinite, non-repeating decimal expansions and cannot be written as a fraction of two integers.
4. **Transcendental Numbers**
   - A subset of irrational numbers, transcendental numbers cannot be the root of any non-zero polynomial equation with integer coefficients.
   - Both π and e are examples of transcendental numbers.

Example:

Which of the following expressions result in irrational numbers?

$\sqrt{18}, \frac{\sqrt{5}}{\sqrt{5}}, \sqrt{49}, \frac{\pi}{\pi}, \pi^2$

Solution:

- $\sqrt{18}$: Irrational: The square root of 18 is not a perfect square, so $\sqrt{18}$ is irrational.
- $\frac{\sqrt{5}}{\sqrt{5}}$: Rational: This simplifies to 1, which is a rational number.
- $\sqrt{49}$: Rational: The square root of 49 is 7, which is a rational number.
- $\frac{\pi}{\pi}$: Rational: This simplifies to 1, which is a rational number.
- π^2 : Irrational: Since π is irrational, any power of π (including π^2) is also irrational.

So, the expressions that result in irrational numbers are: $\sqrt{18}$ and π^2.

Irrational Numbers on Number lines

Rational numbers, such as $\frac{1}{2}, \frac{3}{4}$, or 7, are like well-marked islands. But the irrational numbers, like $\sqrt{2}$ or π, are more mysterious and less easy to pinpoint.

To locate irrational numbers on the number line, think about how we place rational numbers. For instance:

1. **Square Roots**: Suppose you want to find $\sqrt{2}$. It lies between 1 and 2 because 1^2 is 1 and 2^2 is 4. We approximate its position by progressively narrowing the interval. By trying finer divisions, we find $\sqrt{2}$ is around 1.414.
2. **Decimals**: You can use their decimal expansion. For example, π is approximately 3.14159, so you place it slightly beyond 3 but before 4. The more decimal places you use, the closer you get.
3. **Construct Geometrically**: You can also construct them geometrically. For example, to locate $\sqrt{2}$, draw a right triangle with both legs of length 1. The hypotenuse will be $\sqrt{2}$, Now we can draw a circle with the center at zero and the radius of the hypotenuse of the triangle. Therefore, wherever the circle intersects the axis, it will be at a distance of $\sqrt{2}$ from zero.

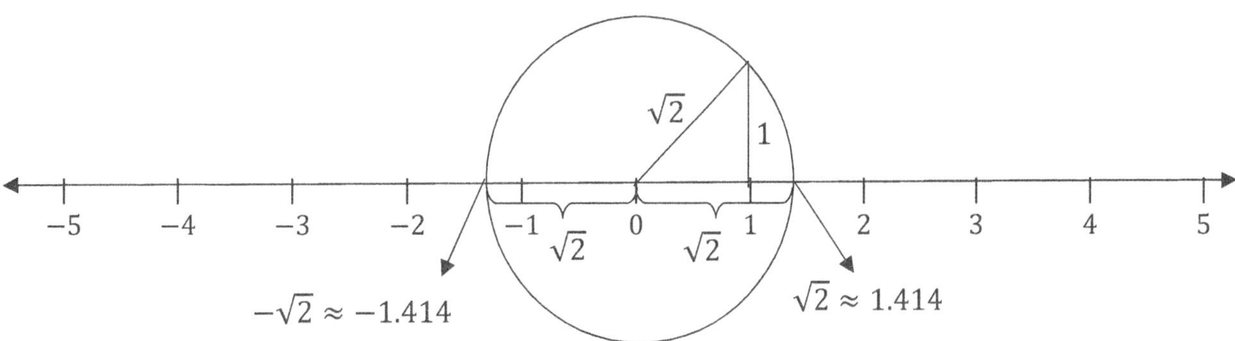

Example:

Find $\sqrt{5}$ on number line.

Solution:

We can draw a right-angled triangle with two legs of 1 and 2, where the hypotenuse, according to the Pythagorean theorem, will be $\sqrt{5}$.

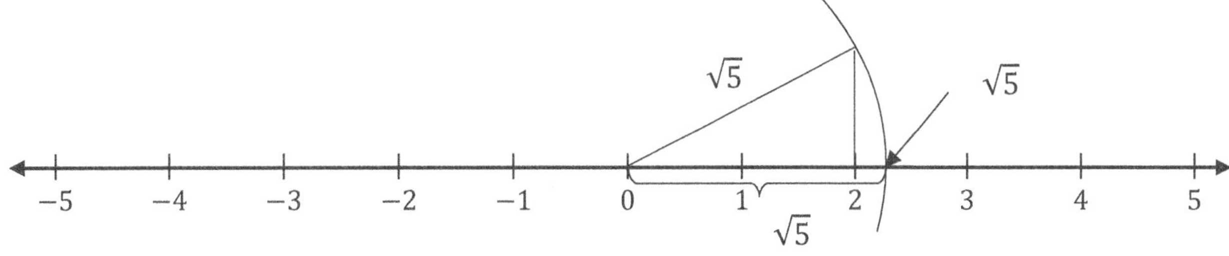

Approximate Irrational Numbers

To find approximate irrational numbers, you typically use mathematical operations or known properties of numbers that lead to irrational results. Here is a common method for finding and approximating irrational non-square numbers:

1. **Identify the Irrational Number**: Let's use $\sqrt{2}$ as an example.
2. **Estimate the Square Root**: Know the squares of integers around the irrational number.
   For $\sqrt{2}$:
   $1^2 = 1$ (less than 2)
   $2^2 = 4$ (greater than 2)
   Therefore, $\sqrt{2}$ lies between 1 and 2.
3. **Refine the Interval**: Narrow down further by squaring decimals if needed. For example:
   $1.4^2 = 1.96$ (less than 2)
   $1.5^2 = 2.25$ (greater than 2)
   So, $\sqrt{2}$ lies between 1.4 and 1.5.
4. **Continue Refining**: Keep this process going to get closer approximations. It helps pinpoint where the irrational number lands on the number line.

Example:

Between which two integers do the following irrational numbers lie?

$\sqrt{28}$, $\sqrt{75}$ and $-\sqrt{14}$

Solution:

$\sqrt{28}$:

- $5^2 = 25$
- $6^2 = 36$ So, $\sqrt{28}$ lies between 5 and 6.

$\sqrt{75}$:

- $8^2 = 64$
- $9^2 = 81$ So, $\sqrt{75}$ lies between 8 and 9.

$-\sqrt{14}$:

- $3^2 = 9$
- $4^2 = 16$ Since it's a negative square root, $-\sqrt{14}$ lies between -4 and -3.

Compare and Order Rational and Irrational Numbers

Comparing and ordering rational and irrational numbers involves placing them on the number line and comparing their positions. Here's a step-by-step guide:

1. **Convert Rational Numbers to Decimals**: Rational numbers can be expressed as decimals (e.g., $\frac{1}{2} = 0.5, \frac{3}{4} = 0.75$). This makes it easier to compare them with irrational numbers.
2. **Approximate Irrational Numbers**: Approximate the value of irrational numbers as decimals. For example:
   - $\sqrt{2} \approx 1.414$
   - $\pi \approx 3.14159$
   - $e \approx 2.718$
3. **Use Number Line Placement**: Place both rational and irrational numbers on a number line. This visual representation helps you see which numbers are larger or smaller.
4. **Compare Values**: Compare the decimal approximations of the numbers:

   Example: Comparing $\frac{1}{2}$ (0.5) and $\sqrt{2}$ (1.414), we see that $\frac{1}{2} < \sqrt{2}$.

   Compare π (3.14159) and 3.14, π is slightly larger than 3.14.
5. **Order the Numbers**: Arrange the numbers in ascending or descending order based on their decimal values.

Example:

Compare and order the numbers:

$\frac{1}{3}, \sqrt{5}, 2$ and π

Solution:

1. Convert rational numbers to decimals:

   $\frac{1}{3} \approx 0.3333 \ldots$ or $0.\overline{3}$

   $2 = 2.000$
2. Approximate irrational numbers:

   $\sqrt{5}$ is between 2 and 3 and is approximately 2.236

   $\pi \approx 3.14159$
3. Place them on the number line and order:

   $0.3333 < 2 < \sqrt{5} < \pi$

Adding and Subtracting Rational Numbers

Adding and subtracting rational numbers involves combining or removing fractions or decimal values. Here's a step-by-step guide:

1. **For Expressions Combining Decimals and Fractions Use One of the Following Methods:**
   - You can convert the fractions to decimals and solve the expression if the denominators can easily be transformed into a power of 10.
   - When the denominators can't be converted to a power of 10, it's more efficient to convert the decimal numbers into fractions to solve the expression.
2. **Convert All Mixed Numbers to Improper Fractions.**
3. **Find a Common Denominator for Fractions:** Find the Least Common Denominator.
4. **Add or Subtract the Numerator:** Add or subtract the numerators while keeping the common denominator.
5. **Align the Decimal Points and Add or Subtract for Decimals.**

Example:

Solve.

a) $-1\frac{2}{5}+\frac{3}{4}+2\frac{1}{6}$

b) $6.1-5.267+7.13$

c) $1\frac{1}{3}-0.25$

Solution a:

- Convert mixed numbers to improper fraction: $-1\frac{2}{5}=-\frac{7}{5}$ and $2\frac{1}{6}=\frac{13}{6}$
- Find a Common Denominator for Fractions: The least common denominator between $5, 4$ and 6 is 60.
- Add the Numerators: $-1\frac{2}{5}+\frac{3}{4}+2\frac{1}{6}=-\frac{7}{5}+\frac{3}{4}+\frac{13}{6}=-\frac{84}{60}+\frac{45}{60}+\frac{130}{60}=\frac{91}{60}$

Solution b:

- First subtract 6.1 by 5.267: $6.1-5.267=0.833$
- Now add 0.833 to 7.13: $0.833+7.13=7.963$
- So, $6.1-5.267+7.13=7.963$

Solution c:

- Convert decimal to Fraction: The denominator of $1\frac{1}{3}$ cannot be converted to power of 10: $0.25=\frac{25}{100}=\frac{1}{4}$
- Convert mixed numbers to improper fraction: $1\frac{1}{3}=\frac{4}{3}$
- Do the subtraction: The least common denominator between 3 and 4 is 12: $1\frac{1}{3}-0.25=\frac{4}{3}-\frac{1}{4}=\frac{16}{12}-\frac{3}{12}=\frac{13}{12}$

Multiplying and Dividing Rational Numbers

Multiplying Rational Numbers

1. **Multiply the Numerators**: Multiply the numerators of the fractions.
2. **Multiply the Denominators**: Multiply the denominators of the fractions.

Dividing Rational Numbers

1. **Reciprocal of the Second Fraction**: Take the reciprocal of the second fraction (invert the numerator and the denominator).
2. **Multiply the Fractions**: Multiply the first fraction by the reciprocal of the second fraction.

Multiplying Rational Numbers in Decimal Format

1. **Multiply as Whole Numbers**: Ignore the decimal points and multiply the numbers as if they were whole numbers.
2. **Count Decimal Places:** Count the total number of decimal places in both original numbers.
3. **Place the Decimal Point:** Place the decimal point in the product so that it has the same number of decimal places as the total you counted.

Dividing Rational Numbers in Decimal Format

1. **Make Divisor a Whole Number**: Move the decimal point to the right in the divisor (the number you're dividing by) to make it a whole number. Move the decimal point in the dividend (the number being divided) the same number of places.
2. **Divide as Whole Numbers:** Divide the numbers as if they were whole numbers.
3. **Place the Decimal Point:** Place the decimal point in the quotient directly above where it now appears in the dividend.

Example: Do the following expressions:

a) $-1.2 \times \frac{3}{4}$ b) $1\frac{5}{9} \div 0.4$

Solution a:

- Convert fraction to decimal or decimal to fraction: In this case we can convert $\frac{3}{4}$ to decimal: $\frac{3}{4} = 3 \div 4 = 0.75$
- Do multiplication: Ignore the decimal points and multiply 12 by 75: $12 \times 75 = 900$
- Count decimal point and place the decimal point: The total number of decimal places is 3: $1.2 \times 0.75 = 0.900 = 0.9$
- Determine the sign of the product: The product of a negative number and positive number is negative: $-1.2 \times 0.75 = -0.900 = -0.9$

Solution b:

- Convert decimal to fraction: $0.4 = \frac{4}{10} = \frac{2}{5}$
- Convert mixed numbers to improper number: $1\frac{5}{9} = \frac{14}{9}$
- Perform the division as described in the steps above:
  $1\frac{5}{9} \div 0.4 = \frac{14}{9} \div \frac{2}{5} = \frac{14}{9} \times \frac{5}{2} = \frac{7}{9} \times \frac{5}{1} = \frac{35}{9}$

Mixed Operations on Rational Numbers

Performing mixed operations on rational numbers involves following the order of operations, commonly remembered by the acronym PEMDAS (Parentheses, Exponents, Multiplication and Division (from left to right), Addition and Subtraction (from left to right). Here's how you can approach it:

Step-by-Step Guide:

1. **Parentheses**: Solve expressions inside parentheses first.
2. **Exponents**: Calculate exponents (powers and roots).
3. **Multiplication and Division**: Perform from left to right.
4. **Addition and Subtraction**: Perform from left to right.

Example:

Evaluate $0.5 \times \left(\frac{1}{3} - \frac{5}{9}\right) - 0.1 \div 0.2^2$

Solution:

1. Convert decimals to fractions: Since the denominators cannot be converted to the power of ten, it is easier to convert decimals to fractions:
   - $0.5 = \frac{5}{10} = \frac{1}{2}$
   - $0.1 = \frac{1}{10}$
   - $0.2 = \frac{2}{10} = \frac{1}{5}$
2. Parentheses: $\frac{1}{3} - \frac{5}{9} = \frac{3}{9} - \frac{5}{9} = \frac{-2}{9}$
3. Exponents: $0.2^2 = (\frac{1}{5})^2 = \frac{1}{25}$
4. Multiplication and division:
   - $0.5 \times \left(\frac{1}{3} - \frac{5}{9}\right) = \frac{1}{2} \times \frac{-2}{9} = \frac{-2}{18} = \frac{-1}{9}$
   - $0.1 \div 0.2^2 = \frac{1}{10} \div \frac{1}{25} = \frac{1}{10} \times \frac{25}{1} = \frac{1}{2} \times \frac{5}{1} = \frac{5}{2}$
5. Addition and subtraction:

   $\frac{-1}{9} - \frac{5}{2} = \frac{-2}{18} - \frac{45}{18} = -\frac{47}{18}$

   So, in summary: $0.5 \times \left(\frac{1}{3} - \frac{5}{9}\right) - 0.1 \div 0.2^2 = -\frac{47}{18}$

Adding and Subtracting Irrational Numbers

When it comes to adding and subtracting irrational numbers, the process is similar to working with rational numbers. Here's a quick guide:

Adding and Subtracting Irrational Numbers:

1. **Simplify Each Term**: Make sure each term is in its simplest form.
2. **Combine Like Terms**: If you have similar irrational terms (like $\sqrt{2}$ and $\sqrt{2}$), you can combine them just like you would with variables.

Notations:

- If the irrational numbers are different (like $\sqrt{2}$ and $\sqrt{3}$), you cannot combine them directly.
- The result of adding or subtracting irrational numbers can still be irrational.

Examples:

1) Evaluate $5\sqrt{3} + 4\sqrt{2} - \sqrt{3} + \frac{\sqrt{2}}{2}$

   Solution:

   1. Combine like terms: group the similar terms together: $(5\sqrt{3} - \sqrt{3}) + (4\sqrt{2} + \frac{\sqrt{2}}{2})$
   2. Simplify within each group and write the result:
      - $5\sqrt{3} - \sqrt{3} = 4\sqrt{3}$
      - $4\sqrt{2} + \frac{\sqrt{2}}{2} = \left(4 + \frac{1}{2}\right)\sqrt{2} = \frac{9}{2}\sqrt{2}$
   3. The final result: $4\sqrt{3} + \frac{9}{2}\sqrt{2}$

2) Evaluate $2\sqrt{12} - \sqrt{75} + 2\sqrt{27}$

   Solution:

   1. Simplify each term:
      - Simplify $\sqrt{12}$: $\sqrt{12} = \sqrt{4 \times 3} = \sqrt{4} \times \sqrt{3} = 2\sqrt{3}$, thus, $2\sqrt{12} = 2 \times 2\sqrt{3} = 4\sqrt{3}$
      - Simplify $\sqrt{75}$: $\sqrt{75} = \sqrt{25 \times 3} = \sqrt{25} \times \sqrt{3} = 5\sqrt{3}$
      - Simplify $\sqrt{27}$: $\sqrt{27} = \sqrt{9 \times 3} = \sqrt{9} \times \sqrt{3} = 3\sqrt{3}$, thus, $2\sqrt{27} = 2 \times 3\sqrt{3} = 6\sqrt{3}$
   2. Substitute simplified terms back into expression:

      $2\sqrt{12} - \sqrt{75} + 2\sqrt{27} = 4\sqrt{3} - 5\sqrt{3} + 6\sqrt{3}$
   3. Combine like terms:

      $4\sqrt{3} - 5\sqrt{3} + 6\sqrt{3} = (4 - 5 + 6)\sqrt{3} = 5\sqrt{3}$

Multiplying and Dividing Irrational Numbers

Multiplying Irrational Numbers:

1. **Multiply the Numbers Inside the Radicals**: If the numbers under the radicals can be combined, do so.
$$\sqrt{a} \times \sqrt{b} = \sqrt{a \times b}$$
2. **Multiply a Coefficient and Radical:** Multiply the coefficients and the radicals separately.
$$a\sqrt{b} \times c\sqrt{d} = a \times c\sqrt{b \times d}$$

Dividing Irrational Numbers:

1. **Divide the Numbers Inside the Radicals:** If the numbers under the radicals can be divided, do so.
$$\frac{\sqrt{a}}{\sqrt{b}} = \sqrt{\frac{a}{b}}$$
2. **Divide a Coefficient and a Radical**: Divide the coefficients and the radicals separately.
$$\frac{a\sqrt{b}}{c\sqrt{d}} = \frac{a}{c}\sqrt{\frac{b}{d}}$$

Notations:

- Make sure to simplify the radicals whenever possible.
- The result of multiplying or dividing irrational numbers can be either rational or irrational.

Examples:

1) Evaluate $\sqrt{0.63} \times \sqrt{7} \times \sqrt{100}$

Solution:

1. Convert decimal to fraction and combine the radicals:
$$\sqrt{0.63} \times \sqrt{7} \times \sqrt{100} = \sqrt{\frac{63}{100}} \times \sqrt{7} \times \sqrt{100} = \sqrt{\frac{63}{100} \times 7 \times 100}$$
2. Simplify the inside of radical:
$$\sqrt{\frac{63}{100} \times 7 \times 100} = \sqrt{\frac{63 \times 7 \times 100}{100}} = \sqrt{63 \times 7} = \sqrt{9 \times 7 \times 7} = \sqrt{9 \times 49} = \sqrt{9} \times \sqrt{49} = 3 \times 7 = 21$$
So, the evaluated expression is: $\sqrt{0.63} \times \sqrt{7} \times \sqrt{100} = 21$

2) Do the expression $\frac{2\sqrt{48}}{\sqrt{7}} \times \frac{\sqrt{28}}{\sqrt{3}}$

Solution:

- Simplify each term: Simplify $\sqrt{48}$: $\sqrt{48} = \sqrt{16 \times 3} = \sqrt{16} \times \sqrt{3} = 4\sqrt{3}$
- Simplify $\sqrt{28}$: $\sqrt{28} = \sqrt{4 \times 7} = \sqrt{4} \times \sqrt{7} = 2\sqrt{7}$

1. Substitute simplified terms back into the expression and simplify fractions:
$$\frac{2\sqrt{48}}{\sqrt{7}} \times \frac{\sqrt{28}}{\sqrt{3}} = \frac{8\sqrt{3}}{\sqrt{7}} \times \frac{2\sqrt{7}}{\sqrt{3}} = 8\sqrt{\frac{3}{7}} \times 2\sqrt{\frac{7}{3}}$$
2. Multiply the simplified fractions: $8\sqrt{\frac{3}{7}} \times 2\sqrt{\frac{7}{3}} = 8 \times 2 \times \sqrt{\frac{3}{7} \times \frac{7}{3}} = 16 \times \sqrt{1} = 16$

Mixed Operations on Irrational Numbers

Similar to rational numbers, when working with mixed operations (addition, subtraction, multiplication, division) involving irrational numbers, it's important to follow the order of operations (**P**arentheses, **E**xponents, **M**ultiplication and **D**ivision, **A**ddition and **S**ubtraction)

Notations:

- **Always simplify radicals first** before performing other operations.
- **Combine like terms** (same radicals) whenever possible.
- The sum or difference of two irrational numbers could be rational or irrational.
- The product of two irrational numbers is usually irrational, but not always.
- Division of irrational numbers can yield a rational or irrational result.

Examples:

1) Evaluate $3\sqrt{6} + 2\sqrt{3} \times \sqrt{2} - \frac{4\sqrt{18}}{3\sqrt{2}}$

Solution:

1. Simplify each term (if possible):
   - $3\sqrt{6}$ is already simplified
   - Simplify $2\sqrt{3} \times \sqrt{2}$: $2\sqrt{3} \times \sqrt{2} = 2\sqrt{3 \times 2} = 2\sqrt{6}$
   - Simplify $\frac{4\sqrt{18}}{3\sqrt{2}}$: $\sqrt{18} = \sqrt{9 \times 2} = \sqrt{9} \times \sqrt{2} = 3\sqrt{2}$, thus, $\frac{4\sqrt{18}}{3\sqrt{2}} = \frac{4\times3\sqrt{2}}{3\sqrt{2}} = 4$
2. Perform multiplication and division first:

   Combine the results: $3\sqrt{6} + 2\sqrt{6} - 4$
3. Perform addition and subtraction:

   Combine like terms: $3\sqrt{6} + 2\sqrt{6} = (3+2)\sqrt{6} = 5\sqrt{6}$

Final answer: $3\sqrt{6} + 2\sqrt{3} \times \sqrt{2} - \frac{4\sqrt{18}}{3\sqrt{2}} = 5\sqrt{6} - 4$

2) Evaluate $(\sqrt{2\sqrt{8} - \sqrt{3}})^2 + \sqrt{\frac{18}{5}} \div \sqrt{\frac{1}{10}}$

Solution:

1. Simplify $\left(\sqrt{2\sqrt{8} - \sqrt{3}}\right)^2$:
2. $(\sqrt{2\sqrt{8} - \sqrt{3}})^2 = \sqrt{2\sqrt{8} - \sqrt{3}} \times \sqrt{2\sqrt{8} - \sqrt{3}} = \sqrt{(2\sqrt{8} - \sqrt{3})^2} = 2\sqrt{8} - \sqrt{3} = 2\sqrt{4} \times \sqrt{2} - \sqrt{3} =$ $4\sqrt{2} - \sqrt{3}$
3. Perform multiplication and division: $\sqrt{\frac{18}{5}} \div \sqrt{\frac{1}{10}} = \sqrt{\frac{18}{5}} \times \sqrt{\frac{10}{1}} = \sqrt{\frac{18}{5} \times \frac{10}{1}} = \sqrt{\frac{18}{1} \times \frac{2}{1}} = \sqrt{36} = 6$
4. Perform addition: $4\sqrt{2} - \sqrt{3} + 6$ (there is no like term)

Final answer: $(\sqrt{2\sqrt{8} - \sqrt{3}})^2 + \sqrt{\frac{18}{5}} \div \sqrt{\frac{1}{10}} = 4\sqrt{2} - \sqrt{3} + 6$

Absolute Value of Rational / Irrational Numbers

Absolute Value of Rational and Irrational Numbers:

As we explained in the previous chapter, the absolute value of all numbers, including rational and irrational numbers, is equal to their distance from zero, for example:

- $\left|-\frac{2}{5}\right| = \frac{2}{5}$
- $|-\sqrt{7}| = \sqrt{7}$

Addition and Subtraction:

When you add or subtract absolute values, you first evaluate the absolute values, then perform the operations. For example:

- $\left|-2\frac{1}{5}\right| + |3| = 2\frac{1}{5} + 3 = 5\frac{1}{5}$
- $\left|-\sqrt{5}\right| - |2| = \sqrt{5} - 2 \approx 2.236 - 2 = 0.236$

Notation: $|a + b| \neq |a| + |b|\ and\ |a - b| \neq |a| - |b|$

Multiplication and Division:

When you multiply or divide absolute values, you first evaluate the absolute values, then perform the operations. For example:

- $\left|\frac{8}{11}\right| \times |-0.3| = \frac{8}{11} \times \frac{3}{10} = \frac{24}{110} = \frac{12}{55}$
- $\left|-\sqrt{3}\right| \times |-\pi| = \sqrt{3} \times \pi \approx 1.732 \times 3.1416 \approx 5.441$

Writing Absolute Value of Irrational Expressions Without Absolute Value Sign:

To write irrational expressions without the absolute value sign, you'll generally look at the context of the expression to determine whether the values are positive or negative and use the following rules:

- $|a| = a\ if\ a \geq 0$
- $|a| = -a\ if\ a < 0$

For example: $\left|\sqrt{2} - \pi\right| \approx |1.41 - 3.14| = |-1.73|$ the inside expression is negative, so $\left|\sqrt{2} - \pi\right| = -\left(\sqrt{2} - \pi\right) = \pi - \sqrt{2}$

Example: Write $\left|-\sqrt{8} + 3\right|$ without absolute value sign.

Solution:

The sign of inside expression is positive (because: $-\sqrt{8} + 3 \approx -2.82 + 3 = 0.18$) so: $\left|-\sqrt{8} + 3\right| = -\sqrt{8} + 3$

Real World Applications

Rational and irrational numbers both have important real-world applications. Here's how they appear in different areas that an eighth grader can relate to:

1. Rational Numbers in Real Life

a. Money and Shopping: When you buy items, prices are often expressed as rational numbers. For example, a shirt may cost \$15.99 or a pack of gum might cost \$2.50. Both \$15.99 and \$2.50 are rational numbers

b. Cooking and Recipes: When cooking, you often need to measure ingredients. If a recipe calls for $\frac{3}{4}$ of a cup of sugar, or 1.5 cups of flour, these are rational numbers. Rational numbers help you measure out exact amounts.

c. Time: Time is often measured using rational numbers, such as when you divide an hour into minutes. For instance, $\frac{1}{2}$ hour = 30 minutes, or 0.25 hours = 15 minutes.

2. Irrational Numbers in Real Life

a. Circles and Geometry (Pi): Pi (π) is an irrational number, approximately 3.14159, and it is used in real-world situations when calculating the area or circumference of a circle. For example, if you want to know the circumference of a circular garden with a radius of 5 meters

c. Architecture and Design: In fields like architecture, irrational numbers are used when designing structures that involve curves or angles, such as domes, arches, and spiral staircases. For example, the golden ratio, which is an irrational number (approximately 1.618), is often seen in architecture, art, and nature. Many famous buildings and artworks have used this ratio to create visually pleasing proportions.

Example:

A landscaper is planning to plant a circular garden in the shape of a flower. The diameter of the garden is 10 meters, and the landscaper wants to install a stone pathway that follows the circumference of the garden.

1. What is the exact length of the pathway around the garden? (Use $\pi \approx 3.1416$ for the approximation.)
2. The landscaper has already laid down 20 meters of stones for the pathway. How much more stone is needed to complete the pathway?

Solution:

In this case, the diameter $d = 10$ meters, and π is irrational. To get an approximate value, multiply $\pi \approx 3.14$ by the diameter: $C = 3.14 \times 10 = 31.4\ meters\ (approx.)$

Remaining Stone Needed: The landscaper already has 31.4 meters of stone. To find out how much more is needed: $31.4 - 20 = 11.4\ meters$

So, the landscaper will need an additional 11.4 meters of stone to complete the pathway.

Worksheets

Convert Repeating Decimal

Convert following repeating decimals to fraction:

1) $0.\overline{4}$
2) $1.\overline{16}$
3) $3.0\overline{2}$
4) $12.\overline{714}$
5) $9.82\overline{1}$

Convert following fractions to repeating decimals:

6) $\frac{4}{11}$
7) $\frac{9}{22}$
8) $\frac{26}{9}$
9) $2\frac{3}{11}$
10) $-1\frac{17}{27}$

Identify Rational Numbers

Identify if these numbers are rational and find their reciprocals.

1) $-6\frac{1}{3}$
2) 0.05
3) $\frac{1}{\pi}$
4) $-\sqrt{\frac{1}{25}}$
5) $\frac{\sqrt{3}}{\sqrt{5}}$

Identify Irrational Numbers

Determine whether the following numbers are irrational or not.

1) $-(\sqrt{2})^{2}$
2) $\frac{\sqrt{3}}{3}$
3) $\pi - 2$
4) $\sqrt{\frac{72}{2}}$
5) $\sqrt{6} - \sqrt{2}$

Irrational Numbers on Number lines

Find the following irrational numbers on the number line.

1) $\sqrt{10}$
2) $\sqrt{13}$
3) $-\sqrt{5}$
4) $1 - \sqrt{2}$
5) $2 + \sqrt{5}$

Approximate Irrational Numbers

Which two integers follow irrational numbers between?

1) $\sqrt{18}$
2) $\sqrt{29}$
3) $-\sqrt{7.5}$
4) $-\sqrt{12} + 1$
5) $\sqrt{61} - 5$

Place the appropriate sign (<=>) in the blank space.

6) $\sqrt{17}\ \square\ 4$
7) $3\frac{1}{3}\ \square\ \sqrt{11}$
8) $\sqrt{27} + 1\ \square\ 6$
9) $\sqrt{4.41}\ \square\ 2.1$
10) $\sqrt{8} + 1\ \square\ \sqrt{10} - 2$

Compare and Order Rational and Irrational Numbers

Order following numbers from smallest to largest.

1) $\sqrt{22}, \pi, -15, \frac{12}{5}$

2) $-2\frac{2}{7}, \sqrt{16}, \sqrt{\frac{9}{2}}, -\frac{7}{8}, 0.14$

3) $-3\frac{1}{3}, -\pi, -3.14, \sqrt{9.5}$

4) $\sqrt{10}, -\frac{1}{10}, 1.089, 3.7, \sqrt{8}+1$

5) $-3\pi, -9.5, -\sqrt{91}, -9\frac{1}{4}$

Adding and Subtracting Rational Numbers

Find the answers.

1) $1\frac{5}{10} + 2\frac{1}{3} - \frac{1}{6}$

2) $0.3 - 2\frac{1}{5} + \frac{1}{4}$

3) $1\frac{4}{9} - 1.2 + \frac{5}{18} - 0.7$

4) $1\frac{3}{5} - \left(-\sqrt{\frac{9}{4}}\right) + 1.5$

5) $-\sqrt{\sqrt{81}} - (-0.24 + 3.16)$

6) $\left(1-\frac{1}{2}\right) - \left(1-\frac{1}{4}\right) + \left(1-\frac{1}{8}\right) - (1-\frac{1}{16})$

7) $-7\frac{24}{36} + 2\frac{51}{85} - 5\frac{39}{65} + 1\frac{19}{38}$

8) $\frac{1}{3} - \left(\frac{1}{6} - \left(\frac{1}{9} + \frac{1}{12}\right) - \frac{1}{15}\right) - \frac{1}{18}$

9) $\frac{1}{100} - \frac{2}{100} + \frac{3}{100} - \cdots + \frac{99}{100} - 1$

10) $\left(\frac{1}{2} + \frac{1}{3} + \frac{1}{4} + \cdots + \frac{1}{100}\right) + \left(\frac{1}{2} + \frac{2}{3} + \frac{3}{4} + \cdots + \frac{99}{100}\right)$

Multiplying and Dividing Rational Numbers

Solve.

1) $-2 \div \frac{5}{7} \div (-4\frac{2}{3})$

2) $2.64 \times \frac{5}{-6} \div (-\frac{5}{7})$

3) $-3 \times [-\frac{4}{6} \div (-5) \times \left(-\frac{3}{8}\right)]$

4) $(-(-0.2) \div 0.04) \times \frac{7}{14}$

5) $-\frac{30}{54} \div \frac{-4\times(-3)\times(-2)}{-6\times(-5)\times(-4)}$

Find missing value.

6) $2.3 \div 0.1 = x \div 10$

7) $\frac{x}{2.5} = \frac{18}{-28} \times \frac{-49}{-15}$

8) $\frac{\frac{2}{3}}{\frac{1}{2} \div 2} = \frac{x}{6}$

9) $\frac{1}{3} \times \left(\frac{1}{5} \div \frac{1}{2}\right) = \frac{3}{x}$

10) $\left(1 \div \left[\frac{6}{5} \times \frac{7}{6} \times \frac{8}{7}\right]\right) \times x = 1$

Mixed Operations on Rational Numbers

Evaluate.

1) $(\frac{4}{7} - \frac{1}{3}) \div (-1\frac{2}{3})$

2) $5 \times 0.01 - 4 \times 0.001 + 2 \times 0.1$

3) $(\frac{1}{6} - \frac{3}{8} + \frac{-8}{9}) \div \frac{79}{-24}$

4) $\frac{-2-[-0.5+2]}{1-0.5-0.75\times 4}$

5) $-\frac{1}{2} - \frac{\frac{1}{2}+\frac{2}{3}}{\frac{1}{5}+\frac{1}{2}}$

6) $\frac{-0.3-1\frac{1}{5}}{1.2+\frac{-3}{5}} \div \frac{0.5}{-1\frac{4}{5}}$

7) $\frac{1}{1\times 2} + \frac{1}{2\times 3} + \frac{1}{3\times 4} + \cdots + \frac{1}{80\times 81}$

8) $\frac{3}{30\times 33} + \frac{3}{33\times 36} + \frac{3}{36\times 39} + \cdots + \frac{3}{117\times 120}$

9) $3.25 - \frac{1}{2-\frac{1}{1-0.25}}$

10) $\frac{\sqrt{6.25}+\sqrt{0.25}}{3-\frac{1}{3+\frac{4}{0.5}}} \div 0.4$

Adding and Subtracting Irrational Numbers

Solve.

1) $4\sqrt{7} + 9\sqrt{7} - 3\sqrt{7}$
2) $19\sqrt{5} - 21\sqrt{5} - \frac{\sqrt{5}}{4}$
3) $-\frac{2}{3}\sqrt{3} + \frac{3}{5}\sqrt{3} + \frac{7}{15}\sqrt{3}$
4) $4\sqrt{13} - \sqrt{11} + 3.4\sqrt{13} + 5\sqrt{11}$
5) $3\sqrt{8} - \sqrt{3} + 0.5\sqrt{2} - 1.5\sqrt{12}$
6) $\sqrt{24} - 3\sqrt{5} + \sqrt{6} - 2\sqrt{20}$
7) $11\sqrt{11} + 10\sqrt{11} + 9\sqrt{11} + \cdots + \sqrt{11}$
8) $2\sqrt{3} - \sqrt{27} + \sqrt{75}$
9) $4\sqrt{50} - 3\sqrt{32} + 2\sqrt{18} - \sqrt{8}$
10) $4\sqrt{28} + 2\sqrt{63} - \sqrt{175}$

Multiplying and Dividing Irrational Numbers

Evaluate.

1) $\sqrt{27} \times \sqrt{3}$
2) $\sqrt{0.2} \times \frac{\sqrt{8}}{\sqrt{2.5}}$
3) $\sqrt{0.3} \times \sqrt{\frac{27}{10}}$
4) $4\sqrt{72} \div 2\sqrt{2}$
5) $\frac{\sqrt{4.9}\times\sqrt{3.6}}{\sqrt{1.21}}$
6) $\sqrt{\frac{54}{4.2}} \times \sqrt{\frac{21}{100}} \times \sqrt{9.6}$
7) $\frac{\sqrt{12}}{3\sqrt{8}} \times \frac{45\sqrt{2}}{\sqrt{27}}$
8) $\sqrt{\frac{0.16}{2.5} \times 4.9 \times 0.36}$
9) $\frac{\sqrt{\frac{9}{4}}}{\sqrt{\frac{17}{36}} \times \sqrt{\frac{9}{17}}}$
10) $\sqrt{8 \times \sqrt{\sqrt{\frac{\sqrt{28}}{\sqrt{7}}} \times 8}}$

Mixed Operations on Irrational Numbers

Find answers.

1) $9\sqrt{110} \div 3\sqrt{4.4}$
2) $\sqrt{2 \times 40 + 1} \div \sqrt{6^2 - \sqrt{121}}$
3) $\frac{\sqrt{(144+256)\times 16}}{\sqrt{1-\frac{3}{4}}}$
4) $\frac{\sqrt{12^2+5^2}}{\sqrt{17^2-8^2}}$
5) $\sqrt{\frac{5}{9}} + \sqrt{\frac{20}{9}} - \sqrt{80}$
6) $\frac{\sqrt{4\sqrt{25}-15}}{\sqrt{3\sqrt{9}+1}} \times \frac{\sqrt{\sqrt{81}-2\sqrt{4}}}{\sqrt{5(\sqrt{100}-\sqrt{4})}}$
7) $\sqrt{\sqrt{8\sqrt{0.25}} + \sqrt{\sqrt{1 + 5\sqrt{\sqrt{81}}}}}$
8) $\sqrt{\sqrt{13 + \sqrt{6 + \sqrt{3 + \sqrt{36}}}}}$
9) $\sqrt{(1 - \frac{1}{2})} \times \sqrt{(1 - \frac{1}{3})} \times \sqrt{(1 - \frac{1}{4})} \times \ldots \times \sqrt{(1 - \frac{1}{100})}$
10) $\sqrt{1 + 8\sqrt{9\sqrt{8\sqrt{5\sqrt{0.64}}}}}$

Absolute Value of Rational / Irrational Numbers and Operation on Them

Evaluate.

1) $|-\sqrt{25}| - |\sqrt{9}|$
2) $-\left|\frac{5}{6}\right| + |-\sqrt{4} + \frac{2}{3}|$
3) $|3\sqrt{10} - 5\sqrt{10}| \div -(|-\sqrt{10}|)$
4) $-|\sqrt{9+16}| \times \left(\left|-\frac{2}{9}\right| - \left|\frac{3}{7}\right|\right)$
5) $\frac{|\sqrt{16}-\sqrt{9}\times 0.3|}{-|0.2\times\sqrt{49}+\frac{\sqrt{36}}{5}|}$

Write the following absolute value expressions without absolute value sign:

6) $|-\sqrt{7} - \sqrt{5}|$
7) $|6 - \sqrt{2} - 3|$
8) $|-\sqrt{3} + 1 + \frac{2}{3}\sqrt{3}|$
9) $|\sqrt{12} - 2\sqrt{2}| + |-\sqrt{2} - 3\sqrt{3}|$
10) $\left|\frac{2}{5} - \sqrt{48}\right| + |-\sqrt{27} + 5|$

Real Word Applications

Do the following word problems about rational and irrational numbers:

1) You are buying flooring for a room that is 12 feet by 18 feet. If the cost of flooring is \$3.50 per square foot, how much will it cost to cover the entire room?
2) A high school has 3 eighth-grade classes, and the number of students in each class is equal. From each class, $\frac{1}{4}$ of the students went on a scientific field trip. What fraction of the total students went on the trip?
3) A satellite orbits the Earth at a height of 600 kilometers. If the Earth's radius is approximately 6371 kilometers, what is the approximate length of the satellite's orbit (circumference)? Use $\pi \approx 3.1416$.
4) A city park is a perfect square measuring 100 meters on each side. The park planners want to build a diagonal walkway straight from one corner to the opposite corner. How long will the walkway be?
5) A wheelchair ramp needs to reach a height of 1 meter over a horizontal distance of 3 meters. What is the length of the ramp surface?
6) A bus has 40 passengers. At one stop, $\frac{1}{5}$ of them got off, and then an amount equal to $\frac{1}{2}$ of the remaining passengers got on. Now, how many passengers are on the bus?
7) We have filled in half of the volume of a water storage tank. A worker fills half of the remaining volume with water every 15 minutes. After how many minutes will $\frac{1}{32}$ of the tank's volume remain empty?
8) We want to sew a tablecloth for a circular table with a diameter of $\sqrt{2}$ meters, which hangs 25 centimeters around the edges. How many meters of fabric do we need at a minimum?
9) Two gears, one with a radius of 3 centimeters and the other with a radius of 8 centimeters, are connected in a toy. If the smaller gear makes 4 rotations, how many rotations does the larger gear make?
10) The weight of a container half-filled with water is 2.4 kilograms. If the weight of the container alone is $\frac{1}{5}$ of the weight of the container fully filled with water, what is the weight of the water in the initial half-filled container?

Answer of Worksheets

Convert Repeating Decimal

1) $\frac{4}{9}$
2) $\frac{115}{99}$
3) $\frac{136}{45}$
4) $\frac{4234}{333}$
5) $\frac{8839}{900}$
6) $0.\overline{36}$
7) $0.4\overline{09}$
8) $2.\overline{8}$
9) $2.\overline{27}$
10) $-1.\overline{629}$

Identify Rational Numbers (Reciprocals and multiplicative inverses)

1) Rational, the reciprocal: $-\frac{3}{19}$
2) Rational, the reciprocal: 20
3) Irrational
4) Rational, the reciprocal: -5
5) Irrational

Identify Irrational Numbers

1) No, rational
2) Yes, irrational
3) Yes, irrational
4) No, rational
5) Yes, irrational

Irrational Numbers on Number lines

1) $\sqrt{10}$:

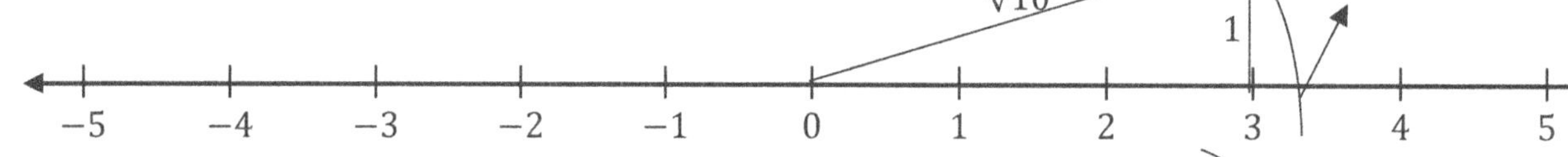

2) $\sqrt{13}$:

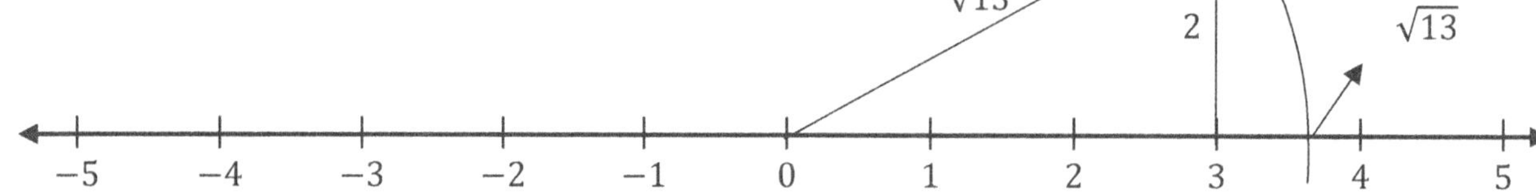

3) $-\sqrt{5}$

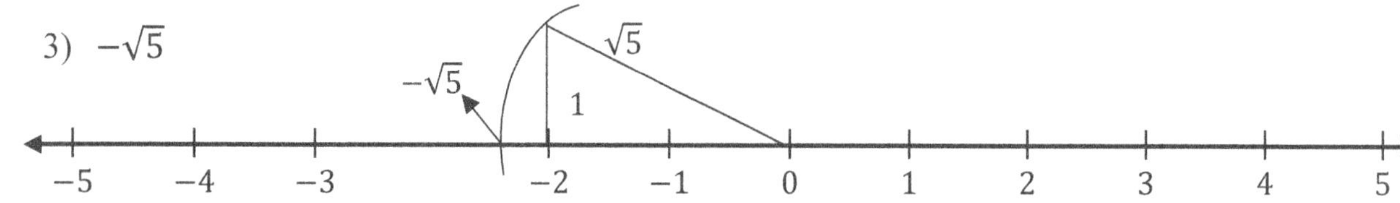

4) $1-\sqrt{2}$:

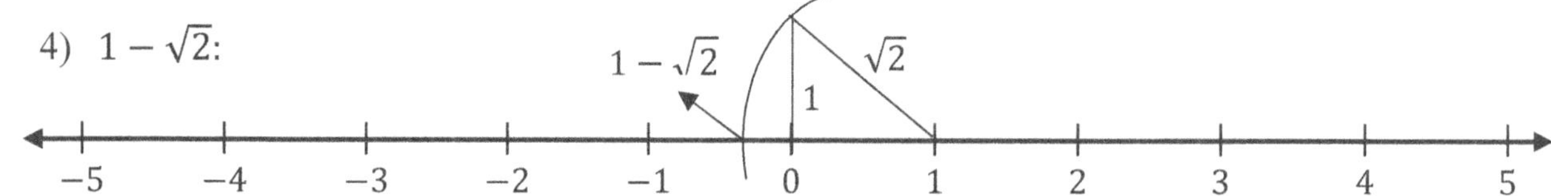

5) $2+\sqrt{5}$:

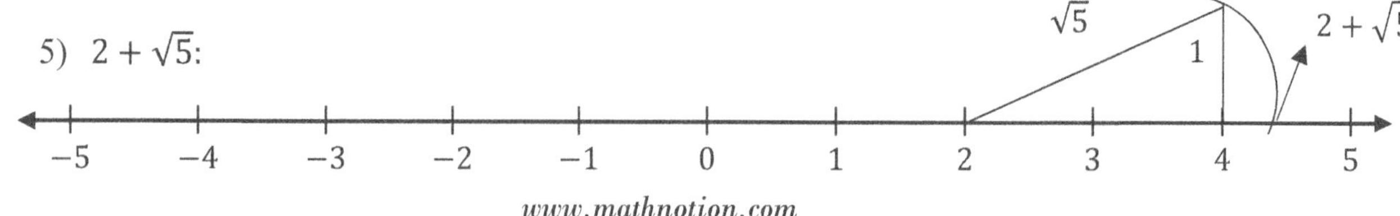

Approximate Irrational Numbers

1) 4 and 5
2) 5 and 6
3) -3 and -2
4) -3 and -2
5) 2 and 3
6) $\sqrt{17}\ \boxed{>}\ 4$
7) $3\frac{1}{3}\ \boxed{>}\ \sqrt{11}$
8) $\sqrt{27}+1\ \boxed{>}\ 6$
9) $\sqrt{4.41}\ \boxed{=}\ 2.1$
10) $\sqrt{8}+1\ \boxed{>}\ \sqrt{10}-2$

Compare and Order Rational and Irrational Numbers

1) $-15 < \frac{12}{5} < \pi < \sqrt{22}$
2) $-2\frac{2}{7} < -\frac{7}{8} < 0.14 < \sqrt{\frac{9}{2}} < \sqrt{16}$
3) $-3\frac{1}{3} < -\pi < -3.14 < \sqrt{9.5}$
4) $-\frac{1}{10} < 1.089 < \sqrt{10} < 3.7 < \sqrt{8}+1$
5) $-\sqrt{91} < -9.5 < -3\pi < -9\frac{1}{4}$

Adding and Subtracting Rational Numbers

1) $\frac{11}{3}$
2) $-\frac{33}{20}$
3) $-\frac{8}{45}$
4) $\frac{23}{5}$
5) -5.92
6) $-\frac{5}{16}$
7) $-9\frac{1}{6}$
8) $\frac{67}{180}$
9) $\frac{-1}{2}$
10) 99

Multiplying and Dividing Rational Numbers

1) $\frac{3}{5}$
2) 3.08
3) $\frac{3}{20}$
4) 2.5
5) $\frac{-25}{9}$
6) $x = 230$
7) $x = -5.25$
8) $x = 16$
9) $x = \frac{45}{2}$
10) $x = \frac{8}{5}$

Mixed Operations on Rational Numbers

1) $\frac{-1}{7}$
2) 0.246
3) $\frac{1}{3}$
4) 1.4
5) $-\frac{13}{6}$
6) 9
7) $\frac{80}{81}$
8) $\frac{1}{40}$
9) 1.75
10) $\frac{165}{64}$

Adding and Subtracting Irrational Numbers

1) $10\sqrt{7}$
2) $-9\frac{\sqrt{5}}{4}$
3) $\frac{2}{5}\sqrt{3}$
4) $7.4\sqrt{13} - 4\sqrt{11}$
5) $6.5\sqrt{2} - 4\sqrt{3}$
6) $-7\sqrt{5} + 3\sqrt{6}$
7) $66\sqrt{11}$
8) $4\sqrt{3}$
9) $12\sqrt{2}$
10) $9\sqrt{7}$

Multiplying and Dividing Irrational Numbers

1) 9
2) 0.8
3) 0.9
4) 12
5) $\frac{42}{11}$
6) $\frac{18}{5}\sqrt{2}$
7) 5
8) $\frac{42}{125}$
9) 3
10) 4

Mixed Operations on Irrational Numbers

1) 15
2) $\frac{9}{5}$
3) 160
4) $\frac{13}{15}$
5) $-3\sqrt{5}$
6) $\frac{\sqrt{10}}{16}$
7) 2
8) 2
9) $\frac{1}{10}$
10) 7

Absolute Value of Rational / Irrational Numbers and Operation on Them

1) 2
2) $\frac{1}{2}$
3) -2
4) $\frac{65}{63}$
5) $\frac{-31}{26}$
6) $+\sqrt{7}+\sqrt{5}$
7) $3-\sqrt{2}$
8) $\frac{-\sqrt{3}}{3}+1$
9) $-\sqrt{2}+5\sqrt{3}$
10) $7\sqrt{3}-\frac{27}{5}$

Real Word Applications

1) $756
2) $\frac{1}{4}$
3) 43,870.9 km
4) $100\sqrt{2}$
5) $\sqrt{10}$ meters
6) 48 passengers
7) 60 minutes
8) ≈ 2.87 square meters
9) The larger gear makes 1.5 rotations,
10) 1.6 kg

Chapter 3: Exponents and Radical Expressions

Topics that you'll learn in this chapter:

- ✓ Integer Exponents
- ✓ Negative, Fraction and Decimal Bases
- ✓ Scientific Notation
- ✓ Adding and Subtracting Scientific Notations
- ✓ Multiplying and Dividing Scientific Notations
- ✓ Scientific Notation (Estimate Large or Small Quantities)
- ✓ Square Roots (Non-Perfect and Perfect)
- ✓ Cube Roots of Positive and Negative Perfect Cubes
- ✓ Approximating Cube Roots
- ✓ Simplifying Cube Root Radical Expressions
- ✓ Adding and Subtracting Radical Expressions
- ✓ Multiplying and Dividing Radical Expressions
- ✓ Real World Applications
- ✓ Worksheets
- ✓ Answer of Worksheets

Integer Exponents

Basic Rules for Integer Exponents

1. **Positive Exponents:** When the exponent is a positive integer, it means you multiply the base by itself as many times as the value of the exponent.

$$a^n = a \times a \times a \times \ldots \times a \ (n\ times)$$

2. **Zero Exponent:** Any non-zero base raised to the power of zero equals 1.

$$a^0 = 1 \ (where\ a \neq 0)$$

3. **Negative Exponents:** A negative exponent represents the reciprocal of the base raised to the positive exponent.

$$a^{-n} = \frac{1}{a^n}$$

Properties of Integer Exponents

1. **Product of Powers** (when multiplying two powers with the same base):

$$a^n \times a^m = a^{n+m}$$

2. **Quotient of Powers:** (when dividing two powers with the same base):

$$\frac{a^n}{a^m} = a^{n-m}$$

3. **Power of a Power Rule** (when raising power to another power):

$$(a^m)^n = a^{m \times n}$$

4. **Power of a Quotient Rule** (when raising a quotient to a power):

$$(\frac{a}{b})^n = \frac{a^n}{b^n}$$

Example: Simplify the following expressions.

a) $\frac{(x^3y^2)^{-2}}{x^{-1}y^4}$

b) $(\frac{3}{4})^2 \times (\frac{8}{6})^{-4}$

Solution a:

1. Apply the power of power rule: $(x^3y^2)^{-2} = x^{3\times(-2)} \times y^{2\times(-2)} = x^{-6} \times y^{-4}$
2. Apply the quotient of power rule:

$$\frac{(x^3y^2)^{-2}}{x^{-1}y^4} = \frac{x^{-6} \times y^{-4}}{x^{-1} \times y^4} = \frac{x^{-6}}{x^{-1}} \times \frac{y^{-4}}{y^4} = x^{-6-(-1)} \times y^{-4-4} = x^{-5} \times y^{-8} = \frac{1}{x^5y^8}$$

Solution b:

Apply power of a quotient rule: $(\frac{3}{4})^2 = \frac{3^2}{4^2}$ and $(\frac{8}{6})^{-4} = (\frac{4}{3})^{-4} = \frac{4^{-4}}{3^{-4}} = \frac{3^4}{4^4}$

Apply product of power and power of quotient rule: $\frac{3^2}{4^2} \times \frac{3^4}{4^4} = \frac{3^6}{4^6} = (\frac{3}{4})^6$

Negative, Fraction and Decimal Bases

Negative Bases: When a negative base is raised to a positive integer exponent, the result depends on whether the exponent is even or odd:

- **Even Exponent:** The result is positive. For example: $(-2)^4 = 16$
- **Odd Exponent**: The result is negative. For example: $(-2)^3 = -8$

Fractional Bases:

- **Positive Exponents:** Similar to power of quotient rule, raising a fractional base to a positive exponent involves raising both the numerator and the denominator to the exponent:

$$(\frac{a}{b})^n = \frac{a^n}{b^n}$$

- **Negative Exponents**: For a fractional base raised to a negative exponent, take the reciprocal and change the exponent to positive:

$$(\frac{a}{b})^{-n} = (\frac{b}{a})^n$$

Decimal Bases: Decimals can be converted to fractions, and the same rules for fractional bases apply.

Examples:

1) Simplify the expression $(-4)^{-2} \times (4)^4$.

   Solution:

   - Simplify $(-4)^{-2}$: The exponent is even so the result is positive: $(-4)^{-2} = 4^{-2}$
   - Apply product of power rule: $(4)^{-2} \times (4)^4 = 4^2$

   So, the final result is 4^2

2) Simplify the following expression and write it in positive power, $\frac{0.2^3 \text{x} 5^{-5} \text{x} 3^8}{(0.6)^{-1}}$.

   Solution:

   1. Simplify the numerator by converting the decimal to fraction and changing the negative exponent to a positive exponent:

      $0.2^3 \times 5^{-5} \times 3^8 = (\frac{2}{10})^3 \times (\frac{1}{5})^5 \times 3^8 = \left(\frac{1}{5}\right)^3 \times (\frac{1}{5})^5 \times 3^8 = (\frac{1}{5})^8 \times 3^8 = \left(\frac{1}{5} \times 3\right)^8 = \left(\frac{3}{5}\right)^8$

   2. Simplify the denominator by converting the decimal to fraction:

      $(0.6)^{-1} = (\frac{6}{10})^{-1} = (\frac{3}{5})^{-1}$

   3. Apply quotient of power rule: $\frac{0.2^3 \text{x} 5^{-5} \text{x} 3^8}{(0.6)^{-1}} = \frac{\left(\frac{3}{5}\right)^8}{(\frac{3}{5})^{-1}} = \left(\frac{3}{5}\right)^{8-(-1)} = \left(\frac{3}{5}\right)^{8-(-1)} = \left(\frac{3}{5}\right)^9$

Scientific Notation

Scientific notation is a way of writing very large or very small numbers in a simplified form. It's especially useful in science and math.

General Format: The general form of scientific notation is: $a \times 10^n$. Where a is a number between 1 and 10.

How to Write in Scientific Notation:

1. **Find the First Non-zero Digit:** For example, for 4,500,000, the first non-zero digit is 4.
2. **Place the Decimal Point:** Position the decimal after the first significant figure: 4.5
3. **Count the Exponent:** Determine how many places the decimal has "moved" to convert the original number to a number between 1 and 10. For 4,500,000, it moves 6 places to the left, so the exponent is 6.

Examples:

1) Convert following numbers to scientific notation.

   a) 0.00000237   b) 8,540,000,000

   Solution a:

   1. Find the first non-zero digit: The first non-zero digit is 2.
   2. Move the decimal point and count the exponent: We move the decimal point 6 places to the right. So 0.00000237 becomes 2.37.
   3. Write it as a power of 10: Since we moved the decimal point to the right, the exponent will be negative. The power of 10 will be -6.

      So, the scientific notation for 0.00000237 is 2.37×10^{-6}

   Solution b:

   1. Find the first non-zero digit: The first non-zero digit is 8.
   2. Move the decimal point and count the exponent: We move the decimal point 9 places to the left. So 8,540,000,000 becomes 8.54.
   3. Write it as a power of 10: Since we moved the decimal point to the left, the exponent will be positive. The power of 10 will be 9.

      So, the scientific notation for 8,540,000,000 is 8.54×10^9

2) Convert 0.0003005×10^{-2} to scientific notation.

   Solution:

   1. Find the first non-zero digit: the first non-zero digit is 3
   2. Move the decimal point and count the exponent: We move the decimal point 4 places to the right so, 0.0003005 becomes 3.005
   3. The scientific notation for 0.0003005: 3.005×10^{-4}

   Finally, the scientific notation for $0.0003005 \times 10^{-2} = 3.005 \times 10^{-4} \times 10^{-2} = 3.005 \times 10^{-6}$

Adding and Subtracting Scientific Notations

To add or subtract numbers in scientific notation, you first need to make sure the exponents of 10 are the same. If they're not, you'll need to adjust one of the numbers so the exponents match.

Steps for Addition/Subtraction:

1. **Make the exponents the same**:
   - If the exponents are different, you'll need to rewrite one or both numbers, so they have the same exponent. To do this, move the decimal point in the coefficient of one number to the left or right, and adjust the exponent accordingly.
   - Example: $4.3 \times 10^6 + 3.5 \times 10^5 \rightarrow$ first, change 3.5×10^5 to 0.35×10^6 so both exponents are 10^6.
2. **Add or subtract the coefficients**:
   - Once the exponents are the same, you can add or subtract the coefficients (the numbers before the times 10^n).
   - Example: $4.3 \times 10^6 + 0.35 \times 10^6 = (4.3 + 0.35) \times 10^6 = 4.65 \times 10^6$
3. **Adjust the result if needed**:
   - After adding or subtracting, you may need to adjust the coefficient to make sure it's still between 1 and 10. If it's not, move the decimal point and adjust the exponent.
   - Example:$1.01 \times 10^7 - 9.78 \times 10^6 = 1.01 \times 10^7 - 0.978 \times 10^7 = (1.01 - 0.978) \times 10^7 = 0.032 \times 10^7 = 3.2 \times 10^{-2} \times 10^7 = 3.2 \times 10^5$

Examples:

1) Subtract $4.0 \times 10^6 - 2.57 \times 10^6$.

   Solution:

   1. Exponents are already the same: Both are 10^6
   2. Subtract the coefficients: $4.0 - 2.57 = 1.43$
   3. Combine the coefficient with the exponent: 1.43×10^6

2) Add $2.5 \times 10^3 + 3.78 \times 10^4$.

   Solution:

   1. Adjust the exponents to be the same: Convert 3.78×10^4 to 37.8×10^3
   2. Add the coefficients: $2.5 + 37.8 = 40.3$
   3. Combine the coefficient with the exponent: 40.3×10^3
   4. Rewrite 40.3×10^3 into proper scientific notation: 4.03×10^4

Multiplying and Dividing Scientific Notations

Steps to Multiply Scientific Notation:

1. **Multiply the coefficients**: Simply multiply the numbers in front.
2. **Add the exponents**: When multiplying, add the exponents of the 10s.
3. **Write the result in scientific notation**: Make sure the final answer is in the proper scientific notation form.

Steps to Divide Scientific Notation:

1. **Divide the coefficients:** Simply divide the numbers in front.
2. **Subtract the exponents**: When dividing, subtract the exponent of the divisor from the exponent of the dividend.
3. **Write the result in scientific notation:** Ensure the final answer is in proper scientific notation form.

Always check:

- **Is the coefficient between 1 and 10?**

  If not, adjust it and change the exponent accordingly.

Examples:

1) Multiply $(3.1 \times 10^5) \times (1.5 \times 10^2)$

   Solution:

   1. Multiply the coefficients: $3.1 \times 1.5 = 4.65$
   2. Add the exponents:$5 + 2 = 7$
   3. Combine the results: 4.65×10^7
   4. Final answer: 4.65×10^7

2) Divide $(7.02 \times 10^5) \div (2.4 \times 10^3)$

   Solution:

   1. Divide the coefficient: $7.02 \div 2.4 = 2.925$
   2. Subtract the exponents: $5 - 3 = 2$
   3. Combine the result: 2.925×10^2

Scientific Notation (Estimate Large or Small Quantities)

Scientific notation is used across various fields to make calculations manageable and to express measurements that are either too large or too small to be conveniently written in decimal form. Here are a few examples:

Astronomy: Distances in space are enormous. For instance, the distance from Earth to the nearest star (other than the Sun) is about 4.22×10^{13} kilometers.

Chemistry: The masses of atoms and molecules are very tiny. For example, the mass of a hydrogen atom is about 1.67×10^{-24} grams.

Physics: Scientific notation is used to express the speed of light (3.00×10^{8}meters per second) or the charge of an electron (-1.602×10^{-19} coulombs).

Biology: In microbiology, the size of bacteria is often measured in micrometers. For instance, an E. coli bacterium is approximately 2×10^{-6}meters long.

Finance: Large financial figures, such as national debts or GDP, can be expressed more clearly. For instance, the GDP of the United States is around 2.2×10^{13}dollars.

Estimation in Scientific Notation:

- To estimate using scientific notation, you can round the coefficient to one significant digit and adjust the exponent accordingly.
- For example: Estimate $3.47 \times 10^{8} + 2.81 \times 10^{8}$:
  Round 3.47 to 3 and 2.81 to 3. Add: $3 \times 10^{8} + 3 \times 10^{8} = 6 \times 10^{8}$

Example:

The distance from Earth to the Sun is approximately 1.496×10^{8}kilometers. Light travels at a speed of 3.00×10^{5}kilometers per second. Estimate how many seconds it takes for light to travel from the Sun to Earth. Express your answer in scientific notation.

Solution:

1. Set up the estimation problem: Time (seconds)= $\frac{Distance}{Speed} \approx \frac{1.496\times10^{8}km}{3.0\times\frac{10^{5}km}{s}} \approx \frac{1.5\times10^{8}}{3.0\times10^{5}}$
2. Divide the coefficients: $\frac{1.5}{3.0} = 0.5$
3. Subtract the exponents: $8 - 5 = 3$
4. Combine the result: 0.5×10^{3}
5. Write in proper scientific notation: $0.5 \times 10^{3} = 5 \times 10^{2}$

Square Roots (Non-Perfect and Perfect)

The Square Root of a number is the value that, when multiplied by itself, gives the original number.

Perfect Square Roots:

A perfect square is a number that is the result of squaring an integer (whole number). So, a perfect square root is the square root of a perfect square. For example, $\sqrt{81} = 9$.

Non-Perfect Square Roots:

A non-perfect square is a number that is not the result of squaring a whole number. When you take the square root of a non-perfect square, the result is usually a decimal or a radical (a number with a square root symbol). For example, $\sqrt{5} \approx 2.236$.

Estimating Square Roots of Non-Perfect Squares:

To estimate the square root of a non-perfect square, you can use the two perfect squares it falls between. For example, to estimate $\sqrt{50}$:

$\sqrt{49} = 7$ and $\sqrt{64} = 8$, Since 50 is closer to 49 than to 64, you can estimate that $\sqrt{50} \approx 7.1$.

Properties of Square Roots:

1. **Square Root of a Product:** The square root of a product is the same as the product of the square roots of the individual numbers:

$$\sqrt{a \times b} = \sqrt{a} \times \sqrt{b}$$

2. **Square Root of a Quotient:** The square root of a fraction (quotient) is the same as the square root of the numerator divided by the square root of the denominator.

$$\sqrt{\frac{a}{b}} = \frac{\sqrt{a}}{\sqrt{b}}$$

3. **Square Root of a Square:** The square root of a number squared is the original number (but the answer will be positive, because the square root is always positive).

$$\sqrt{a^2} = |a|$$

Example:

Estimate the square root of 72 and then simplify its square root.

Solution:

Estimate $\sqrt{72}$: $\sqrt{64} = 8$ and $\sqrt{81} = 9$, Since 72 is closer to 64 than 81, we can find the square of $8.1, 8.2, 8.3, 8.4$ and 8.5 to find the closest to 72. So $\sqrt{72} \approx 8.4$

Simplify $\sqrt{72}$: $\sqrt{72} = \sqrt{36 \times 2} = \sqrt{36} \times \sqrt{2} = 6 \times \sqrt{2} = 6\sqrt{2}$

Cube Roots of Positive and Negative Perfect Cubes

The cube root of a number is a value that, when multiplied by itself three times (cubed), gives the original number. We represent it as $\sqrt[3]{x}$.

Positive Perfect Cubes:

A positive perfect cube is a positive number that results from cubing an integer. For example: $1^3 = 1$, $2^3 = 8$, $3^3 = 27$, $4^3 = 64$ and so on.

To find the cube root of a positive perfect cube, we determine which number, when cubed, equals the original number. For instance: $\sqrt[3]{8} = 2$ because $2^3 = 8$, $\sqrt[3]{27} = 3$ because $3^3 = 27$.

Negative Perfect Cubes:

A negative perfect cube is a negative number resulting from cubing a negative integer. For example: $(-1)^3 = -1$, $(-2)^3 = -8$, $(-3)^3 = -27$, $(-4)^3 = -64$ and so on.

To find the cube root of a negative perfect cube, we follow the same principle but consider the negative sign. For instance: $\sqrt[3]{-8} = -2$ because $(-2)^3 = -8$, $\sqrt[3]{-27} = -3$ because $(-3)^3 = -27$.

Properties of Cube Roots:

Single Real Value: Unlike square roots, which have both positive and negative solutions, cube roots of real numbers have a single real value.

Cube Root of a Product: The cube root of a product is the product of the cube roots.

$$\sqrt[3]{a \times b} = \sqrt[3]{a} \times \sqrt[3]{b}$$

Cube Root of a Cube: The cube root of a power is the power of the cube root.

$$\sqrt[3]{a^3} = (\sqrt[3]{a})^3 = a$$

Cube Root of a Quotient: The cube root of a quotient follows the principle that the cube root of a fraction is the fraction of the cube roots.

$$\sqrt[3]{\frac{a}{b}} = \frac{\sqrt[3]{a}}{\sqrt[3]{b}}$$

Example:

Simplify $\sqrt[3]{-125}$.

Solution:

Rewrite -125 as a cube: $-125 = (-5) \times (-5) \times (-5)$ therefore, $-125 = (-5)^3$

Simplify the cube root: $\sqrt[3]{-125} = \sqrt[3]{(-5)^3} = -5$ (consider $\sqrt[3]{a^3} = a$ property)

Approximating Cube Roots

Similar to approximating non-perfect square roots, estimating cube roots for non-perfect cubes involves using a few strategies to get close to the actual value.

Steps to Approximate Cube Roots:

1. **Identify Nearest Perfect Cubes**
   - Find the two perfect cubes between which your number lies.
   - Example: To estimate $\sqrt[3]{50}$, note that 27 (3^3) and 64 (4^3) are the nearest perfect cubes.
2. **Estimate the Interval**
   - Establish the interval: $3 < \sqrt[3]{50} < 4$.
3. **Linear Interpolation**
   - Use linear interpolation to make a more accurate guess.
   - Example: Since 50 is closer to 64 than to 27, estimate a value closer to 4.
4. **Refine Using Trial and Error**
   - Try numbers within your interval and adjust as needed.
   - Example:
     - $3.5^3 = 42.875$ (less than 50)
     - $3.6^3 = 46.656$ (less than 50)
     - $3.7^3 = 50.653$ (slightly more than 50)
   - Conclude that $\sqrt[3]{50} \approx 3.7$

Example:

Approximate $\sqrt[3]{150}$.

Solution:

1. Identify perfect cubes: $125(5^3)$ and $216(6^3)$.
2. Estimate interval: $5 < \sqrt[3]{150} < 6$.
3. Trial and error:

   Since 150 is closer to 125 than to 216, estimate a value closer to 5:
   - $5.2^3 = 140.608$ (less than 150)
   - $5.3^3 = 148.877$ (less than 150)
   - $5.4^3 = 157.464$ (more than 150)
   - Conclude that $\sqrt[3]{150} \approx 5.3$

Simplifying Cube Root Radical Expressions

To simplify cube root radical expressions (similar to square root radical expressions), you need to break down the number inside the cube root into its prime factors and identify perfect cubes. Here's a step-by-step guide to simplifying cube root expressions:

Step-by-Step Guide to Simplifying Cube Roots:

For a simple cube root: If the number inside the cube root is a perfect cube, take the cube root. For example ($\sqrt[3]{1000} = 10$).

For Non-Perfect Cube:

1. Break down the number inside the cubic root into its prime factors.

   For example: $72 = 2^3 \times 3^2$.

2. Identify any perfect cubes (a perfect cube is a number that can be written as a^3).

   2^3 is a perfect cube.

3. Take the cube root of the perfect cube part and leave the rest inside the cube root.

   $\sqrt[3]{72} = \sqrt[3]{2^3 \times 3^2} = 2 \times \sqrt[3]{9}$.

4. The simplified form is $2\sqrt[3]{9}$.

For a Cube Root of a Fraction:

1. Separate the cube root of the numerator and denominator.

   For example: $\sqrt[3]{\frac{27}{8}} = \frac{\sqrt[3]{27}}{\sqrt[3]{8}}$

2. Simplify the cube roots of the numerator and denominator separately.

   $\sqrt[3]{27} = 3$ and $\sqrt[3]{8} = 2$

3. The simplified expression is $\frac{3}{2}$.

Example: Simplify $\sqrt[3]{\frac{96}{343}}$.

Solution:

1. Simplify $\sqrt[3]{343}$: 343 is a perfect cube, since $343 = 7^3$. Therefore, $\sqrt[3]{343} = 7$.
2. Simplify $\sqrt[3]{96}$: Break 96 into its prime factors: $96 = 2^5 \times 3$.

   Take out the perfect cube from 2^5: $\sqrt[3]{96} = \sqrt[3]{2^3 \times 2^2 \times 3} = 2 \times \sqrt[3]{4 \times 3} = 2 \times \sqrt[3]{12}$

3. Put it all together: Now, substitute these simplified cube roots back into the original expression: $\frac{\sqrt[3]{96}}{\sqrt[3]{343}} = \frac{2 \times \sqrt[3]{12}}{7}$

Adding and Subtracting Radical Expressions

Adding and subtracting radical expressions (including square roots and cube roots) follow specific rules. The key idea is to look for like terms — terms that have the same radicand (the number inside the radical). Here's a guide for both square roots and cube roots:

General Process for Adding/Subtracting Radical Expressions:

1. **Simplify Each Radical Expression:**
   - Before adding or subtracting, simplify each radical expression if possible (e.g., reduce square roots or cube roots to their simplest form).
2. **Look for Like Terms:**
   - You can only add or subtract radical expressions with the same radicand and the same index. For square roots, this means the numbers inside the square roots must be the same. For cube roots, the numbers inside the cube roots must be the same.
   - If the terms have the same radicand and index, you can combine them by adding or subtracting their coefficients (the numbers outside the radical).
3. **If the Radicals Are Not Like Terms:**
   - If the terms don't have the same radicand, you cannot combine them. The expression will remain separate.

Examples:

1) Simplify the expression: $\sqrt[3]{16} + 3\sqrt[3]{16} - 5\sqrt[3]{16}$.

   Solution:

   Both terms have the same radicand ($\sqrt[3]{16}$), so we can add and subtract the coefficients:

   $(1 + 3 - 5)\sqrt[3]{16} = -\sqrt[3]{16}$

2) Simplify the expression: $4\sqrt[3]{81} - \sqrt{12} + 2\sqrt{3} - \sqrt[3]{3}$.

   Solution:

   1. Simplify $\sqrt[3]{81}$ (since $81 = 3^4$): $\sqrt[3]{81} = \sqrt[3]{3^4} = 3\sqrt[3]{3}$
   2. Simplify $\sqrt{12}$ (since $12 = 3 \times 2^2$): $\sqrt{12} = \sqrt{3 \times 2^2} = 2\sqrt{3}$
   3. Rewrite the expression: $4\sqrt[3]{81} - \sqrt{12} + 2\sqrt{3} - \sqrt[3]{3} = 4(3\sqrt[3]{3}) - 2\sqrt{3} + 2\sqrt{3} - \sqrt[3]{3}$
   4. Simplify like terms:

      $4(3\sqrt[3]{3}) - 2\sqrt{3} + 2\sqrt{3} - \sqrt[3]{3} = (12 - 1)\sqrt[3]{3} + (-2 + 2)\sqrt{3} = 11\sqrt[3]{3}$

   So, the final result is: $4\sqrt[3]{81} - \sqrt{12} + 2\sqrt{3} - \sqrt[3]{3} = 11\sqrt[3]{3}$

Multiplying and Dividing Radical Expressions

Multiplying Radical Expressions

Same Index (Same Root Type): As we learned in the previous sections, when multiplying two radicals with the same index (like square roots or cube roots), you multiply the numbers inside the radicals and then take the radical of the product.

Different Indices: If the radicals have different indices, this may involve rationalizing the radical or rewriting them to have the same root. For example, if you wanted to multiply $\sqrt{2}$ and $\sqrt[3]{4}$ you'd need to rewrite them with a common index (perhaps using exponents), But in general, direct multiplication across different roots doesn't apply unless they share the same index.

Dividing Radical Expressions

Same Index: As we discussed in the earlier sections, when dividing radicals with the same index, you divide the numbers inside the radicals and then simplify the result.

Rationalizing the Denominator: If you have a radical in the denominator, it's often necessary to rationalize the denominator by multiplying both the numerator and denominator by the conjugate (if it's a binomial) or by the radical itself (if it's a single radical).

- Example with square roots: $\frac{1}{\sqrt{2}} \times \frac{\sqrt{2}}{\sqrt{2}} = \frac{\sqrt{2}}{2}$
- Example with cube roots: $\frac{1}{\sqrt[3]{2}} \times \frac{\sqrt[3]{4}}{\sqrt[3]{4}} = \frac{\sqrt[3]{4}}{\sqrt[3]{8}} = \frac{\sqrt[3]{4}}{2}$

Examples:

1) Simplify the expression $\sqrt[3]{15} \times \sqrt[3]{24}$.

   Solution:

   Using the multiplying property, we mentioned before ($\sqrt[3]{a} \times \sqrt[3]{b} = \sqrt[3]{a \times b}$) we have:

   $\sqrt[3]{15} \times \sqrt[3]{24} = \sqrt[3]{360} = \sqrt[3]{2^3 \times 3^2 \times 5} = \sqrt[3]{2^3} \times \sqrt[3]{3^2 \times 5} = 2 \times \sqrt[3]{3^2 \times 5}$

   So, the final result is: $\sqrt[3]{15} \times \sqrt[3]{24} = 2\sqrt[3]{45}$

2) Simplify by rationalizing the denominator: a) $\frac{1}{\sqrt{7}}$ b) $\frac{2}{\sqrt[3]{9}}$

   Solution a:

   Multiply both the numerator and the denominator by $\sqrt{7}$: $\frac{1}{\sqrt{7}} \times \frac{\sqrt{7}}{\sqrt{7}} = \frac{\sqrt{7}}{7}$, so $\frac{1}{\sqrt{7}}$ is $\frac{\sqrt{7}}{7}$

   Solution b:

   To rationalize $\frac{2}{\sqrt[3]{9}}$, we must eliminate the cube root from the denominator. Since $9 = 3^2$

   We multiply both the numerator and the denominator by $\sqrt[3]{3}$: $\frac{2}{\sqrt[3]{9}} \times \frac{\sqrt[3]{3}}{\sqrt[3]{3}} = \frac{2\sqrt[3]{3}}{\sqrt[3]{27}} = \frac{2\sqrt[3]{3}}{3}$

Real World Applications

Exponents and radical expressions might seem abstract at first, but they have many real-world applications. Here are some examples:

1. **Population Growth**: Exponential growth models describe how populations grow over time. For instance, the bacteria population doubling every hour is an example of exponential growth.
2. **Technology**: Computer memory and processing power often grow exponentially, meaning they double in capacity over regular intervals.
3. **Science**: Scientific notation, which involves exponents, is used to express very large or very small numbers, like the distance between stars or the size of microscopic organisms.
4. **Physics**: Calculating the distance traveled by an object under uniform acceleration often involves square roots.
5. **Architecture and Engineering**: Determining the dimensions of materials, such as the diagonal length of a piece of wood or metal, can involve square roots.
6. **Engineering**: Understanding the relationships between different dimensions in three-dimensional objects, like in design and manufacturing processes, often involves cube roots.
7. **Chemistry**: In calculations involving molar volumes and gas laws, cube roots can be used to determine specific dimensions and properties.

Example:

A city has been experiencing rapid population growth. The population of the city is modeled by a cubic growth formula, which states that the population P (in thousands) of the city at time t years can be expressed as: $P(t) = 0.5t^3$

where t is the number of years since the city was founded.

After 10 years, the city plans to build a new community center, but it needs to ensure that the population is at least 40,000 to justify the project. The city planners are trying to determine when the population will reach 40,000. How many years after the city's founding will the population reach 40,000 people?

Solution:

1. To find when the population will reach 40,000, we set $P(t) = 40$ (since the population is measured in thousands): $40 = 0.5 \times t^3$
2. Solve for t^3: $\frac{40}{0.5} = t^3 \rightarrow 80 = t^3$
3. No, take the cube root of both sides: $t = \sqrt[3]{80} \approx 4.31$ years.

So, the population will reach 40,000 people approximately 4.31 years after the city's founding.

Worksheets

Integer Exponents

Simplify following expressions:

1) $7^{10} \times 7^0 \times 7^{-9} \times 7^4$
2) $\frac{a^{27} \div a^{17}}{a^{12} \times a^{-2}}$
3) $x^4 \times (x^2y^3)^2 \times y^{-1}$
4) $15^{-7} \times 3^{10} \times 4^{-7} \times 20^{10}$
5) $\frac{(ab)^4 \div (ab^2)^4}{(b^{-2})^{-1}}$
6) $(64)^{4^2} \times (128)^{5^2} \times (16^4)^{10}$

If $2^a = k$, find the result of the following expressions in terms of k:

7) 4^{a+2}
8) 8^{a-1}
9) 2^{5a+3}
10) $(\frac{1}{32})^{2-a}$

Negative, Fraction and Decimal Bases

Simplify.

1) $5^3 \times 0.2^{-4} \times (\frac{1}{25})^{-2}$
2) $(\frac{3}{7})^{-9} \times (-\frac{9}{49})^4 \times (\frac{14}{6})^5$
3) $(\frac{4}{5})^7 \times (\frac{16}{25})^8 \times (-\frac{12}{15})^9$
4) $\frac{(0.125)^{11} \times (0.0625)^{-20}}{(0.25)^{-15} \times (-2)^{-50}}$
5) $\frac{x^3(-y)^5(xy^2)^{-3}}{(xy)^{-3}x^{-2}}$
6) $\frac{(0.25)^{-3} \times (-16)^3 \times 3^3}{(\frac{3}{4})^{-5} \times 3^8}$

If $2^x = 3$ and $3^y = 2$ find the result of the following expressions:

7) $(2^{3x})^y$
8) 4^{xy+2}
9) 24^{xy+1}
10) $\frac{2^{x+2}}{9^{xy-3}}$

Scientific Notation

Convert numbers to scientific notation:

1) 582000000
2) 0.00002316
3) 82.36×10^{-5}
4) $7700 \div 10^3 \times 0.1^2$
5) $1400000 \div 0.001^5 \times 100$

Adding and Subtracting Scientific Notations

Write the result of the following expression in scientific notation:

1) $2.346 \times 10^{-3} + 1.3447 \times 10^{-3}$
2) $7.81 \times 10^8 - 5.002 \times 10^7$
3) $8.03 \times 10^3 + 9.066 \times 10^4$
4) $3.01 \times 10^5 - 2.98 \times 10^4$
5) $4.1 \times 10^9 + 1.054 \times 10^{-7} - 2.14 \times 10^3$

Multiplying and Dividing Scientific Notations

Write the result of the following expression in scientific notation:

1) $1.31 \times 10^{-3} \times 6.87 \times 10^8$
2) $9.18 \times 10^7 \div 3.06 \times 10^{-5}$
3) $4.002 \times 10^2 \times 7.12 \times 10^4 \times 1.1 \times 10$
4) $12.03 \times 10^{-7} \div 2.005 \times 10^{-1}$
5) $3.12 \times 10^{12} \times 5.78 \times 10^3 \div 1.80336 \times 10^2$

Scientific Notation (Estimate Large or Small Quantities)

Do following word problem about scientific notation:

1) The distance from Earth to the nearest star, Proxima Centauri, is about 424×10^{11} kilometers. Convert this number into scientific notation.
2) The mass of a proton is approximately $0.00000000000000000000000000167$ kilograms. Express this mass in scientific notation.
3) The Earth's mass is approximately 5.972×10^{24} kilograms, and the mass of a hydrogen atom is roughly 1.67×10^{-27} kilograms. Estimate how many hydrogen atoms would be needed to equal the mass of the Earth. Provide your answer in scientific notation.
4) A grain of salt has a mass of approximately 0.1 grams, and it is composed mostly of sodium chloride ($NaCl$). Using the atomic masses of sodium ($23\ u$) and chlorine ($35.5\ u$), estimate how many atoms are in a single grain of salt. Express your answer in scientific notation.
5) The elementary charge (the charge of a single electron) is approximately 1.6×10^{-19}coulombs. If you have a total charge of 1 coulomb, estimate how many electrons would be needed to make up this charge. Express your answer in scientific notation.

Square Roots (Non-Perfect and Perfect)

Estimate the square root of following numbers and then simplify their square root.

1) $\sqrt{125}$
2) $\sqrt{800}$
3) $\sqrt{5^2 + 12^2}$
4) $\sqrt{(1-\sqrt{2})^2}$
5) $\sqrt{\frac{0.0004}{10^6}}$

How many natural numbers satisfy in the following inequalities?

6) $6 < \sqrt{x} < 13$
7) $2 < \sqrt{x} < 7$
8) $-6 < -\sqrt{x} < -1$
9) $1 < \sqrt{x-1} < 10$
10) $10 < \sqrt{x^2+2} < 30$

Cube Roots of Positive and Negative Perfect Cubes

Calculate.

1) $\sqrt[3]{10^6}$
2) $\sqrt[3]{\sqrt{64}}$
3) $\sqrt[3]{x^9 \times 27}$
4) $\sqrt[3]{-y} \times \sqrt[3]{y^5}$
5) $\sqrt[3]{\frac{-1}{125 \times 0.008}}$
6) $\sqrt[3]{\frac{z^6 \times 0.27}{-10}}$
7) $\frac{\sqrt[3]{-4} \times \sqrt[3]{10}}{\sqrt[3]{320000}}$
8) $\frac{\sqrt[3]{0.729}}{-\sqrt[3]{0.027}}$
9) $\sqrt[3]{\frac{343 \times x^3}{-y^{12} \times z^9}}$
10) $\sqrt[3]{\frac{-a^{-3} \times 64}{(xy^6)^{-3}}}$

Approximating Cube Roots

Estimate the cube roots of the following numbers to one decimal place:

1) 180
2) 200
3) 66
4) 41
5) 350

Simplifying Cube Root Radical Expressions (Including Fractions)

Simplify following cube root expressions:

1) $\sqrt[3]{625}$
2) $\sqrt[3]{-80}$
3) $\sqrt[3]{3000}$
4) $\sqrt[3]{-\frac{81}{125}}$
5) $\sqrt[3]{-7^8}$
6) $\frac{\sqrt[3]{32}}{\sqrt[3]{-4000}}$
7) $\sqrt[3]{0.343 \times 2^5}$
8) $\sqrt[3]{\frac{-8}{7^{10}}}$
9) $\frac{\sqrt[3]{108}}{\sqrt[3]{0.008}}$
10) $\frac{\sqrt[3]{75}}{\sqrt[3]{-81}}$

Adding and Subtracting Radical Expressions

Simplify following expressions:

1) $2\sqrt[3]{2} - \sqrt[3]{2} + 5\sqrt[3]{2}$
2) $4\sqrt[3]{16} - 3\sqrt[3]{54} + \sqrt[3]{108}$
3) $-4\sqrt[3]{3} + 2\sqrt[2]{3} - 8\sqrt[3]{3} - 7\sqrt[2]{3}$
4) $3\sqrt{12} - 2\sqrt[3]{16} - \sqrt[3]{32}$
5) $\sqrt{50} - 2\sqrt[3]{125} - \sqrt[3]{54}$
6) $\sqrt[3]{24} - 5\sqrt[3]{3} + 10\sqrt[3]{81}$
7) $\sqrt[3]{2000} - 5\sqrt{4000}$
8) $\sqrt[3]{128} - 4\sqrt{45} + 2\sqrt{125}$
9) $\sqrt{98} - \sqrt[3]{108} + \sqrt[3]{64} + \sqrt[3]{32}$
10) $4\sqrt[3]{54} - \sqrt{24} + \sqrt[3]{81} - \sqrt[3]{375} + \sqrt{54}$

Multiplying and Dividing Radical Expressions

Simplify following expressions:

1) $\sqrt[3]{-2^7} \times \sqrt[3]{5^4}$
2) $\frac{\sqrt[3]{18} \times \sqrt[3]{6}}{\sqrt[3]{-9000}}$
3) $\sqrt{45} \times \sqrt{0.05}$
4) $\frac{\sqrt{27}+\sqrt{12}}{\sqrt{3}}$
5) $\frac{\sqrt[3]{75} \times \sqrt[3]{-9}}{\sqrt[3]{-81}}$

Simplify following expressions by rationalizing the denominator.

6) $\frac{1}{\sqrt{6}}$
7) $\frac{5}{\sqrt[3]{3}}$
8) $\frac{9\sqrt{5}}{\sqrt{10}}$
9) $\frac{4}{\sqrt[3]{9}}$
10) $\sqrt[3]{\frac{7}{25}}$

Real World Applications

Do the following problems:

1) If a cubic bottle can hold 8 liters of water, what is the length of each side of the cube in centimeters?
2) A square garden has an area of 144 square meters, calculating the side length.

3) A computer represents integers in binary format. The maximum integer that can be stored in an n-bit signed integer format is given by $2^{n-1} - 1$.
   - Calculate the maximum integer that can be stored in a 32-bit signed integer format.
   - If you have a 256-bit signed integer format, express the maximum integer that can be stored using radical expressions.
4) The distance to stars is often measured in light years. The nearest star system, Alpha Centauri, is approximately 4.367×10^{13} kilometers away from Earth.
   - Convert this distance to meters.
   - Express the distance in terms of square root and cube root.
5) The power (P) dissipated in an electrical resistor is given by $P = \frac{V^2}{R}$, where V is the voltage across the resistor and R is the resistance. If a resistor with a resistance of $50\,\Omega$ has a voltage drop of $230\,V$ across it, calculate the power dissipated by the resistor.
6) The rate of mutation in a particular species of frog is linked to environmental radiation levels. Researchers have found that the mutation rate M increases with the square of the radiation level R, according to the equation: $M = kR^2$
   If the mutation rate is 16 when the k is 8, find the value of R.
7) The safety of the school playground is being tested. The amount of rubber mulch needed to cover the circular playground can be approximated by the formula: $V = \pi r^2 h$
   where r is the radius of the playground and h is the depth of mulch. If the volume of mulch needed is 2,500 m^3 and the mulch is to be spread to a depth of 0.3 m, find the radius of the playground.
8) The value of an investment grows exponentially according to the formula $V = P(1 + r)^t$, where P is the initial amount, r is the annual growth rate, and t is the number of years. If an initial investment of 1,000 dollars grow at an annual rate of 5%, calculate the value of the investment after 10 years.

Answer of Worksheets

Integer Exponents

1) 7^5
2) 1
3) x^8y^5
4) 60^3
5) $\frac{1}{b^6}$
6) 2^{431}
7) $16k^2$
8) $\frac{k^3}{8}$
9) $8k^5$
10) $\frac{k^5}{2^{10}}$

Negative, Fraction and Decimal Bases

1) 5^{11}
2) $(\frac{7}{3})^6$
3) $-(\frac{4}{5})^{32}$
4) 2^{67}
5) $-y^2x^5$
6) -2^8
7) 8
8) 64
9) 576
10) 972

Scientific Notation

1) 5.82×10^8
2) 2.316×10^{-5}
3) 8.236×10^{-4}
4) 7.7×10^{-2}
5) 1.4×10^{23}

Adding and Subtracting Scientific Notations

1) 3.6907×10^{-3}
2) 7.3098×10^8
3) 9.869×10^4
4) 2.712×10^5
5) $\approx 4.1 \times 10^9$

Multiplying and Dividing Scientific Notations

1) 9.0037×10^5
2) 3×10^{12}
3) $\approx 3.14 \times 10^8$
4) 6×10^{-6}
5) 1.001×10^{14}

Scientific Notation (Estimate Large or Small Quantities)

1) 4.24×10^{13} km
2) 1.67×10^{-24}kg
3) 3.57×10^{51}
4) 2.06×10^{21}
5) 6.25×10^{18}electrons.

Square Roots (Non-Perfect and Perfect)

1) $\sqrt{125} = 5\sqrt{5} \approx 11.2$
2) $\sqrt{800} = 20\sqrt{2} \approx 28.3$
3) 13
4) $|1-\sqrt{2}| = \sqrt{2} - 1 \approx 0.4$
5) $\sqrt{\frac{0.0004}{10^6}} = 2 \times 10^{-5}$
6) 132 natural numbers.
7) 44 natural numbers.
8) 34 natural numbers.
9) 98 natural numbers.
10) 20 natural numbers.

Cube Roots of Positive and Negative Perfect Cubes

1) 10^2
2) 2
3) $3x^3$
4) $-y^2$
5) -1

6) $-\frac{3z^2}{10}$
7) $\frac{-1}{20}$
8) -3
9) $-\frac{7x}{y^4 \times z^3}$
10) $-\frac{4xy^6}{a}$

Approximating Cube Roots

1) 5.6
2) 5.8
3) 4.0
4) 3.4
5) 7.0

Simplifying Cube Root Radical Expressions (Including Fractions)

1) $5 \times \sqrt[3]{5}$
2) $-2 \times \sqrt[3]{10}$
3) $10 \times \sqrt[3]{3}$
4) $-\frac{3 \times \sqrt[3]{3}}{5}$
5) $49\sqrt[3]{49}$
6) $-\frac{1}{5}$
7) $1.4\sqrt[3]{4}$
8) $\frac{-2}{343 \times \sqrt[3]{7}}$
9) $15 \times \sqrt[3]{4}$
10) $\frac{\sqrt[3]{25}}{3}$

Adding and Subtracting Radical Expressions

1) $6\sqrt[3]{2}$
2) $-\sqrt[3]{2} + 3\sqrt[3]{4}$
3) $-12\sqrt[3]{3} - 5\sqrt[2]{3}$
4) $6\sqrt{3} - 4\sqrt[3]{2} - 2\sqrt[3]{4}$
5) $5\sqrt{2} - 10 - 3\sqrt[3]{2}$
6) $27\sqrt[3]{3}$
7) $10\sqrt[3]{2} - 100\sqrt{10}$
8) $4\sqrt[3]{2} - 2\sqrt{5}$
9) $7\sqrt{2} - \sqrt[3]{4} + 4$
10) $12\sqrt[3]{2} + \sqrt{6} - 2\sqrt[3]{3}$

Multiplying and Dividing Radical Expressions

1) $-20\sqrt[3]{10}$
2) $\frac{-3}{10} \times \sqrt[3]{\frac{4}{9}}$
3) $\frac{3}{2}$
4) 5
5) $\sqrt[3]{\frac{25}{3}}$
6) $\frac{\sqrt{6}}{6}$
7) $\frac{5\sqrt[3]{9}}{3}$
8) $\frac{9\sqrt{2}}{2}$
9) $\frac{4\sqrt[3]{3}}{3}$
10) $\frac{\sqrt[3]{35}}{5}$

Real World Applications

1) 20 centimeters.
2) 12 meters.
3) The maximum integer that can be stored in a 32-bit signed integer format: $2^{31} - 1$. And the maximum integer that can be stored in a 256-bit using radical expressions: $\sqrt[3]{2^{765}} - 1$
4) 4.367×10^{16} meters. Square root: $(2.09 \times 10^8)^2$, cube root: $(1.62 \times 10^{5.33})^3$
5) 1058 watts.
6) $\sqrt{2}$
7) $\frac{50}{\sqrt{0.3\pi}}$ meters.
8) $\approx$ 1,628.89 dollars.

Chapter 4: Proportions and Percents

Topics that you'll learn in this chapter:

- ✓ Equations of Proportional Relationships
- ✓ Proportional Relationships in Tables
- ✓ Proportional Relationships on Graphs
- ✓ Proportional Relationships in Real-World Problems
- ✓ Percents Using Proportions
- ✓ Simple Interest
- ✓ Percent Increases and Decreases
- ✓ Calculating Percent Change
- ✓ Calculating the Original Amount
- ✓ Real World Applications
- ✓ Worksheets
- ✓ Answer of Worksheets

Equations of Proportional Relationships

Definition to Proportional Relationship:

A proportional relationship is a relationship between two quantities where the ratio (or quotient) between them is constant. In other words, when one quantity changes, the other changes at a consistent rate. For example, "If you buy 2 apples for $1, 4 apples would cost $2. The cost is proportional to the number of apples"

The proportional relationship can be represented with an equation, a graph, or a table.

Equation of Proportional Relationships:

Definition: A proportional relationship between two variables x and y can be expressed by the equation $y = kx$,

Where: y is the dependent variable (the one that depends on the other), x is the independent variable (the one you control) and k is the constant of proportionality.

For example, $y = 3x$ is a proportional relationship because it's in the form $y = kx$, where $k = 3$. But $y = 2x + 1$ is not proportional because the equation is not in the form $y = kx$; it has an additional constant $(+1)$.

Examples:

1) Write an equation that represents the proportional relationship where y is 4 times x.

   Solution:

   This means that for any value of x, y will be four times that value. Using the formula for proportional relationships $y = kx$, substitute the value of k (which is 4): $y = 4x$

   For example:

   - If $x = 1$, then $y = 4 \times 1 = 4$.
   - If $x = 2$, then $y = 4 \times 2 = 8$.

   Both cases satisfy the condition that y is 4 times x. Therefore, the equation that represents the proportional relationship where y is 4 times x is: $y = 4x$.

2) The speed of a car is proportional to the time it takes to travel a fixed distance. If the car travels 60 miles in 0.5 hour, write an equation that represents the relationship between speed s and time t.

   Solution:

   The speed of a car (s) is proportional to the time (t) it takes to travel a fixed distance. The car travels 60 miles in 0.5 hour. Speed (s) can be calculated using the formula:

   $s = \frac{distance}{time} = \frac{60\ miles}{0.5\ hour} = 120$ mile per hour, so the constant of proportionality k is 120.

   Using the formula for proportional relationships, where $s = k \cdot t$: $s = 120t$

Proportional Relationships in Tables

If you're given a table of values for x and y, you can check if the relationship is proportional by calculating the ratio of y to x for each pair. In a proportional relationship, the ratio $\frac{y}{x}$ (also called the constant of proportionality k) will be the same for all pairs of values.

Steps to Check:

1. For each row in the table, divide y by x (i.e., $\frac{y}{x}$).
2. If the ratio is the same for every row, the relationship is proportional.
3. If the ratio changes between rows, the relationship is not proportional.

For example:

x	y
1	2
2	4
3	6
4	8

For this table $\frac{2}{1} = 2$, $\frac{4}{2} = 2$, $\frac{6}{3} = 2$ and $\frac{8}{4} = 2$, Since the ratio is the same (2) for each pair, the relationship is proportional.

Example:

Determine which of the following tables is a proportion table and which one is not.

a)

x	3	12	1.5	9
y	5	20	2.5	15

b)

x	1	2	3	4
y	1	4	9	16

Solution a:

For each column in the table, divide y by x:

$\frac{5}{3}, \frac{20}{12} = \frac{5}{3}, \frac{2.5}{1.5} = \frac{5}{3}, \frac{15}{9} = \frac{5}{3}$, since the ratio is same $(\frac{5}{3})$ for each pair, the relationship is proportional.

Solution b:

For each column in the table, divide y by x:

$\frac{1}{1}, \frac{4}{2} = \frac{2}{1}, \frac{9}{3} = \frac{3}{1}, \frac{16}{4} = \frac{4}{1}$, Since all the ratios are different $(\frac{1}{1} \neq \frac{2}{1} \neq \frac{3}{1} \neq \frac{4}{1})$, this relationship is not proportional.

Proportional Relationships on Graphs

As we mentioned before, a proportional relationship between two quantities means that they change at a constant rate. When you graph these relationships on a coordinate plane, you'll see a straight line that passes through the origin (0,0).

Key Characteristics of Proportional Relationships on Graphs:

1. **Straight Line Through the Origin:**
   - The graph of a proportional relationship will always be a straight line.
   - This line will always pass through the point (0,0). This is because if one quantity is zero, the other must also be zero in a proportional relationship.
2. **Constant Rate of Change:**
   - The slope of the line represents the constant rate of change or the "unit rate".
   - For example, if you have a proportional relationship where $y = 2x$, the slope is 2. This means for every increase of 1 in x, y increases by 2.
3. **Ratio of Coordinates:**
   - Similar to 2 different previous proportional relationships, the ratio of the y-coordinate to the x-coordinate $(\frac{y}{x})$ will be the same for every point on the line. This ratio is the same as the slope or unit rate.
   - For example, if one point on the line is $(1, 3)$, then the ratio is $\frac{3}{1} = 3$. If another point is (2,6), the ratio is $\frac{6}{2} = 3$, maintaining consistency.

Example:

Consider the proportional relationship described by the equation $y = \frac{3}{5}x$. Determine the slope of the line and graph the relationship.

Solution:

1. Determine the slope of the line: The equation of the line $y = \frac{3}{5}x$ is in the form $y = mx$, where m is the slope. The slope m is $\frac{3}{5}$. This means for every increase of 5 in x, y increase by 3.
2. Graph the relationship: You can plot a few points:
   - When $x = 0, y = 0$ (the origin)
   - When $x = 5, y = \frac{3}{5} \times 5 = 3$.
   - When $x = -5, y = \frac{3}{5} \times (-5) = -3$.

   Plot the points on a coordinate plane.

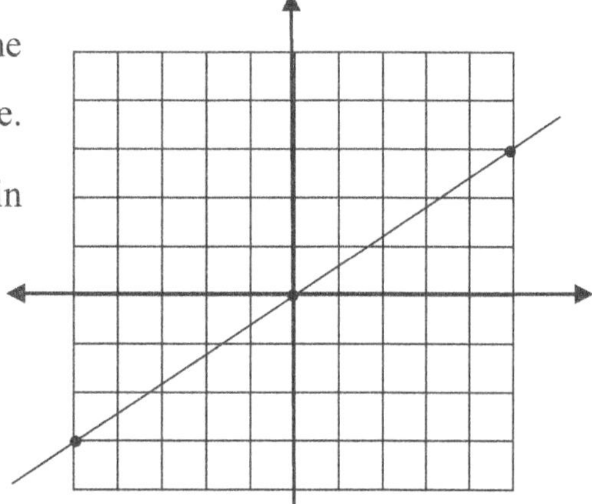

Proportional Relationships in Real-World Problems

Proportional relationships show up everywhere in real life! They help us understand how two quantities change in a predictable way. For example, when you're shopping, cooking, or even traveling, you'll find proportional relationships in action.

How to Recognize a Proportional Relationship in Real Life:

- Constant rate or ratio: In any real-world situation, if there's a consistent rate of change (like price per item, miles per hour, or ingredients per recipe), that's a good sign you're dealing with a proportional relationship.
- Straight line on a graph: If you were to graph the situation, it would be a straight line that passes through the origin $(0, 0)$, because when one quantity is zero, the other is also zero (e.g., no apples, no cost).

How You Could Solve Problems with Proportional Relationships:

When solving real-world problems with proportional relationships, you can:

- Use the constant of proportionality (the rate or ratio) to set up an equation.
- Use the equation to solve unknown quantities.

Example:

You are planning a hiking trip and looking at a map. The map has a scale of 1 centimeter representing 5 kilometers in real life. You want to hike from Point A to Point B. On the map, the distance between Point A and Point B is 7.2 centimeters.

- What is the actual distance between Point A and Point B in real life?
- If your average hiking speed is 4 kilometers per hour, how long will it take you to hike from Point A to Point B?

Solution:

1. Calculate the actual distance:
    - The scale of the map is 1 centimeter = 5 kilometers.
    - The distance on the map between Point A and Point B is 7.2 centimeters.
    - To find the actual distance, multiply the distance on the map by the scale factor:

      Actual distance = $7.2cm \times \frac{5km}{cm} = 36\ km$
2. Calculate the hiking time:
    - The actual distance between Point A and Point B is 36 kilometers.
    - Your average hiking speed is 4 kilometers per hour.
    - To find the hiking time, divide the actual distance by your average speed:

      Hiking time $= \frac{36km}{4km/hour} = 9\text{hours}$

Percents Using Proportions

As we know, a percent is a way of expressing a number as a fraction of 100. The word "percent" means "per hundred," so when we say 50%, it means 50 out of 100.

Using Proportions to Solve Percent Problems:

We can use proportions to solve problems involving percents. Here's a simple step-by-step approach:

Step 1: Set Up the Proportion: For any percent problem, we can set up a proportion in the form: $\frac{part}{whole} = \frac{percent}{100}$

Step 2: Fill in the Known Values: Identify the known values in the problem and fill them into the proportion. One of these values will be the percent, and the other values will be part and whole.

Step 3: Solve for the Unknown: Cross-multiply and solve for the unknown value.

Examples:

1) 30 is what percent of 50?

   Solution:

   1. Set up the proportion: $\frac{part}{whole} = \frac{percent}{100}$
   2. Fill in the known values: Part: 30, whole: 50 and percent:?

      So, the proportion becomes: $\frac{30}{50} = \frac{x}{100}$
   3. Solve for the unknown cross-multiply:

      $50x = 30 \times 100 \rightarrow 50x = 3{,}000 \rightarrow x = \frac{3{,}000}{50} = 60$; So, 30 is 60% of 50.

2) In a class of 40 students, 30 percent of the class is very interested in painting. How many students are not interested in painting?

   Solution:

   1. Set Up the Proportion: $\frac{part}{whole} = \frac{percent}{100}$
   2. Fill in the known values: Percent interested in painting: 30%, whole (total number of students): 40, so, the proportion becomes: $\frac{x}{40} = \frac{30}{100}$
   3. Solve for the Unknown (number of students interested in painting) Cross-multiply and solve for x: $100x = 30 \times 40 \rightarrow 100x = 1200 \rightarrow x = \frac{1200}{100} = 12$
   4. Calculate the number of students not interested in painting:

      Total number of students $= 40$; number of students interested in painting $= 12$

      Number of students not interested in painting: $40 - 12 = 28$

Simple Interest

Simple interest is a quick way to calculate interest on a loan or investment based on the original amount of money (principal), the interest rate, and the time period involved.

Formula for Simple Interest: The formula to calculate simple interest is:

$$Simple\ interest = P \times R \times T$$

Where:

- P is the principal amount (the initial amount of money)
- R is the Rate of interest per year (expressed as a decimal)
- T is the Time period the money is invested or borrowed for (in years)

Examples:

1) Let's say you invest $\$500$ in a savings account with a simple interest rate of 4% per year for 3 years. How much interest will you earn?

   Solution:

   1. Identify the values:
      - Principal (P): $\$500$
      - Rate (R): $4\%\ or\ 0.04$ (as a decimal)
      - Time (T): 3 years
   2. Plug the values into the formula:

      $Simple\ interest = P \times R \times T = 500 \times 0.04 \times 3 = 60$
   3. Calculate total amount: To find the total amount of money after the interest is added, simply add the interest to the principal:

      $Total\ amount = principal + simple\ interest = 500 + 60 = 560$

      Therefore, after 3 years, you will have $\$560$ in total.

2) John borrowed $\$1,200$ to buy a new laptop. After 2 years, he had to pay back $\$1,320$ in total. What was the annual interest rate?

   Solution:

   1. Identify the values:
      - Principal (P): $\$1,200$
      - Time (T): 2 years
      - Total amount (A): $\$1,320$
      - Simple interest: (I): $Total\ amount - principal = 1,320 - 1,200 = 120$
   2. Plug the values into the formula:

      $Simple\ interest = P \times R \times T \rightarrow 120 = 1,200 \times R \times 2 \rightarrow R = \frac{120}{2,400} = 0.05$

      S0, the annual interest rate is 5%.

Percent Increases and Decreases

Percent increases and decreases is all about understanding how to calculate changes in quantities using percentages.

Understanding Percent Increases: A percent increase happens when a quantity grows by a certain percentage. To calculate the new amount after a percent increase, you can follow these steps:

1. **Calculate the Increase:** Use the formula:

$$Increase = \frac{Percent\ Increase}{100} \times Original\ Amount$$

2. **Find the New Amount:** Add the increase to the original amount:

$$New\ Amount = Original\ Amount + Increase$$

Understanding Percent Decreases: A percent decrease happens when a quantity reduces by a certain percentage. To calculate the new amount after a percent decrease, you can follow these steps:

1. **Calculate the Decrease:** Use the formula:

$$Decrease = \frac{Percent\ Decrease}{100} \times Original\ Amount$$

2. **Find the New Amount:** Subtract the from the original amount:

$$New\ Amount = Original\ Amount - Decrease$$

Examples:

1) A shirt originally costs $\$50$. The price increases by 10%. What is the new price of the shirt?

   Solution:

   1. Calculate the increase: $Increase = \frac{10}{100} \times 50 = 0.1 \times 50 = 5$
   2. Finde the new amount: $New\ price = 50 + 5 = 55$

   So, the new price of the shirt is $\$55$.

2) A pair of shoes originally costs $\$80$. The price decreases by 15%. What is the new price of the shoes?

   Solution:

   1. Calculate the decrease: $Decrease = \frac{15}{100} \times 80 = 0.15 \times 80 = 12$
   2. Find the new amount: $New\ price = 80 - 12 = 68$

   So, the new price of the shoes is $\$68$.

Calculating Percent Change

Percent change is a way to express how much a quantity has increased or decreased in comparison to its original value. It's commonly used in various real-world contexts like finance, population studies, and sales.

Formula for Percent Change:

$$Percent\ Change = \frac{New\ Value - Original\ Value}{Original\ Value} \times 100$$

Here's a breakdown of the steps to calculate percent change:

Find the Difference:

- Subtract the original value from the new value to determine the amount of change.
- $Difference = New\ Value - Original\ Value$

Divided by the Original Value:

- Divide the difference by the original value to find the proportional change.
- $Proportional\ Change = \frac{Difference}{Original\ Value}$

Multiply by 100:

- Multiply the proportional change by 100 to convert it to a percentage.
- $Percent\ Change = Proportional\ Change \times 100$

Examples:

1) A laptop's price increased from \$800 to \$900. What is the percent increase?

   Solution:

   1. Find the difference: $Difference = 900 - 800 = 100$
   2. Divide by the original value: Proportional cjange $= \frac{100}{800} = 0.125$
   3. Multiply by 100: $Percent\ change = 0.125 \times 100 = 12.5\%$

   So, the laptop's price increased by 12.5%.

2) The population of a town decreased from 1,200 to 1,000. What is the percent decrease?

   Solution:

   1. Find the difference: $Difference = 1{,}000 - 1{,}200 = -200$
   2. Divide by the original value: Proportional cjange $= \frac{-200}{1{,}200} = -0.1667$
   3. Multiply by 100: $Percent\ change = -0.1667 \times 100 = -16.67\%$

   So, the laptop's price decreased by 16.67%.

Calculating the Original Amount

This typically involves working backwards from a final amount to find out what you started with. This can come up in a variety of contexts, like discounts, taxes, or interest rates. Here are a couple of common scenarios:

1. **Discounts:**

Imagine you bought something at a discounted price, and you want to know the original price before the discount was applied.

If you know the final price (F) after a discount and the discount percentage (D), you can use this formula: $Original\ Amount = \frac{F}{1-\frac{D}{100}}$

2. **Taxes:**

If you have the final price (F) including tax, and you know the tax rate (T), you can find the original price before tax. The formula is: $Original\ Amount = \frac{F}{1+\frac{T}{100}}$

Examples:

1) You bought a gadget for $80 after a 20% discount. What's the original price?

   Solution:

   1. Identify the values:
      - Final price (F): $80
      - Discount percentage (D): 20%
   2. Find the original value: Using the discount formula:

      $Original\ Amount = \frac{80}{1-\frac{20}{100}} = \frac{80}{\frac{80}{100}} = 80 \div \frac{80}{100} = 80 \times \frac{100}{80} = 100$

   So, the gadget originally cost $\$100$.

2) Sarah bought a piece of furniture and the final price, including tax, was $\$120$. The sales tax rate is 8%. Sarah wants to find out what the original price of the furniture was before the tax was added.

   Solution:

   3. Identify the values:
      - Final price (F): $120
      - Tax rate (T): 8%
   4. Find the original value: Using the discount formula:

      $Original\ Amount = \frac{120}{1+\frac{8}{100}} = \frac{120}{\frac{108}{100}} = 120 \div \frac{108}{100} = 120 \times \frac{100}{108} \approx 111.11$

   So, the original price of the furniture before tax was approximately $\$111.11$.

Real World Applications

Proportions and percents are incredibly useful in everyday life. Here are some real-world applications:

Shopping Discounts: Understanding how to calculate discounts during sales helps us find the best deals. For example, if an item cost $50 and is on sale for 30% off, students can use percents to determine the sale price.

Cooking and Recipes: When adjusting recipes, we can use proportions to scale ingredients up or down. If a recipe designed for 4 people need to serve 6, proportions help in calculating the right amount of each ingredient.

Travel and Maps: When planning trips, we can use proportions to estimate travel time and distance. For example, if a map scale shows that 1 inch represents 10 miles, they can use this proportion to find the real distance between two points.

Financial Literacy: Understanding interest rates and loan terms is essential for financial planning. we can use percents to calculate interest on savings accounts or loans.

Sports Statistics: Analyzing player performance in sports often involves proportions and percents. For example, calculating the shooting percentage in basketball helps in understanding a player's effectiveness.

Architecture and Design: When designing a space, proportions are used to create models and ensure that dimensions are accurate. This skill is essential for future architects and engineers.

Health and Fitness: Tracking progress in fitness goals can involve proportions and percents. For example, if a student wants to increase their running distance by 25%, they can use these concepts to set realistic goals.

Example:

You have a job during summer where you earn $12 per hour. You decide to save 40% of your earnings each week. If you work for 8 weeks and work 25 hours per week, how much money will you have saved by the end of the summer?

Solution:

1. Calculate your weekly earnings: $Weekly\ earnings = 12 \times 25 = \$300$
2. Calculate your total earnings for 8 weeks: $Total\ earnings = 300 \times 8 = \$2{,}400$
3. Calculate the amount you save each week: You save 40% of your weekly earnings. $40\% = 0.40 \rightarrow weekly\ savings = 0.4 \times 300 = \$120$
4. Calculate the total amount saved by the end of summer: You save $120 each week for 8 weeks. $Total\ savings = 120 \times 8 = \$960$

Worksheets

Equations of Proportional Relationships

Write an equation that represents the relationship between two values:

1) If a car travel 150 miles in 3 hours at a constant speed, write the equation that represents the proportional relationship between distance traveled (d) and time (t).
2) A painting company charges \$3 per square foot for painting a wall. Write the equation that represents the proportional relationship between the cost (C) and the area (A) of the wall.
3) A company produces widgets, and the cost of production is directly proportional to the number of widgets produced. If the cost to produce 250 widgets is \$1,500 write the equation that represents the relationship between the cost (C) and the widgets (W).
4) A bakery sells cookies, and the revenue generated from selling cookies is directly proportional to the number of cookies sold. If the bakery generates \$60 in revenue by selling 200 cookies, write the equation that represents the relationship between the revenue (R) and the number of cookies sold (C).
5) A chemist is preparing a solution and needs to maintain a constant concentration of a substance. If 5 grams of the substance are dissolved in 200 milliliters of solvent. Write the equation that represents the relationship between the gram of substance (g) and the milliliter of solvent (m).

Proportional Relationships in Tables

Determine which of the following tables is a proportion table and which one is not.

1)

x	1	3	2	10
y	5	15	10	50

2)

x	12	22	1.2	2
y	16	26	1.6	6

3)

x	18	6	12	27
y	-20	$-\frac{20}{3}$	$-\frac{40}{3}$	-30

4)

x	2.5	3.5	2	5
y	5.5	6.5	5	11

5)

x	1	2	3	4
y	1	8	27	64

Proportional Relationships on Graphs

Determine the slope of the following lines and graph the relationships.

1) $y = -2x$
2) $y = \frac{2}{3}x$
3) $4y = 3x$
4) $x = -\frac{5}{4}y$

Write an equation that represents the relationship between x and y in following graphs:

5)

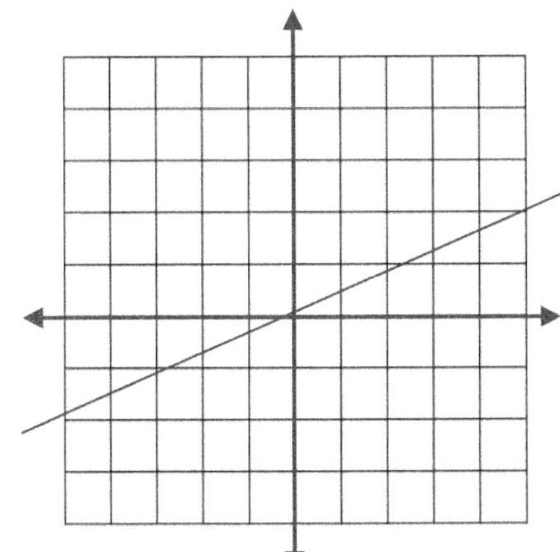

6)

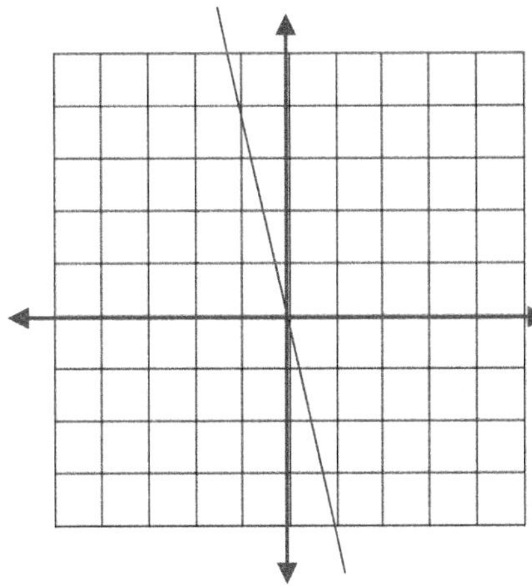

7)

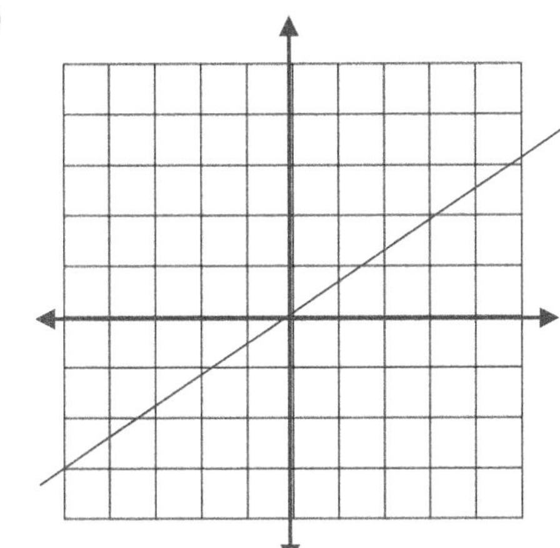

8)

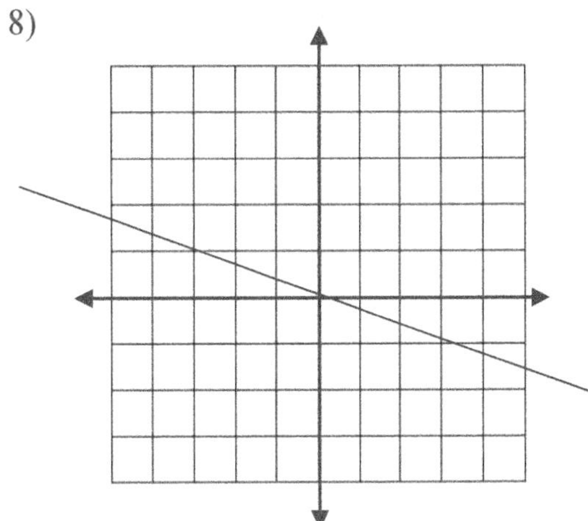

Proportional Relationships in Real-World Problems

Find the answers.

1) An artist is creating a scale model of a building. If the building is 120 feet tall and the model is to be built at a scale of 1 inch to 10 feet, how tall will the model be in inches?
2) A worker earns $15 per hour. If they worked 35 hours in a week, how much would they earn in 45 hours?
3) A map has a scale of 1: 50,000, meaning 1 centimeter on the map represents 50,000 centimeters in reality. If the distance between two cities on the map is 5 centimeters, what is the actual distance between the cities in kilometers?
4) To make a type of orange paint, mix yellow and red paint in a 3 to 7 ratio. If we need 80 grams of orange paint, how much of each yellow and red paint should we mix?
5) A sawmill machine can cut a 10 −meter log into 6 pieces in half an hour. How many cuts can this machine make in 1.5 hours?

Percents Using Proportions

Do following percent problems using proportion:

1) What percent is $\frac{2.8}{7}$?

2) About 0.05% of the people have AB blood type. If the population of a country is 10 million, approximately how many of them have AB blood type?
3) What is 0.25% of a 1-kilometer distance in meters?
4) The distance from John's house to the school is about 1.2 kilometers. He travels 800 meters of this distance with his father every day. What percentage of this distance does John travel alone?
5) In a multiple-choice test, each correct answer is worth 3 points, and 1 point is deducted for each incorrect answer. If in a 20 −question test, you answer 12 questions correctly and 6 questions incorrectly, what will be the test score as a percentage?

Simple Interest

Do the problems.

1) You invest $500 in a savings account with a simple interest rate of 4% per year. How much interest will you earn in 1 year?
2) Sam invested $4,000 in a savings account. After 3 years, the interest earned was $720. What was the annual simple interest rate?
3) John invests $1,000 in a fixed deposit for 3 years at a simple interest rate of 5% per year. How much interest will he earn by the end of the 3 years, and what will be the total amount in the account?
4) Maria takes out a loan of $2,500 at a simple interest rate of 6% per year. If she repays the loan after 2 years, how much interest will she have paid, and what is the total amount she needs to repay?
5) A loan of $5,500 was taken out, and after 5 years, the total amount repaid was $7,150. Calculate the annual simple interest rate.

Percent Increases and Decreases

Solve problems.

1) A shirt originally costs $20. If the price is increased by 10%, what is the new price of the shirt?
2) A town's population was 12,000 people. After a year, the population decreased by 8%. What is the new population of the town?
3) A worker earns $40,000 per year. After a 12% raise, what is their new annual salary?
4) The value of a house was $250,000 last year. This year, its value increased by 12%. Next year, it is expected to decrease by 5%. What will be the value of the house next year?
5) An investment portfolio was worth $50,000 at the beginning of the year. During the first half of the year, it increased by 15%. However, during the second half of the year, it decreased by 8%. What is the value of the investment portfolio at the end of the year?

Calculating Percent Change

Do the following problems.

1) The price of a book increased from $15 to $18. Calculate the percent increase in the price of the book.
2) Last year, a smartphone model was sold for $600. This year, the price dropped to $540. Calculate the percent decrease in the price of smartphone.
3) The city's population was 150,000 five years ago. Due to migration, the population decreased to 135,000. Calculate the percent decrease in the city's population over the five years.

4) An investor's portfolio was valued at $50,000 at the beginning of the year. By mid-year, the portfolio's value increased to $57,500. However, by the end of the year, it decreased to $52,000. Calculate the percent change from the beginning to mid-year and from mid-year to the end of the year.
5) A person weighed 180 pounds at the beginning of the year. After six months, they now weigh 162 pounds. What is the percent decrease in their weight?

Calculating the Original Amount

Find the answers.

1) Sarah bought a dress on sale for $36, which was 20% off the original price. What was the original price of the dress?
2) After a 10% increase, the cost of a book is $22. What was the original price of the book?
3) After losing 12% of their body weight, a person weighs 140 pounds. What was their original weight?
4) A city's population decreases by 12% in the first year and another 10% in the second year. The population after two years is 72,000. What was the original population before the decrease?
5) An investment grows to $12,000 after 3 years with an annual interest rate of 8%, compounded yearly. What was the original investment amount?

Real World Applications

Do the following problems:

1) If we mix 100 liters of 96% alcohol with 140 liters of 72% alcohol, what will be the percentage of the resulting alcohol?
2) If the price of an item has tripled compared to last year, what is the percent increase in the price?
3) David and Mary's money is equal. If David gives $\frac{3}{4}$ of his money to Mary, what is the ratio of Mary's money to David's money?
4) An employee's salary decreased by 20%. By what percentage should the salary be increased to return to the original salary?
5) If we mix 20 kilograms of $20 tea with 30 kilograms of $30 tea, what will be the price per kilogram of the mixed tea?
6) To 10 grams of an alloy of copper and gold in a 1 to 3 ratios, how many grams of copper should be added to change the alloy ratio to 2 to 5?
7) What single percentage gain is equivalent to three consecutive gains of 10%, 20%, and 50%?
8) A library had 40 members, and 60% of them were girls. After some time, 10 more boys joined the library. What percentage of the members are girls now?
9) A company that manufactures children's pools reduced each side of a square-shaped pool by 25% to make the side 12 meters long. What was the area of this pool in its original state?
10) What percentage of 75% of the volume of a container is equal to 25% of its volume?

Answer of Worksheets

Equations of Proportional Relationships

1) $d = 50t$
2) $C = 3A$
3) $C = 6W$
4) $R = \frac{3}{10}C$
5) $g = \frac{1}{40}m$

Proportional Relationships in Tables

1) Proportional, $y = 5x$
2) Disproportional
3) Proportional, $y = -\frac{10}{9}x$
4) Disproportional
5) Disproportional

Proportional Relationships on Graphs

1) $y = -2x$, slope$= -2$

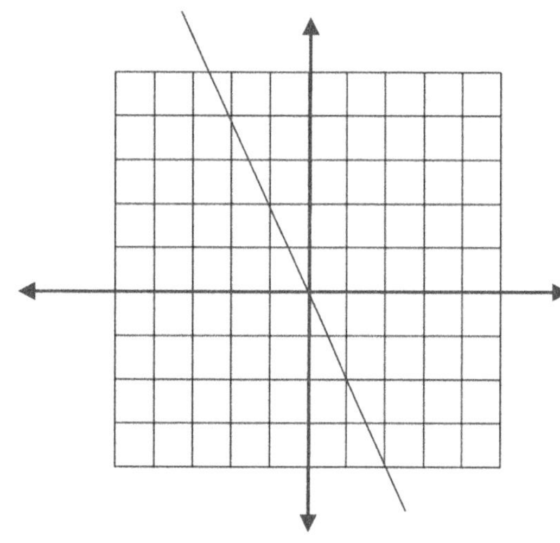

2) $y = \frac{2}{3}x$, slope$= \frac{2}{3}$

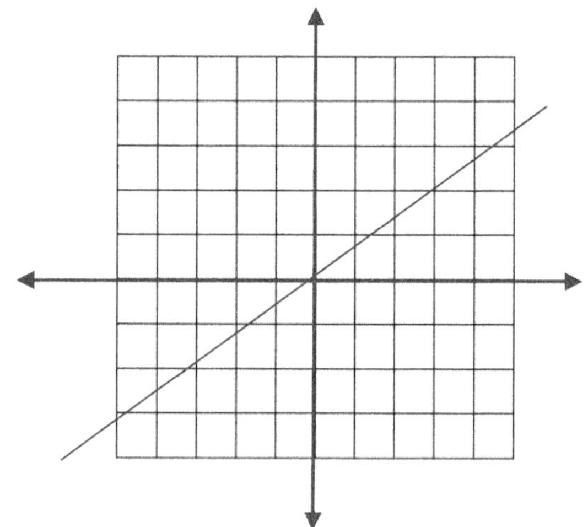

3) $4y = 3x$, slope$= \frac{3}{4}$

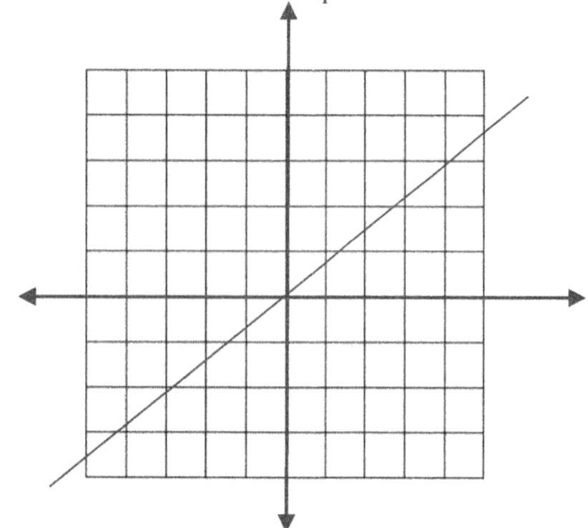

4) $x = -\frac{5}{4}y$, slope$= -\frac{5}{4}$

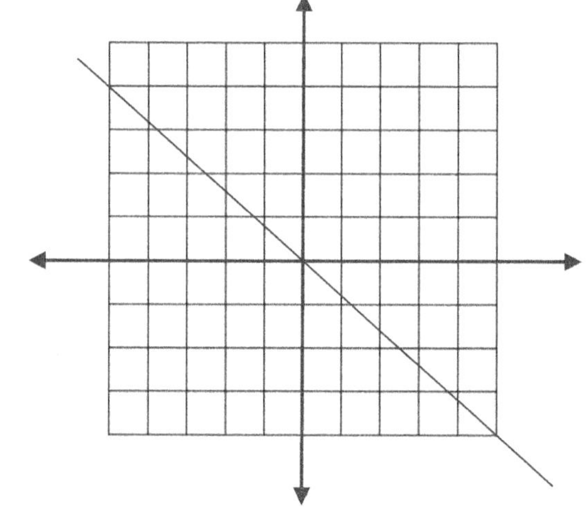

5) $y = \frac{1}{2}x$
6) $y = -4x$
7) $y = \frac{2}{3}x$
8) $y = -\frac{1}{3}x$

Proportional Relationships in Real-World Problems

1) 12 inches tall.
2) $675.
3) 2.5 km.
4) 26 grams.
5) 15 cuts.

Percents Using Proportions

1) 40%.
2) 5,000 people.
3) 2.5 meters.
4) Approximately 33.33%.
5) 50%.

Simple Interest

1) $20.
2) 6%.
3) $1,150.
4) $2,800.
5) 6%.

Percent Increases and Decreases

1) $22.
2) 11,040 people.
3) $44,800.
4) $266,000.
5) $52,900.

Calculating Percent Change

1) 20%
2) 10%
3) 10%
4) Beginning to mid-year: 15% increase, and mid-year to end: 9.75% decrease
5) 10%

Calculating the Original Amount

1) 45$
2) 20$
3) Approximately 159.09 pounds
4) Approximately 90,909 people
5) Approximately $9,528.32

Real World Applications

1) Approximately 82% alcohol
2) 200%
3) 7: 1
4) 25%
5) $26/kg
6) 0.5 grams
7) 98%
8) 48%
9) 256 square meters
10) Approximately 33.33%

Chapter 5: Algebraic Expressions

Topics that you'll learn in this chapter:

- ✓ Write Variable Expressions
- ✓ Simplifying and Combining Like Terms
- ✓ Distributive Property
- ✓ Distributive Property - Area Models
- ✓ Multiplying Binomials
- ✓ Multiplying Other Polynomials
- ✓ Factoring Using Distributive Property
- ✓ Factoring by Grouping
- ✓ Factoring Quadratics
- ✓ Factoring by Sorting GCF
- ✓ Evaluate Radical Expressions
- ✓ Evaluate Rational Expressions
- ✓ Real World Applications
- ✓ Worksheets

Write Variable Expressions

Variable expressions are mathematical phrases that can include numbers, variables (letters that represent unknown values), and operations (addition, subtraction, multiplication, and division). Understanding how to write and interpret these expressions is essential for solving algebraic problems. Here's how to do it step-by-step for phrases with two or three operations:

Key Steps:

1. **Identify the Variables**: Look for words that describe an unknown quantity (e.g., "a number," "some quantity").
2. **Recognize the Operations**: Find the mathematical operations described in the phrase (e.g., addition, subtraction, multiplication, division).
3. **Translate the Words to Symbols**: Convert the words into algebraic symbols and operations.
4. **Follow the Order of Operations**: Ensure that the expression follows the correct mathematical order (PEMDAS rules: Parentheses, Exponents, Multiplication and Division, Addition and Subtraction).

Examples:

1) Write an expression for "twice a number y plus five."

   Solution:

   1. Identify the variable: A number y
   2. Recognize the operations: Multiplication and addition.
   3. Translate the words to symbols: Multiply the variable y by 2 $(2y)$ then add 5 to the result $(2y + 5)$
   4. Write the expressions: $2y + 5$

2) Write an expression for "Nine divided by the difference of three times a number and seven."

   Solution:

   1. Identify the variable: Let's use x to represent "a number."
   2. Recognize the operations: Multiplication, division and subtraction
   3. Translate the words to symbols:
      - Three times a number: $3x$
      - The difference between three times a number and seven: $3x - 7$
      - Nine divided by the difference: $9 \div (3x - 7)$
   4. Write the expression: $\frac{9}{3x-7}$

Simplifying and Combining Like Terms

Simplifying and combining like terms is an essential skill in algebra that helps make expressions easier to work with.

Key Concepts:

1. **Terms:** In an algebraic expression, the components that are divided by addition $(+)$ or subtraction $(-)$ symbols are called terms. For example, in the expression $5y^2 - 2x + 1$, the terms are $5y^2$, $-2x$ and 1.
2. **Coefficients:** A coefficient refers to a number that is used to multiply a variable. For example, in the terms $5y^2$ the coefficient is 5 and the variable is y^2.
3. **Like Terms:** Terms that have the same variable(s) raised to the same power(s). The coefficients can be different.
   - Example of like terms: $3x$ and $5x$; $\frac{a^3}{4}$ and $-a^3$.
   - Example of unlike terms: $2x$ and $5y$; $4y^2$and $4y$

Steps to Simplify and Combine Like Terms:

1. **Identify Like Terms:** Look for terms that have the same variable(s) and exponent(s).
2. **Group Like Terms Together:** Rearrange the expression, if necessary, so that like terms are next to each other.
3. **Combine the Coefficients**: Add or subtract the coefficients of the like terms while keeping the variable part the same.
4. **Write the Simplified Expression:** Combine all the like terms to get the simplified expression.

Example:

Simplify the expression $-2x^3 + y^2 - 4.5x^2 - 6y^2 + \frac{2}{5}x^3$.

Solution:

- Identify like terms: $-2x^3$ and $\frac{2}{5}x^3$; y^2 and $-6y^2$; $-4.5x^2$.
- Group like terms: $-2x^3 + \frac{2}{5}x^3, y^2 - 6y^2$, and $-4.5x^2$.
- Combine coefficients and write simplified expression:

  $-2x^3 + y^2 - 4.5x^2 - 6y^2 + \frac{2}{5}x^3 = -2x^3 + \frac{2}{5}x^3 + y^2 - 6y^2 - 4.5x^2 = \left(-2 + \frac{2}{5}\right)x^3 + (1 - 6)y^2 - 4.5x^2 = -\frac{8}{5}x^3 - 5y^2 - 4.5x^2$

Distributive Property

The distributive property is a fundamental concept in algebra that allows you to multiply a single term by each term inside a set of parentheses. This property is very useful when simplifying expressions and solving equations and in general, this property states that:

$$a(b + c) = ab + ac$$

Distributive Property with Numbers: This distributive property means that you multiply the number outside the parentheses by each term inside the parentheses and then add the results. For example, $-3(2a - b) = -3 \times 2a + (-3) \times -b = -6a + 3b$

Distributive Property with Variables: This property allows you to distribute the term outside the parentheses to each term inside the parentheses.

For example, $2x(x - 4y) = 2x \cdot x - 2x \cdot y = 2x^2 - 8xy$

Examples:

1) Simplify the expression $\frac{2}{3}(-6a^2 + 2a - 9) + 2(a + 3a^2)$.

   Solution:

   - Distribute $\frac{2}{3}$ to each term inside the first set of parentheses:

     $\frac{2}{3}(-6a^2 + 2a - 9) = \frac{2}{3} \times -6a^2 + \frac{2}{3} \times 2a + \frac{2}{3} \times -9 = -4a^2 + \frac{4}{3}a - 6$
   - Distribute 2 to each term inside the first set of parentheses: $2(a + 3a^2) = 2a + 6a^2$
   - Combine the two expressions: $-4a^2 + \frac{4}{3}a - 6 + 2a + 6a^2$
   - Combine like terms and write the simplified expression:

     $-4a^2 + 6a^2 + \frac{4}{3}a + 2a - 6 = 2a^2 + \frac{10}{3}a - 6$

2) Simplify the expression $-3x^2(4x + 2x^2 - 2) + x(-x^3 + x^2)$.

   Solution:

   - Distribute $-3x^2$ to each term inside the first set of parentheses:

     $-3x^2(4x + 2x^2 - 2) = -12x^3 - 6x^4 + 6x^2$
   - Distribute $x$to each term inside the first set of parentheses:

     $x(-x^3 + x^2) = -x^4 + x^3$
   - Combine the two expressions: $= -12x^3 - 6x^4 + 6x^2 - x^4 + x^3$
   - Combine like terms and write the simplified expression:

     $-12x^3 + x^3 - 6x^4 - x^4 + 6x^2 = -7x^4 - 11x^3 + 6x^2$

Distributive Property - Area Models

The area model is a visual way to represent the distributive property. Here are the general steps for using the area model to apply the distributive property to variable expressions:

General Steps for Using Area Model for Distributive Property (Variable Expressions):

1. **Break Down the Expression:** Write the expression inside the parentheses as a sum of two or more terms. For example, for the expression $a(b+c)$, break down $b+c$.
2. **Draw a Rectangle:** Draw a large rectangle to represent the entire product. Then, divide it into smaller rectangles that correspond to each term inside the parentheses.
3. **Label the Sides:**
   - Label one side of the large rectangle with the term outside the parentheses (e.g., a).
   - Label the other sides with each of the terms inside the parentheses (e.g., b and c).
4. **Calculate the Areas of Smaller Rectangles:**
   - Multiply the term outside the parentheses by each term inside the parentheses to get the areas of the smaller rectangles.
   - For the example $a(b + c)$, calculate $a \times b$ and $a \times c$.
5. **Sum the Areas:**

   Add up the areas of the smaller rectangles to get the final expression.

   For the example $a(b+c)$, the sum is $ab + ac$.
6. **Simplify (if needed):** Combine like terms, if any, to simplify the expression.

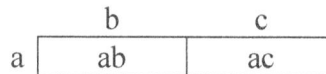

	b	c
a	ab	ac

Example:

Simplify $-2x(3x^2 + 4x)$ using an area model.

Solution:

1) Break down the expression: The expression $3x^2 + 4x$ can be thought of as two separate parts: $3x^2$ and $4x$.
2) Set up the area model: Draw a rectangle and split it into two parts: one for $3x^2$ and one for $4x$. Then label the widths: $-2x \times 3x^2$ and $-2x \times 4x$.
3) Calculate the areas of the smaller rectangles: First part: $-2x \times 3x^2 = -6x^3$. Second part: $-2x \times 4x = -8x^2$.
4) Add the areas together: $= -6x^3 - 8x^2$.

	$3x^2$	$4x$
$-2x$	$-6x^3$	$-8x^2$

Multiplying Binomials

Multiplying binomials is a key concept in algebra that involves multiplying two binomial expressions together. A binomial is a polynomial with exactly two terms.

Multiplying Binomials:

When you multiply binomials, you're multiplying two expressions that each have two terms. It can feel tricky at first, but we can break it down step by step.

1. **Identify the Binomials:** Let's say you have two binomials: $(a + b)$ and $(c + d)$.
2. **Apply the Distributive Property:** You'll distribute each term in the first binomial to each term in the second binomial. For $(a + b)(c + d)$, it looks like this:
   - $a \times c$
   - $a \times d$
   - $b \times c$
   - $b \times d$
3. **Write Out the Product:** Combine all the products: $ac + ad + bc + bd$
4. **Combine Like Terms (if any):** Check if there are any like terms to combine and simplify the expression if necessary.

The FOIL Method:

Another way to remember how to multiply binomials is by using the FOIL method, which stands for First, Outer, Inner, Last. This method helps you remember the steps of multiplying binomials.

- First: Multiply the first terms in each binomial.
- Outer: Multiply the outer terms in the binomials.
- Inner: Multiply the inner terms in the binomials.
- Last: Multiply the last terms in each binomial.

Example: Multiply the binomials $(2x + 3)(x - 4)$:

Solution:

- First: $2x \times x = 2x^2$
- Outer: $2x \times -4 = -8x$
- Inner: $3 \times x = 3x$
- Last: $3 \times -4 = -12$

Combine these results: $2x^2 - 8x + 3x - 12$

Simplify by combining like terms: $2x^2 - 5x - 12$

So, $(2x + 3)(x - 4) = 2x^2 - 5x - 12$

Multiplying Other Polynomials

Multiplying other polynomials involves extending the distributive property to expressions that have more than two terms. This includes multiplying binomials with trinomials, trinomials with trinomials, and so on.

General Method for Multiplying Polynomials:

1. **Distribute Each Term:** Start by distributing each term in the first polynomial to every term in the second polynomial. You'll multiply each term in the first polynomial by each term in the second polynomial.
2. **Write Out Each Product:** Write out the product of each pair of terms. Ensure you keep track of all the terms separately.
3. **Combine Like Terms:** Identify and combine any like terms in the resulting expression to simplify it. Like terms are terms that have the same variables raised to the same powers.
4. **Arrange in Standard Form (optional):** It's often useful to arrange the final expression in standard form, which means ordering the terms by descending powers of the variables.

Example: Multiply $(-x+2-x^2)$ by (x^2+3x+4).

Solution:

1. Distribute each term:
   - $-x$ distribute to x^2, $3x$, and 4.
   - 2 distribute to x^2, $3x$, and 4.
   - $-x^2$ distribute to x^2, $3x$, and 4.
2. Write out each product:
   - $-x \times x^2 = -x^3$
   - $-x \times 3x = -3x^2$
   - $-x \times 4 = -4x$
   - $2 \times x^2 = 2x^2$
   - $2 \times 3x = 6x$
   - $2 \times 4 = 8$
   - $-x^2 \times x^2 = -x^4$
   - $-x^2 \times 3x = -3x^3$
   - $-x^2 \times 4 = -4x^2$
3. Combine like terms: $-x^3 - 3x^2 - 4x + 2x^2 + 6x + 8 - x^4 - 3x^3 - 4x^2$
4. Simplify and arrange in standard form: $-x^4 - 4x^3 - 5x^2 + 2x + 8$

Factoring Using Distributive Property

Factoring using the distributive property is essentially the reverse of the distributive process. It involves identifying a common factor in the terms of a polynomial and then rewriting the expression as a product of that common factor and another polynomial. Here are the general steps to factor using the distributive property:

Steps to Factor Using Distributive Property

1. **Identify the Greatest Common Factor (GCF):** Find the greatest common factor among the terms in the polynomial.
2. **Rewrite the Expression:** Rewrite the polynomial as a product of the GCF and a polynomial. This often involves dividing each term by the GCF and writing the results within parentheses.
3. **Use the Distributive Property:** Apply the distributive property in reverse to factor out the GCF.

Examples:

1) Factor the polynomial $12x^3 - 18x^2 + 24x$.

   Solution:

   1. Identify the GCF: The GCF of $12x^3$, $-18x^2$ and $24x$ is $6x$.
   2. Rewrite the expression: Divide each term by $6x$:
      - $12x^3 \div 6x = 2x^2$
      - $-18x^2 \div 6x = -3x$
      - $24x \div 6x = 4$
   3. Factor using distributive property: Rewrite the expression as $6x(2x^2 - 3x + 4)$

   So, $12x^3 - 18x^2 + 24x$ factors to $6x(2x^2 - 3x + 4)$.

2) Simplify the expression $\frac{10a^2b^3-15b^2a^3}{25a^2+20a^3}$ by factoring.

   Solution:

   1. Identify the GCF:
      - The GCF of $10a^2b^3 - 15b^2a^3$ is $5a^2b^2$
      - The GCF of $25a^2 + 20a^3$ is $5a^2$
   2. Rewrite the numerator and denominator:
      - Numerator: Divide each term by $5a^2b^2$: $10a^2b^3 - 15b^2a^3 = 5a^2b^2(2b - 3a)$
      - Denominator: Divide each term by $5a^2$: $25a^2 + 20a^3 = 5a^2(5 + 4a)$
   3. Simplify: $\frac{10a^2b^3-15b^2a^3}{25a^2+20a^3} = \frac{5a^2b^2(2b-3a)}{5a^2(5+4a)} = \frac{b^2(2b-3a)}{(5+4a)}$

Factoring by Grouping

Definition: Factoring by grouping involves rearranging and grouping terms in a polynomial so that each group has a common factor and then factoring out the common factors from each group.

Steps for Factoring by Grouping:

Group Terms: Split the polynomial into two pairs of terms.

$ax + ay + bx + by$ can be grouped as $(ax + ay) + (bx + by)$

Factor Each Group: Factor out the GCF from each group.

$$(ax + ay) + (bx + by) \rightarrow a(x + y) + b(x + y)$$

Factor out the Common Binomial Factor: After factoring out the GCF from each group, you should have a common binomial factor.

$$a(x + y) + b(x + y) \rightarrow (a + b)(x + y)$$

Examples:

1) Factor $12x^3 + 18x^2 + 8x + 12$.

   Solution:

   1. Group terms:

      $(12x^3 + 18x^2) + (8x + 12)$

   2. Factor each group:

      $6x^2(2x + 3) + 4(2x + 3)$

   3. Factor out the common binomial factor:

      $(2x + 3)(6x^2 + 4)$

   So, the factored form of $12x^3 + 18x^2 + 8x + 12$ is $(2x + 3)(6x^2 + 4)$.

2) Factor $-x^4 - x^3 + 3x^2 + 3x$.

   Solution:

   1. Group terms:

      $(-x^4 - x^3) + (3x^2 + 3x)$

   2. Factor each group:

      $-x^3(x + 1) + 3x(x + 1)$

   3. Factor out the common binomial factor:

      $(x + 1)(-x^3 + 3x)$

      So, the factored form of $-x^4 - x^3 + 3x^2 + 3x$ is $(x + 1)(-x^3 + 3x)$

Factoring Quadratics

Factoring quadratics is the process of breaking down a quadratic expression into simpler binomial expressions. A quadratic expression is typically in the form:

$$ax^2 + bx + c$$

The goal of factoring is to rewrite this quadratic expression as a product of two binomials.

Steps to Factor a Quadratic:

1. **Identify the coefficients:** For a quadratic expression $ax^2 + bx + c$, identify a, b, and c.
2. **Find two numbers that multiply to ac (the product of a and c) and add up to b:** Look for two numbers that multiply to ac (if $a \neq 1$) and add to b.
3. **Split the middle term**: Rewrite the middle term bx using the two numbers found in step 2.
4. **Factor by grouping:** Group the terms in pairs and factor out the greatest common factor (GCF) from each pair.
5. **Final factoring:** You should now be able to factor the quadratic into two binomials.

Examples:

1) Factor $x^2 + 5x + 6$.

Solution:

1. Identify $a = 1, b = 5$ and $c = 6$.
2. Find two numbers that multiply to $ac = 1 \times 6 = 6$ and add up to $b = 5$. These numbers are 2 and 3.
3. Split the middle term: $x^2 + 5x + 6 = x^2 + 2x + 3x + 6$.
4. Group: $(x^2 + 2x) + (3x + 6)$.
5. Factor out the GCF from each group: $x(x + 2) + 3(x + 2)$.
6. Factor out the common binomial factor: $(x + 2)(x + 3)$.

So, the factorization of $x^2 + 5x + 6$ is $(x + 2)(x + 3)$.

2) Factor $2x^2 + 9x - 5$.

Solution:

1. Identify $a = 2, b = 9$ and $c = -5$.
2. Find two numbers that multiply to $ac = 2 \times -5 = -10$ and add up to $b = 9$. These numbers are 10 and -1.
3. Split the middle term: $2x^2 + 9x - 5 = 2x^2 + 10x - x - 5$.
4. Group: $(2x^2 + 10x) - (x + 5)$.
5. Factor out the GCF from each group: $2x(x + 5) - 1(x + 5)$.
6. Factor out the common binomial factor: $(x + 5)(2x - 1)$.

So, the factorization of $2x^2 + 9x - 5$ is $(x + 5)(2x - 1)$.

Factoring by Sorting GCF

As we learned about factoring, in algebra, factors are the numbers or expressions that multiply together to give a product. For example, the factors of 12 are $1, 2, 3, 4, 6, 12$. When we work with variable expressions, we look for factors in terms of variables and constants. For example, the factors of $6x^2$ are 6 and x^2.

Steps to Sort Factors of an Expression:

1. **Look for the greatest common factor (GCF):** To sort the factors, we first identify the greatest common factor (GCF) of the terms in the expression. For example, for $6x^2 + 9x$ the GCF is $3x$.
2. **Factor out the GCF:** Now, we can factor out the GCF from the expression:
$$6x^2 + 9x = 3x(2x + 3)$$
3. **Sort the factors:** So, the factors of are $3x$ and $(2x + 3)$.

☑ **Tip for Sorting Factors of an Expression:**

"Always look for what's common first!"

1. Find the GCF of all terms – numbers and variables.
2. Factor it out by dividing each term.
3. Rewrite the expression as GCF × (remaining terms).

Example: Sort factors of $3x^2 - x - 2$.

Solution:

1. Identify the coefficient: The expression is in the form $ax^2 + bx + c$, where: $a = 3, b = -1$ and $c = -2$.
2. Find two numbers that multiply to $a \times c$ and add up to b: $a \times c = -6$, the two numbers that work are 2 and -3 because $2 \times -3 = -6$ and $2 + (-3) = -1$.
3. Split the middle term: We can rewrite the middle term $-x$ as $2x - 3x$. So, the expression becomes: $3x^2 + 2x - 3x - 2$.
4. Factor by grouping: Now, group the terms in pairs and factor out the greatest common factor (GCF) from each pair:
$(3x^2 + 2x) - (3x + 2) = x(3x + 2) - 1(3x + 2)$
5. Final factoring: Now, factor out the common binomial factor
$(3x + 2)$: $(3x + 2)(x - 1)$ So, the factors of $3x^2 - x - 2$ are $(3x + 2)$ and $(x - 1)$

Evaluate Radical Expressions

Radical expressions are expressions that contain a radical symbol (√), which is used to denote the square root (or higher roots) of a number or variable.

Steps to Evaluate Radical Expressions with Variables:

Step 1: Identify the radical expression: A radical expression looks like this: $\sqrt{x-2}$ or $\sqrt[3]{a^2+5a}$.

Step 2: Substitute the value of the variable: When you are given an expression like $\sqrt{x+5}$ and told that $x = 4$, you substitute the value of x into the expression. For example, $\sqrt{x+5}$ Substituting $x = 4$: $\sqrt{4+5} = \sqrt{9}$.

Step 3: Simplify the radical: Now, simplify the square root. Since $\sqrt{9} = 3$, the value of the expression when $x = 4$ is 3.

Example: Evaluate $\sqrt{2x^3 - x^2 + 3}$ when $x = 1$.

Solution:

1. Substitute $x = 1$: $\sqrt{2(1)^3 - (1)^2 + 3}$
2. Simplify inside the square root: $\sqrt{2-1+3} = \sqrt{4} = 2$

Evaluate Rational Expressions

A rational expression is simply an expression that involves fractions where the numerator (top part) and/or the denominator (bottom part) is an algebraic expression.

Steps to Evaluate Rational Expressions:

Step 1: Identify the rational expression: A rational expression may look like: $\frac{3x+4}{2x-1}$.

Step 2: Substitute the values of the variables: When you are given a rational expression like $\frac{3x+4}{2x-1}$ and told that $x = 5$, you substitute the value of x into the expression. For example: $\frac{3x+4}{2x-1}$, Substituting $x = 5$: $\frac{3(5)+4}{2(5)-1} = \frac{19}{9}$.

Step 3: Simplify the Fraction if possible.

Example: Evaluate $\frac{x^2+5}{1-x}$ when $x = -2$.

Solution:

1. Substitute $x = 2$: $\frac{(-2)^2+5}{1-(-2)}$
2. Simplify the fraction: $\frac{4+5}{3} = \frac{9}{3} = 3$

Real World Applications

Algebraic expressions play a crucial role in various real-world applications. Here are some examples:

Shopping and Budgeting: Understanding discounts, sales tax, and total cost can be simplified using algebraic expressions.

Cooking and Baking: Recipes often need to be adjusted based on serving sizes. Algebra helps in scaling ingredients proportionally. For example, if a recipe calls for x cups of flour to serve 4 people, to serve 8 people you'd need $2x$ cups of flour.

Sports and Fitness: Calculating scores, averages, and statistics in sports often involve algebraic expressions.

Travel and Distance: Planning trips involves calculating distances, travel time, and fuel costs using algebra. For example, if a car travels at $x\ km/h$ for 3 hours, the distance covered is $3x\ km$.

Construction and Engineering: Algebra is used to calculate areas, volumes, and material requirements in construction projects.

Personal Finance: Managing savings, interest rates, and investments requires understanding algebraic expressions. For example, if you have x dollars in savings with an annual interest rate of 5%, after one year you'll have $x + 0.05x = 1.05x$ dollars.

Example:

Your school is organizing a fundraiser event. The cost of renting a venue is a fixed amount of \$500. Additionally, each ticket to the event costs \$8, and your school plans to sell tickets to raise money for the event. Write an algebraic expression for the total money raised from selling tickets and then solve it for 100 numbers of tickets sold.

Solution:

1. Define the variables:
    - Let x represent the number of tickets sold.
    - The fixed cost of renting the venue is \$500.
    - Each ticket costs \$8, so the revenue from selling x tickets is $8x$.
2. Write an algebraic expression for the total money raised: To find the total amount of money raised, we have to subtract the fixed cost from the revenue generated by selling tickets. Total money raised= $8x - 500$
3. Solve for 100 numbers of tickets sold: Substitute $x = 100$ into the expression:
   Total money raised= $8(100) - 500 = 300$

So, if 100 tickets are sold, the total money raised will be \$300.

Worksheets

Write Variable Expressions

Write a variable expression.

1) Three times a number increased by 5.
2) The difference between the square of a number and 10.
3) Half of the sum of a number and 12.
4) The product of 7 and the difference between twice a number and 3.
5) The quotient of the sum of a number and 9 divided by the product of 4 and the number.

Simplifying and Combining Like Terms

Simplify following algebraic expressions:

1) $4x^2 + 0.5x - x^2 + \frac{x}{2}$
2) $-xy + xy^2 + 2y^2x + 5yx - yx^2$
3) $m^2 - \frac{2nm}{5} + 1.5m^2 + mn$
4) $\frac{1}{4}(ab)^2 - ab^2 + 3a^2b^2 - 0.5b^2a + \frac{4}{5}ba^2$
5) $\frac{6x+12xy}{3} - \frac{-4yx+14y+8x}{2}$

Distributive Property (With Numbers, Variables)

Simplify following expressions using distributive property:

1) $-4(2xy + 5x) + 2(x - xy)$
2) $a(-2a + b) - 3ab + a^2 + 0.5b^2$
3) $-3x\left(-y + 2x + \frac{1}{3}x^2\right) - 1.2(5xy - \frac{5}{6}x^2)$
4) $-5xy^2 + x^2 - 4(0.2x^2 + xy^2 - \frac{3}{8}y^2)$
5) $a^2(-a + 3b - 2) - 4b(a^2 + 2) - (ba^2 - 2a^3)$

Distributive Property with Area Models

Simplify following expressions using area model:

1) $8(2a + b)$
2) $2x(y^3 - 5x)$
3) $-ab(a + 2b^2)$
4) $xy^2(3x - y + 5)$
5) $(-2b + 3)(a + 1)$

Multiplying Binomials

Simplify following expressions:

1) $(a - 1)(a + 1)$
2) $(-3x + 1)(x - 4)$
3) $(a^2 + 5a)(a - 5)$
4) $(-4x + y^2)(2y - x^2)$
5) $(2xy^2 + 5)(x - 2y^2)$
6) $-(3 - m^2)(3 + m^2)$
7) $(2x - 1)(x + 3) - (x - 3)(3x + 2)$
8) $(4x - y)(xy^2 - 1)$
9) $3ab^3 - (a - 3b^2)(2b - a)$
10) $(1 - x)(1 + x^2 - x) + x^3$

Multiplying Other Polynomials

Multiply following polynomials:

1) $(x + 1)(x^2 - 3x + 2)$
2) $(2x^2 + x + 1)(x - 1)$
3) $(2x + 3)(x^2 + x + 1)$
4) $(x^2 - 4)(x + 2)$
5) $(x + 5)(x^3 - 2x^2 + x - 4)$
6) $(3x^2 - x + 2)(2x^2 + x - 2)$

7) $(x^4 - x^3 + x^2)(x^2 - 1)$
8) $(2x^3 + x^2 - 2x + 1)(x^2 - 3x + 4)$
9) $(x^2 - 1)(x^2 + 1)(x^4 + 1)$
10) $(2x + 1)(4x^2 + 2x + 1)(x^3 - 1)$

Factoring Using Distributive Property

Factor following polynomials using distributive property:

1) $3ab + 2a$
2) $x^4 - x^3 + x^2$
3) $4a^4b^5 + 8a^5b^4$
4) $a^5 - a^4 + a^3 - a^2$
5) $16x^2y^3 - 24x^3y^2$
6) $\frac{45z^2x^4 - 9x^3z^3}{15x^2z^3}$
7) $6a^2b^2 + 8a^2 - 12b^3a^3 + 14b^2a^4$
8) $\frac{4x^2y^4z^6 - 6x^6y^4z^2 + 2x^4y^6z^2}{8x^3y^3z^2}$
9) $\frac{2a^3b - 6a^2b^2}{ab^3 - 3b^4}$
10) $\frac{15x^3y^2 - 3x^3y^3 + 6x^2y^2}{5x^2y^2 - x^2y^3 + 2xy^2}$

Factoring by Grouping

Factor following polynomials by grouping:

1) $ac - bc + ad - bd$
2) $4x - 4 + 2xy - 2y$
3) $a^2 + ab - ba - b^2$
4) $x^2 + xy - xz - yz$
5) $ba + 2a + 3b + 6$
6) $-a^2b^2 + a^3b + ab^3 - a^2b^2$
7) $6x^2 + 3x - 4x - 2$
8) $15a^2b - 5a^2 + 6b^2 - 2b$
9) $4xy - 4x - 2y + 2 + y^3 - y^2$
10) $-8x + 2 + 20x^2 - 5x - 4x^3 + x^2$

Factoring Quadratics

Factor.

1) $x^2 + x - 6$
2) $y^2 - 2y - 35$
3) $a^2 + 13a + 36$
4) $b^2 + 2b - 99$
5) $2x^2 + 3x - 2$
6) $9x^2 + 12x - 5$
7) $5a^2 + 19a - 4$
8) $9x^2 + 6x - 8$
9) $9y^2 + 9xy - 10x^2$
10) $16a^4 + 8a^2 - 35$

Factoring by Sorting GCF

Factor by sorting.

1) $6x^2 + 9x$
2) $10a^3b + 15a^2b^2$
3) $2x^2 + 5x - 3$
4) $x^3 + 6x^2 + x + 6$
5) $3x^3 - 6x^2 - 9x$

Evaluate Radical Expressions

Simplify the following radical expressions based on the given value.

1) $\sqrt{6x - x^2}, x = 2$
2) $\sqrt{7a^2 - 3a + 10}, a = -1$
3) $\sqrt{5x + 6} + \sqrt{2x}, x = 2$
4) $\sqrt{-y^3 + 5y^2 - 4}, y = -2$
5) $\sqrt[3]{\frac{2x+3}{-4x+4}}, x = -1$

Evaluate Rational Expressions

Simplify the following rational expressions based on the given value.

1) $\frac{3a-1}{2}, a = 5$
2) $\frac{4x+1}{3-2x}, x = -2$
3) $\frac{2x^2+1}{2-x^2}, x = 1$
4) $\frac{b^2-2b^3+1}{3b^2+b^4}, b = -1$
5) $\frac{8x^2+2x-1}{6x-5}, x = \frac{1}{2}$

Real World Applications

Do the following problems.

1) The shirt originally costs c dollars. The store is offering a 20% discount. Write an algebraic expression for the price of the shirt after the discount.
2) A cell phone plan charges a fixed monthly fee of \$25 and \$0.10 per text message sent. Write an expression for the total monthly cost if t text messages are sent.
3) A bookstore sells notebooks for \$4 each and pens for \$1.50 each. Write an expression for the total cost if you buy n notebooks and p pens.
4) You want to fence a rectangular garden. The length of the garden is l meters, and the width is w meters. The fencing costs \$5 per meter. Write an expression for the total cost of fencing the garden.
5) A school is holding a fundraising event where they sell raffle tickets for \$2 each and baked goods for \$1.50 each. If they sell r raffle tickets and b baked goods, write an expression for the total amount of money raised.
6) A book has 250 pages. If we read 10 pages per day, write an algebraic expression for the number of remaining pages after m days.
7) An architect is designing a rectangular room. The length of the room is $3x + 5$ feet, and the width is $2x + 4$ feet. Write an expression for the area of the room in terms of x.
8) A speeding fine consists of a base fee of 50 dollars plus 5 dollars for every mile per hour over the speed limit. Write an expression for the total fine f if someone is caught speeding x miles per hour over the speed limit.
9) A savings account earns compound interest. The balance B after t years is given by the formula $B = P(1 + r)^t$, where P is the principal, r is the annual interest rate, and t is the number of years. If $P = 1000$ dollars, and the interest rate is $r = 0.05$, how much will the balance be after 3 years?
10) A car travels at a speed of 60 miles per hour for the first part of a trip and then slows down to 40 miles per hour for the second part of the trip. If the car spends t hours on the first part and $t + 1$ hours on the second part. Write an expression for the total distance traveled.

Answer of Worksheets

Write Variable Expressions (Two or Three Operations)

1) $3x+5$
2) x^2-10
3) $\frac{1}{2}(x+12)$
4) $7(2x-3)$
5) $\frac{x+9}{4x}$

Simplifying and Combining Like Terms

1) $3.5x^2+x$
2) $4xy+3xy^2-yx^2$
3) $2.5m^2+0.6mn$
4) $\frac{13}{4}(ab)^2-1.5ab^2+\frac{4}{5}ba^2$
5) $-2x+6xy-7y$

Distributive Property (With Numbers, Variables)

1) $-10xy-18x$
2) $-2ab-a^2+0.5b^2$
3) $-3xy-5x^2-x^3$
4) $-9xy^2+0.2x^2+\frac{3}{2}y^2$
5) $a^3-2ba^2-2a^2-8b$

Distributive Property with Area Models

1) $16a+8b$
2) $2xy^3-10x^2$
3) $-a^2b-2ab^3$
4) $3x^2y^2-xy^3+5xy^2$
5) $-2ab-2b+3a+3$

Multiplying Binomials

1) a^2-1
2) $-3x^2+13x-4$
3) a^3-25a
4) $-8xy+4x^3+2y^3-x^2y^2$
5) $2x^2y^2-4xy^4+5x-10y^2$
6) m^4-9
7) $-x^2+12x+3$
8) $4x^2y^2-4x-xy^3+y$
9) $a^2+3ab^3-2ab-3ab^2+6b^3$
10) $2x^2-2x+1$

Multiplying Other Polynomials

1) x^3-2x^2-x+2
2) $2x^3-x^2-1$
3) $2x^3+5x^2+5x+3$
4) x^3+2x^2-4x-8
5) $x^4+3x^3-9x^2+x-20$
6) $6x^4+x^3-3x^2+4x-4$
7) $x^6-x^5+x^3-x^2$
8) $2x^5-5x^4+3x^3+11x^2-11x+4$
9) x^8-1
10) $8x^6+8x^5+4x^4-7x^3-8x^2-4x-1$

Factoring Using Distributive Property

1) $a(3b+2)$
2) $x^2(x^2-x+1)$
3) $4a^4b^4(b+2a)$
4) $a^2(a^3-a^2+a-1)$
5) $8x^2y^2(2y-3x)$
6) $\frac{3x(5x-z)}{5z}$
7) $2a^2(3b^2+4-6b^3a+7b^2a^2)$
8) $\frac{y(2z^4-3x^4+x^2y^2)}{4x}$
9) $\frac{2a^2}{b^2}$
10) $3x$

Factoring by Grouping

1) $(a-b)(c+d)$
2) $2(x-1)(2+y)$
3) $(a+b)(a-b)$
4) $(x+y)(x-z)$
5) $(a+3)(b+2)$
6) $(a-b)^2ab$
7) $(3x-2)(2x+1)$
8) $(3b-1)(5a^2+2b)$
9) $(y-1)(4x-2+y^2)$
10) $(4x-1)(-2+5x-x^2)$

Factoring Quadratics

1) $(x+3)(x-2)$
2) $(y-7)(y+5)$
3) $(a+9)(a+4)$
4) $(b+11)(b-9)$
5) $(2x-1)(x+2)$
6) $(3x+5)(3x-1)$
7) $(5a-1)(a+4)$
8) $(3x+4)(3x-2)$
9) $(3y-2x)(3y+5x)$
10) $(4a^2+7)(4a^2-5)$

Factoring by Sorting GCF

1) The factors are: $3x$ and $(2x+3)$
2) The factors are: $5a^2b$ and $(2a+3b)$
3) The factors are: $(2x-1)$ and $(x+3)$
4) The factors are: $(x+6)$ and (x^2+1)
5) The factors are: $3x$, $(x-3)$ and $(x+1)$

Evaluate Radical Expressions

1) $2\sqrt{2}$
2) $2\sqrt{5}$
3) 6
4) $2\sqrt{6}$
5) $\frac{1}{2}$

Evaluate Rational Expressions

1) 7
2) -1
3) 3
4) 1
5) -1

Real World Applications

1) $0.80c$
2) $25+0.10t$
3) $4n+1.50p$
4) $5(2l+2w)$
5) $2r+1.50b$
6) $250-10m$
7) $6x^2+22x+20$
8) $50+5x$
9) $\approx 1157.63\$$
10) $100t+40$

Chapter 6: Equation and Inequalities

Topics that you'll learn in this chapter:

- ✓ Solving Equations
- ✓ One-Step equations
- ✓ Two/Multi-Step Equations
- ✓ Equations with One Solution
- ✓ Equations with No Solution
- ✓ Equations with Infinitely Many Solutions
- ✓ One-Step Inequalities
- ✓ Two/Multi Step Inequalities
- ✓ Graphing Inequalities
- ✓ Classifying Systems of Equations
- ✓ Systems of Equations Using Graphing
- ✓ Systems of Equations Using Substitutions
- ✓ Systems of Equations Using Elimination
- ✓ Real World Applications
- ✓ Worksheets
- ✓ Answer of Worksheets

Solving Equations

An equation is a mathematical sentence that shows that two expressions are equal. It contains a variable (like x or y) that we want to solve for. The general form is:

$$ax + b = c$$

Where:

- a is the coefficient (the number multiplying the variable),
- b is a constant (a fixed number),
- c is another constant (another fixed number),
- and x is the variable we want to solve for.

Diagram Representing an Equation

A diagram is a visual representation of a situation that can be turned into an equation. These diagrams often use shapes or blocks to represent unknowns, quantities, or operations (like addition, subtraction, multiplication, etc.). Writing equations based on diagrams means converting those visual representations into mathematical statements.

Steps for Writing Equations from Diagrams:

1. Look at the diagram and identify what is being represented (addition, subtraction, multiplication, division, etc.).
2. Translate the visual information into a mathematical expression or equation: For addition, use $+$. For subtraction, use $-$. For multiplication, use $\times$ or dot. For division, use $\div$ or /.
3. Write the equation that matches the diagram. If there are multiple parts to the diagram (like several blocks), combine them appropriately.

Example:

Write an equation that says that the length of the red line is equal to the length of black line minus the length of the blue line.

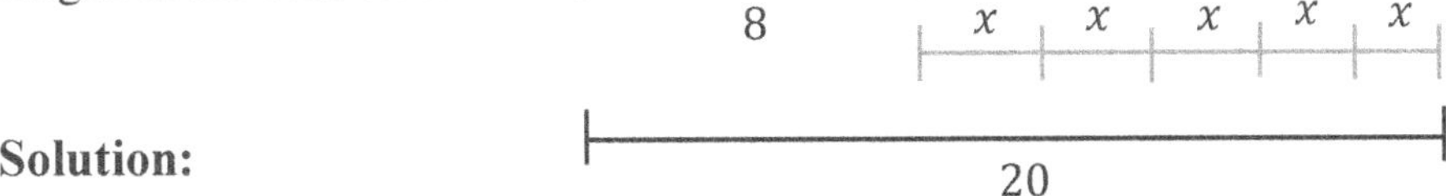

Solution:

Our goal is to substitute both sides of the equation below with numerical values and algebraic expressions.

Length of red line = length of black line − length of blue line

The length of red line is 8, the length of the black line is 20 and the blue line is the sum of its segments ($x + x + x + x + x = 5x$). Putting these together gives: $8 = 20 - 5x$

One-Step equations

A one-step equation is an equation where you only need to perform one operation (addition, subtraction, multiplication, or division) to isolate the variable and solve it.

The goal:

The goal is to find the value of the variable (usually represented by x, y or another letter) that makes the equation true.

How to solve a one-step equation:

1. **Identify the operation:** First, look at the equation and identify what operation is being performed on the variable (addition, subtraction, multiplication, or division).
2. **Undo the operation:** To isolate the variable, you perform the opposite operation to both sides of the equation.
3. **Solve for the variable:** After performing the opposite operation, you'll have the value of the variable.

☑ **Easy Trick:**

"Do the opposite to get the answer!"

If something is added, you subtract. If something is multiplied, you divide.

Example: Solve the equations:

a) $x + 5 = 12$

b) $y - 3 = 8$

c) $4z = 16$

d) $\frac{a}{5} = 2$

Solution:

a) To solve this, you need to get x by itself on one side of the equation. Subtract 5 from both sides: $x + 5 - 5 = 12 - 5 \rightarrow x = 7$

b) To solve this, you need to get y by itself on one side of the equation. Add 3 to both sides: $y - 3 + 3 = 8 + 3 \rightarrow y = 11$

c) To solve this, you need to get z by itself on one side of the equation. Divide both sides by 4: $\frac{4z}{4} = \frac{16}{4} \rightarrow z = 4$

d) To solve this, you need to get a by itself on one side of the equation. Multiply both sides by 5: $\frac{a}{5} \times 5 = 2 \times 5 \rightarrow a = 10$

Two/Multi-Step Equations

Two-step equations: These are equations that require two operations to isolate the variable.

Multi-step equations: These are equations that require more than two steps to isolate the variable. This may involve combining like terms, using the distributive property, and performing multiple operations.

Steps to Solve Two-Step and Multi-Step Equations:

Step 1: Simplify the equation: If the equation has parentheses or like terms, simplify it first.

Step 2: Perform inverse operations: Use inverse operations to get the variable by itself.

- The Inverse of addition is subtraction.
- The Inverse of subtraction is addition.
- The inverse of multiplication is division.
- The Inverse of division is multiplication.

Step 3: Solve step by step: Start with the operation that is farthest away from the variable and work your way in.

Step 4: Check your solution: Once you find the value of the variable, substitute it back into the original equation to make sure it works.

Examples:

1) Solve $2x + 3 = 11$.

Solution:

1. Start by subtracting 3 from both sides (to get rid of the $+3$):
   $2x + 3 - 3 = 11 - 3 \rightarrow 2x = 8$
2. Now divide both sides by 2 (because the opposite of multiplying by 2 is dividing by 2): $\frac{2x}{2} = \frac{8}{2} \rightarrow x = 4$

2) Solve $3(x - 2) + 4 = 19$.

Solution:

1. Start by distributing the 3 to both terms inside the parentheses:
   $3x - 6 + 4 = 19 \rightarrow 3x - 2 = 19$
2. Add 2 to both sides to get rid of the -2:
   $3x - 2 + 2 = 19 + 2 \rightarrow 3x = 21$
3. Divide both sides by 3 to isolate x: $\frac{3x}{3} = \frac{21}{3} \rightarrow x = 7$

Equations with One Solution

When we solve an equation, we are looking for a value (or values) for the variable that makes the equation true. An equation with one solution has exactly one value that satisfies it.

When you solve these equations, you will end up with a single number for the variable, and when you substitute that number back into the original equation, both sides of the equation will be equal.

Example of One-Solution Equations:

- $2x - 1 = -x \rightarrow 2x + x = 1 \rightarrow 3x = 1 \rightarrow x = \frac{1}{3}$
- $2(x + 3) = 14 \rightarrow 2x + 6 = 14 \rightarrow 2x = 14 - 6 \rightarrow 2x = 8 \rightarrow x = 4$

Equations with No Solution

An equation with no solution means there is no possible value for the variable that will satisfy the equation. In other words, no matter what value you choose for the variable, the equation will never be true.

Identifying Equations with No Solution:

Equations with no solution typically result in a contradiction, where the equation simplifies to a false statement, such as $0 = 5$ or any other false numerical equality.

Examples of Equations with No Solution:

- $2x + 3 = 2x - 5 \rightarrow 2x + 3 - 2x = 2x - 5 - 2x \rightarrow 3 = -5$, Since 3 is not equal to -5, the equation is false, and there is no solution.
- $3x - 4 = 2x + x + 6 \rightarrow 3x - 2x - x = 6 + 4 \rightarrow 0 = 10$, Since we end up with a false statement, this means there is no solution.

Equations with Infinitely Many Solutions

An equation with infinitely many solutions means that any value you substitute for the variable will satisfy the equation. This often happens when both sides of the equation are essentially the same or can be made identical through simplification.

Identifying Equations with Infinitely Many Solutions:

Equations with infinitely many solutions typically result in a true statement that does not involve the variable, such as $0 = 0$ or any other true numerical equality.

Examples of Equations with Infinitely Many Solutions:

- $x - 2 = x - 2$, Since the lines are the same, there are infinitely many solutions.
- $2(x + 3) = 2x + 6 \rightarrow 2x + 6 = 2x + 6$, Since both sides are identical, any value of x will satisfy the equation, resulting in infinitely many solutions.

One-Step Inequalities

Solving inequalities is similar to solving equations, but instead of finding a single value for the variable, you find a range of values that make the inequality true.

One-step inequalities are inequalities that can be solved with just one operation, similar to one-step equations. The goal is to isolate the variable on one side of the inequality by performing a single operation, such as addition, subtraction, multiplication, or division.

Steps to Solve One-Step Inequalities:

1. **Identify the Operation**: Determine which operation (addition, subtraction, multiplication, or division) is being used.
2. **Perform the Inverse Operation:** Use the inverse operation to isolate the variable on one side of the inequality. Remember to perform the same operation on both sides to maintain balance.
3. **Reverse the Inequality Sign (if necessary):** If you multiply or divide both sides by a negative number, reverse the direction of the inequality sign.

☑ **Easy Trick:**

"Solve like an equation, flip the sign if negative!"

If you multiply or divide both sides by a negative number, remember to turn the inequality sign around.

Example:

Solve.

a) $x - 9 \leq -2$

b) $2 + x > -10$

c) $-3x \geq 15$

d) $\frac{x}{4} < 3$

Solution:

a) Add 9 to both sides: $x - 9 + 9 \leq -2 + 9 \rightarrow x \leq 7$

b) Subtract 2 from both sides: $2 + x - 2 > -10 - 2 \rightarrow x > -12$

c) Divide both sides by -3 (and reverse the inequality sign): $\frac{-3x}{-3} \leq \frac{15}{-3} \rightarrow x \leq -5$

d) Multiply both sides by 4: $4 \times \frac{x}{4} < 3 \times 4 \rightarrow x < 12$

Two/Multi Step Inequalities

Two-Step Inequalities:

To solve two-step inequalities, you perform two inverse operations to isolate the variable. Here's a step-by-step process:

1. **Undo Addition or Subtraction:** First, move the constant term to the other side of the inequality.
2. **Undo Multiplication or Subtraction:** Next, divide or multiply both sides of the inequality by the coefficient of the variable.

Multi-Step Inequalities:

Multi-step inequalities involve more than two steps and may require combining like terms and using the distributive property. Here's the process:

1. **Simplify Both Sides:** Use the distributive property and combine like terms.
2. **Undo Addition or Subtraction:** Move the constant term to the other side.
3. **Undo Multiplication or Division:** Divide or multiply both sides of the inequality by the coefficient of the variable.

Examples:

1) Solve $2x + 3 > 7$.

   Solution:

   1. Subtract 3 from both sides: $2x + 3 - 3 > 7 - 3 \rightarrow 2x > 4$
   2. Divide both sides by 2: $\frac{2x}{2} > \frac{4}{2} \rightarrow x > 2$

2) Solve $-3(x - 2) + 4 \leq 2x + 5$.

   Solution:

   1. Distribute -3: $-3x + 6 + 4 \leq 2x + 5$
   2. Combine like terms: $-3x + 10 \leq 2x + 5$
   3. Subtract $2x$ from both sides:
      $-3x + 10 - 2x \leq 2x + 5 - 2x \rightarrow -5x + 10 \leq 5$
   4. Subtract 10 from both sides:
      $-5x + 10 - 10 \leq 5 - 10 \rightarrow -5x \leq -5$
   5. Divide both sides by -5 (and reverse the inequality sign):
      $\frac{-5x}{-5} \geq \frac{-5}{-5} \rightarrow x \geq 1$

Graphing Inequalities

Graphing inequalities on a number line is a useful skill to visualize the solutions. Here are the steps to graph inequalities:

Steps to Graph Inequalities:

1. **Solve the Inequality:** First, solve the inequality to find the value of the variable.
2. **Draw the Number Line:** Draw a horizontal line and mark appropriate numbers on it, including the value you found in the solution.
3. **Plot the Solution:**
   - If the inequality is strict (e.g., $x > a$ or $x < a$), use an open circle to indicate that the number itself is not included.
   - If the inequality includes the number (e.g., $x \geq a$ or $x \leq a$), use a closed circle to indicate that the number itself is included.
4. **Shade the Solution Area:**
   - For $x > a$ (or $x \geq a$), shade the number line to the right of a, indicating all numbers greater than a.
   - For $x < a$ (or $x \leq a$), shade the number line to the left of a, indicating all numbers less than a.

Example: Solve $2x - (4x + 1) < -x$, then graph on a number line:

Solution:

1. Distribute: $2x - 4x - 1 < -x$
2. Combine like terms: $-2x - 1 < -x$
3. Add x to both sides: $-2x - 1 + x < -x + x \rightarrow -x - 1 < 0$
4. Add 1 to both sides: $-x - 1 + 1 < 0 + 1 \rightarrow -x < 1$
5. Divide both sides by -1 and reverse the inequality sign: $\frac{-x}{-1} > \frac{1}{-1} \rightarrow x > -1$
6. Draw a number line and use an open circle:

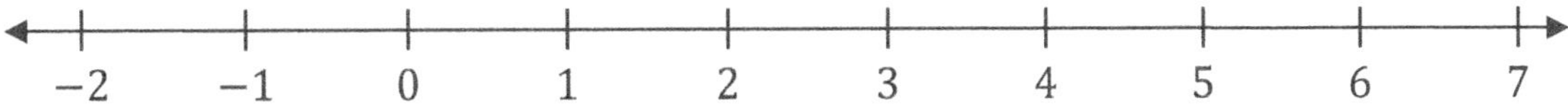

7. Shade the solution area:

Classifying Systems of Equations

Classifying systems of equations is an important skill that helps us understand how different equations relate to each other.

Classifications of Systems of Equations:

1. **Consistent Systems:**
   - **Independent:** These systems have exactly one solution where the two lines intersect at a single point.
     - **Example:** $\begin{cases} y = 2x + 1 \\ y = -x + 4 \end{cases}$, The lines intersect at one point, giving exactly one solution.
   - **Dependent:** These systems have infinitely many solutions because the equations represent the same line.
     - **Example:** $\begin{cases} y = 2x + 1 \\ 2y = 4x + 2 \end{cases}$, Both equations describe the same line, so every point on the line is a solution.
2. **Inconsistent Systems:**
   - **These systems have no solutions because the lines are parallel and never intersect.**
     - **Example:** $\begin{cases} y = 2x + 1 \\ 2y = 4x - 3 \end{cases}$, The lines have the same slope but different y-intercepts, so they never meet.

Example: Classify the types of the following systems of equations.

a) $\begin{cases} y = -5x + 2 \\ 3y = -15x + 6 \end{cases}$ b) $\begin{cases} y = -4x - 4 \\ y = 3x + 7 \end{cases}$ c) $\begin{cases} y = 6x - 1 \\ y = 6x + 6 \end{cases}$

Solution:

a) Rewrite the second equation in a similar form to the first equation to compare them more easily: $3y = -15x + 6$, Divide both sides by 3 to isolate y: $y = -5x + 2$

Since both equations represent the same line, the system is consistent and dependent.

b) Look at the slopes and y-intercepts of the lines: The first equation $y = -4x - 4$ has a slope of -4. The second equation $y = 3x + 7$ has a slope of 3.

Since the slopes are different (-4 and 3), the lines will intersect at exactly one point. This means the system is consistent independent.

c) Look at the slopes and y-intercept of the lines: The first equation $y = 6x - 1$ has a slope of 6 and a y-intercept of -1. The second equation $y = 6x + 6$ has a slope of 6 and a y −intercept of 6.

Since the slopes are the same (both are 6) but the y-intercepts are different (one is -1 and the other is 6), the lines are parallel and will never intersect. This means the system is inconsistent.

Systems of Equations Using Graphing

Graphing systems of equations is a visual way to find the solution to a system by identifying where the lines intersect on a graph. Here's a step-by-step guide to how to do it:

Steps to Graph Systems of Equations

1. **Write Each Equation in Slope-Intercept Form:** The slope-intercept form is $y = mx + b$, where m is the slope and b is the y-intercept. If the equations are not already in this form, rearrange them.
2. **Plot the y-Intercept:** For each equation, start by plotting the y-intercept (the value of b) on the y-axis.
3. **Use the Slope to Plot Another Point:** From the y-intercept, use the slope (m) to find another point on the line. The slope is the ratio of the rise (change in y) to the run (change in x).
4. **Draw the Line:** Connect the points with a straight line. Extend the line across the graph.
5. **Repeat for the Second Equation:** Follow the same steps to graph the second equation on the same set of axes.
6. **Identify the Intersection Point:** The solution to the system of equations is the point where the two lines intersect.

Example: Graph the system of equations: $\begin{cases} y = 2x + 1 \\ y = -x + 4 \end{cases}$

Solution:

1. First equation: $y = 2x + 1$
   - y-Intercept (b) $= 1$ (Plot the point $(0, 1)$ on the y-axis)
   - Slope (m) $=$ 2 (which means rise 2 units up and run 1 unit to the right)
   - From $(0, 1)$, plot another point $(1, 3)$
2. Second Equation: $y = -x + 4$
   - y-Intercept (b) $= 4$ (Plot the point $(0, 4)$ on the y-axis)
   - Slope (m) $= -1$ (which means fall 1 unit down and run 1 unit to the right)
   - From $(0, 4)$, plot another point $(1, 3)$

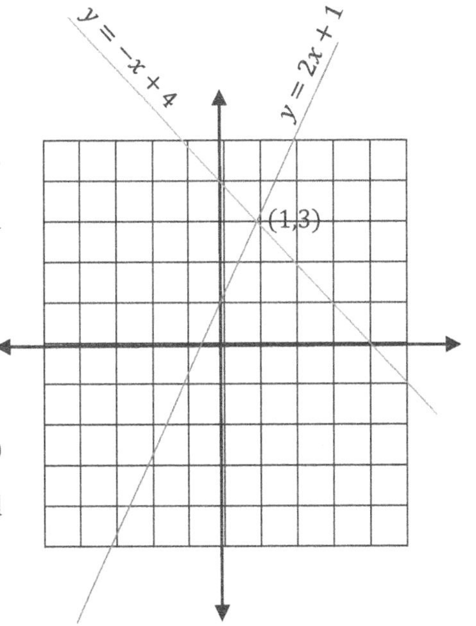

Systems of Equations Using Substitutions

Solving systems of equations using substitution is a method where you solve one of the equations for one variable and then substitute that expression into the other equation. This allows you to solve for one variable at a time. Here's a step-by-step guide:

Steps to Solve Systems of Equations Using Substitution

1. **Solve One Equation for One Variable:** Choose one of the equations and solve it for one of the variables (either x or y). This should be the easiest variable to isolate.
2. **Substitute the Expression:** Substitute the expression obtained in step 1 into the other equation. This will result in an equation with one variable.
3. **Solve for the Remaining Variable:** Solve the equation obtained in step 2 for the remaining variable.
4. **Substitute Back:** Substitute the value obtained in step 3 back into the expression from step 1 to find the value of the other variable.
5. **Check Your Solution:** Plug the values of both variables into the original equations to verify that they satisfy both equations.

Example: Solve the system of equations using substitution: $\begin{cases} y = -3x - 2 \\ y - 2x = +3 \end{cases}$

Solution:

1. Solve one equation for one variable: The first equation is already solved for y: $y = -3x - 2$
2. Substitute the expression: Substitute $y = -3x - 2$ into the second equation:
   $-3x - 2 - 2x = 3$
3. Solve for the remaining variable: Simplify and solve for x:
   $-3x - 2 - 2x = 3 \to -5x - 2 = 3 \to -5x = 3 + 2 \to -5x = 5 \to x = \frac{5}{-5} = -1$
4. Substitute back: Substitute $x = -1$ back into the expression $y = -3x - 2$:
   $y = -3 \times -1 - 2 = 3 - 2 = 1$
5. Check your solution: Plug $x = -1$ and $y = 1$ into the original equations to verify:

$$\begin{cases} y = -3x - 2 \to 1 = -3 \times -1 - 2 \to 1 = 1 \\ y - 2x = +3 \to 1 - 2 \times -1 \to 1 + 2 = 3 \to 3 = 3 \end{cases}$$

So, the solution to the system is $x = -1$ and $y = 1$

Systems of Equations Using Elimination

Solving systems of equations using elimination involves eliminating one of the variables by adding or subtracting the equations. This method helps you find the values of both variables. Here's a step-by-step guide:

Steps to Solve Systems of Equations Using Elimination

1. **Align the Equations:** Make sure the equations are in standard form ($Ax + By = C$) and align them vertically so that the variables and constants are lined up.
2. **Multiply if Necessary:** If the coefficients of one of the variables are not the same or opposite, you may need to multiply one or both equations by a constant to create coefficients that are the same or opposite.
3. **Add or Subtract the Equations:** Add or subtract the equations to eliminate one of the variables. This will leave you with a single equation with one variable.
4. **Solve for the Remaining Variable:** Solve the resulting equation for the remaining variable.
5. **Substitute Back:** Substitute the value of the solved variable back into one of the original equations to find the value of the other variable.
6. **Check Your Solution:** Plug the values of both variables into the original equations to verify that they satisfy both equations.

Example: Solve the system of equation using elimination: $\begin{cases} 2x + 3y = 16 \\ 4x - 3y = 8 \end{cases}$

Solution:

- Align the equations: $\begin{cases} 2x + 3y = 16 \\ 4x - 3y = 8 \end{cases}$
- Add the equations: Notice that $3y$ and $-3y$ will cancel out when we add the equations:

$$\begin{array}{r} \begin{cases} 2x + 3y = 16 \\ 4x - 3y = 8^{+} \end{cases} \\ \hline 6x = 24 \end{array}$$

- Solve for x: $6x = 24 \rightarrow x = \frac{24}{6} = 4$
- Substitute back: Substitute $x = 4$ back into one of the original equations to find y.

  $2(4) + 3y = 16 \rightarrow 8 + 3y = 16 \rightarrow 3y = 16 - 8 = 8 \rightarrow y = \frac{8}{3}$
- Check your solution: Plug $x = 4$ and $y = 8$ into the original equations to verify:

  For $2x + 3y = 16$: $2(4) + 3\left(\frac{8}{3}\right) = 8 + 8 = 16$ (True).

  For $4x - 3y = 8$: $4(4) - 3\left(\frac{8}{3}\right) = 16 - 8 = 8$ (True).

So, the solution to the system is $x = 4$ and $y = \frac{8}{3}$

Real World Applications

Equations and inequalities are not just abstract mathematical concepts; they have numerous real-world applications that are relevant and useful in everyday life. Here are some examples:

1. **Budgeting and Money Management**
   When you need to manage your money, you might use an equation to figure out how much you can spend or save.
2. **Speed and Distance**
   Equations are used to calculate speed, time, or distance, which is really helpful when planning trips.
3. **Temperature**
   Inequalities can help when you're dealing with temperature ranges, like for weather forecasts or knowing when something will freeze.
4. **Sports and Team Scoring**
   In sports, equations and inequalities help teams figure out how many points they need to win or how many they need to keep up with their opponent.
5. **Shopping Discounts**
   Inequalities also come up when you're shopping and looking for sales. Let's say you have a budget and want to make sure you don't spend too much.

Example: You go to a store to buy apples and bananas. Apples cost $2 each, and bananas cost $1 each. You bought a total of 10 fruits and spent $15. How many apples and bananas did you buy?

Solution:

1. Define variables:
   - Let x be the number of apples
   - Let y be the number of bananas.
2. Set up the equations:
   - Total number of fruits: $x + y = 10$
   - Total cost: $2x + y = 15$
3. Solve the System: $\begin{cases} x + y = 10 \\ 2x + y = 15 \end{cases}$:
   - Solve the first equation for y: $y = 10 - x$
   - Substitute y in the second equation:
     $2x + (10 - x) = 15 \rightarrow 2x + 10 - x = 15 \rightarrow x + 10 = 15 \rightarrow x = 5$
   - Substitute $x = 5$ back into the first equation: $5 + y = 10 \rightarrow y = 5$

So, you bought 5 apples and 5 bananas.

Worksheets

Solving Equations

Write an equation for each diagram:

1) The length of the black line is equal to the length of the blue line.

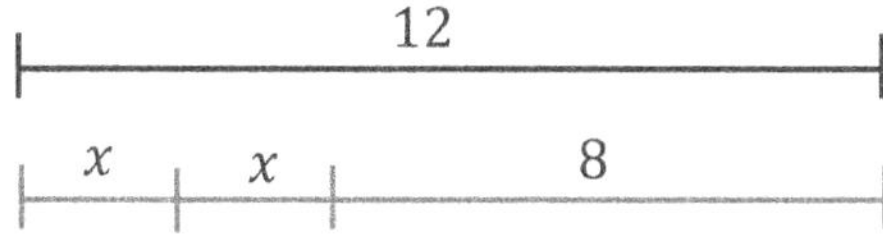

2) The length of the red line is equal to the length of the black line.

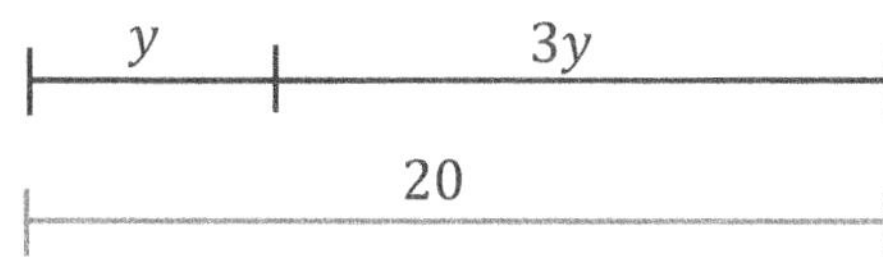

3) The length of the green line is equal to the length of the black line minus the length of the blue line.

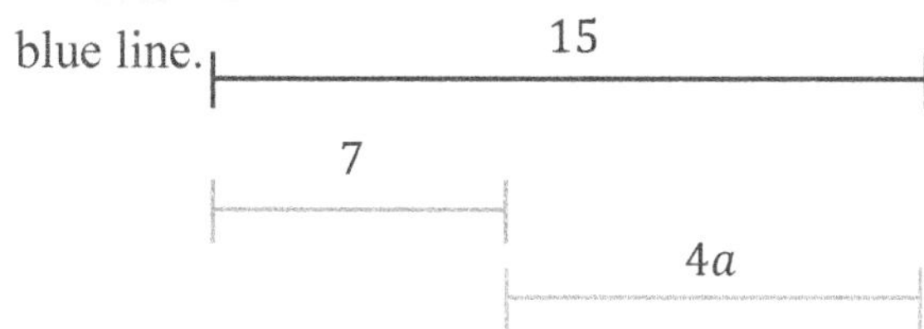

4) The length of the black line is equal to the sum of the length of the orange line and the length of the blue line.

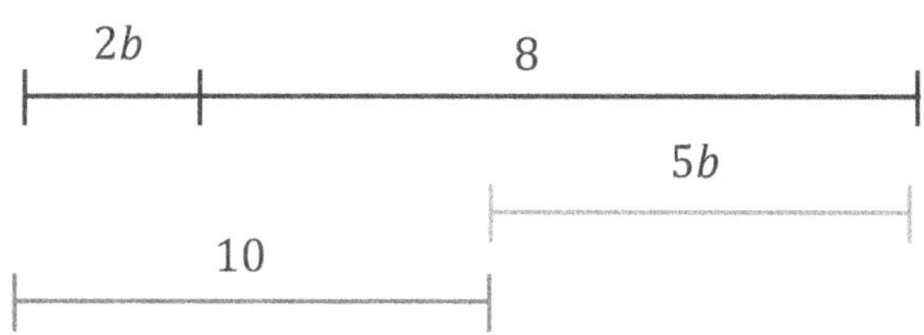

5) The length of the blue line is equal to the length of the black line minus the length of the red line.

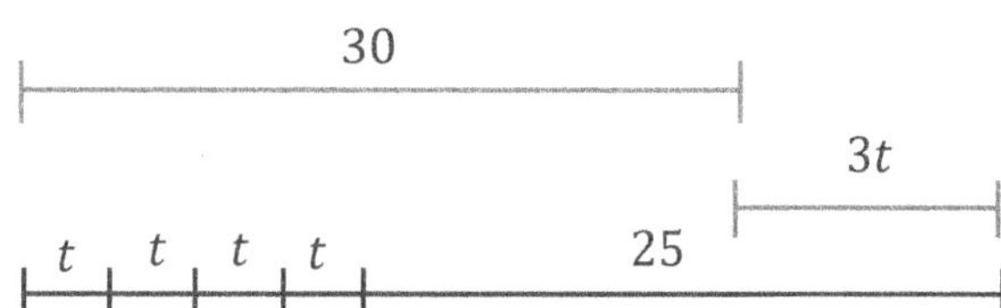

One-Step equations

Solve.

1) $x - 2 = \frac{2}{5}$
2) $11 = y + 1.3$
3) $\frac{x}{-5} = 0.3$
4) $3x = \frac{-2}{3}$
5) $x - 0.2 = -4\frac{4}{5}$

Two/Multi-Step Equations

Solve equations.

1) $12 - 4x = -\frac{1}{2}$
2) $5 = -\frac{3}{5} - 2.3x$
3) $-3(x - 4) = 2$
4) $0.2(-10x + 5) = -3x - 4$
5) $-7x + 3(-x - 15) = -2(4 - 5x)$
6) $\frac{0.4x-2}{3} = \frac{-1+x}{4}$
7) $0.4(2 - 3x) + 2x = -\frac{1}{4}x - 2(2x - 0.3)$
8) $\frac{5x-9}{-5} + \frac{2x}{10} = \frac{3x-8}{20}$
9) $\frac{4x-3}{3} = -2x + 7(2x - 1)$
10) $\frac{-2x+3}{4x} = -\frac{2}{5}$

Equations with One Solution

Determine whether the following equations have a solution or not. Then find the solution of "one-solution" equations.

1) $3x - 2 = 3x$
2) $-2x + 5 = 1 - x$
3) $4 - 2x = -x + 4 - x$
4) $-(x + 2) = 2x$
5) $2x - 7 = -(3 - 2x) - 4$

Equations with No Solution

Determine whether the following equations have no solution or not.

1) $-4x + 2(x - 3) = -2x$
2) $\frac{-3x+5}{2} = -1$
3) $8 - 2x = 4 - x$
4) $6(5x - 3) - 25x = 5x - 18$
5) $-7 - 3x = -3(x + 2) + 4$

Equations with Infinitely Many Solutions

Determine whether the following equations have infinitely many solutions or not.

1) $5x - 3(2 + x) + 3 = -3 + 2x$
2) $\frac{-1+3x}{2} = 1.5x$
3) $5x + 7 = -2 + x$
4) $-3(2x - 7) = -16x$
5) $4x - 2 = 2(x - 3) + 2x + 4$

One-Step Inequalities

Solve inequalities.

1) $2x < -5$
2) $x + 2 > -9$
3) $\frac{-5}{3} \le -5x$
4) $-3.8 + x \ge -1.5$
5) $-2.5 \le \frac{x}{0.3}$

Two/Multi Step Inequalities

Solve following two/multi-step inequalities:

1) $0.4 - 4x \ge -1.2$
2) $12 > -2x + 3$
3) $\frac{-5x}{3} \le \frac{1}{4}$
4) $3x - 2(x - 5) < 6x$
5) $-4.2(5x - 1) + 2x \le -9$
6) $\frac{2}{5}x - 0.1 + \frac{3}{4}x > -\frac{3}{5}x - 2$
7) $1 - \frac{2x-3}{5} + 4x \ge -0.2x$
8) $\frac{4}{5}(-3x - 2) - 0.1x < -(2x + \frac{1}{2})$
9) $\frac{-6x+2}{2} \ge \frac{4x+12}{-4}$
10) $-5x - 1.2(10x - 5) < \frac{-3x}{10}$

Graphing Inequalities

Graph following inequalities:

1) $\frac{-3x}{4} \le -2$
2) $3(x-2) < \frac{1}{2}$
3) $-2x + 5(x-1) > -2$
4) $2x - (3x+5) \le \frac{-6x+1}{3}$
5) $-\frac{3}{4}x + 2 \ge \frac{1}{4}x$

Classifying Systems of Equations

Classify following systems of equations (Consistent (independent or dependent)/ Inconsistent systems).

1) $\begin{cases} y = x - 3 \\ y = x + 10 \end{cases}$
2) $\begin{cases} y = 2x + 4 \\ 2y = 2x - 6 \end{cases}$
3) $\begin{cases} y - 4x = 5 \\ y = -4x + 7 \end{cases}$
4) $\begin{cases} 3y = 3x + 12 \\ y - x - 4 = 0 \end{cases}$
5) $\begin{cases} 5y + 2x = -1 \\ y = -\frac{4}{10}x + 1 \end{cases}$

Systems of Equations Using Graphing

Solve the following system of equations using graphing:

1) $\begin{cases} y = 2x - 2 \\ y = \frac{x}{4} + 1 \end{cases}$
2) $\begin{cases} x + y = 3 \\ 2x - y = 1 \end{cases}$
3) $\begin{cases} 3x + y = 5 \\ 4x - y = 2 \end{cases}$
4) $\begin{cases} y = 0.5x + 2 \\ y = -3.5x - 2 \end{cases}$
5) $\begin{cases} y = -2x + 2 \\ y = x + 1 \end{cases}$

Systems of Equations Using Substitutions

Solve the following system of equations using substitutions:

1) $\begin{cases} y = 3x + 5 \\ y = 1 + 4x \end{cases}$
2) $\begin{cases} y = \frac{1}{2}x - \frac{1}{2} \\ y = -\frac{3}{4}x + \frac{11}{4} \end{cases}$
3) $\begin{cases} x + y = 7 \\ 2x - y = 4 \end{cases}$
4) $\begin{cases} 3x + 2y = 8 \\ x - y = 1 \end{cases}$
5) $\begin{cases} 4x - y = 13 \\ x + 2y = 6 \end{cases}$

Systems of Equations Using Elimination

Solve the following system of equations using elimination:

1) $\begin{cases} x + y = 5 \\ x - y = 1 \end{cases}$
2) $\begin{cases} 3x + 4y = 18 \\ 2x - y = 5 \end{cases}$
3) $\begin{cases} 4x - y = 7 \\ 3x + 2y = -3 \end{cases}$
4) $\begin{cases} 5x + 6y = -9 \\ 3x - 4y = 2 \end{cases}$
5) $\begin{cases} 2x - 3y = -2 \\ 5x + 2y = 4 \end{cases}$

Real World Applications

Do thc following problems.

1) John wants to buy a book that costs \$15. He has \$7 and earns \$5 per hour walking dogs. Write an inequality to determine how many hours he needs to work to afford the book.
2) Sarah has twice as many dimes as nickels. If the total value of her coins is \$3.25, how many dimes and nickels does she have?
3) Chloe spent half of her savings on a gift and then \$20 on a meal. If she has \$40 left, how much did she originally have?
4) Lily wants to buy a new dress that costs \$35. She has \$20 saved and plans to save \$5 each week. How many weeks will it take for her to have enough money to buy the dress?
5) A pizza parlor charges \$10 for a pizza and \$2 for each topping. John wants to spend no more than \$20. How many toppings can he get at most?
6) A rectangle's length is 5 units more than twice its width. If the perimeter of the rectangle is 34 units, find its dimensions.
7) A movie theater sells tickets for \$8 each for adults and \$4 each for children. Last night, 100 tickets were sold, and the total revenue was \$700. How many adults and how many children's tickets were sold?
8) We wanted to distribute some money among several people, with each person receiving \$1,000. However, when it came time to distribute, 2 people were absent, and as a result, each remaining person received \$1,200. Calculate the total amount of money and the initial number of people
9) Two workers with different wages were assigned to do a job. After completing the job, the first worker received \$2700, and the second worker, who worked 6 hours less, received \$1800. If the first worker worked twice as much as the second worker, and the second worker worked one-third as much as the first worker, they would have received the same amount in wages. Determine how many days each of the two workers worked and their daily wages.
10) Lucy asked David, 'How old are you?' David replied, 'When you reach my age, I will be twice your current age'. If the sum of their ages is currently 30 years, how old is each of them?

Answer of Worksheets

Solving Equations

1) $12 = 2x + 8$
2) $20 = 4y$
3) $7 = 15 - 4a$
4) $2b + 8 = 10 + 5b$
5) $3t = 4t + 25 - 30$

One-Step equations

1) $x = 2.4$
2) $y = 9.7$
3) $x = -1.5$
4) $x = \frac{-2}{9}$
5) $x = -4.6$

Two/Multi-Step Equations

1) $x = \frac{25}{8}$
2) $x = -\frac{56}{23}$
3) $x = \frac{10}{3}$
4) $x = -5$
5) $x = \frac{-37}{20}$
6) $x = \frac{-25}{7}$
7) $x = \frac{-4}{505}$
8) $x = \frac{44}{19}$
9) $x = \frac{9}{16}$
10) $x = \frac{15}{2}$

Equations with One Solution

1) No solution
2) One solution, $x = 4$
3) Infinitely many solutions
4) One solution, $x = \frac{-2}{3}$
5) Infinitely many solutions

Equations with No Solution

1) No solution
2) One solution
3) One solution
4) Infinitely many solutions
5) No solution

Equations with Infinitely Many Solutions

1) Infinitely many solutions
2) No solution
3) One solution
4) One solution
5) Infinitely many solutions

One-Step Inequalities

1) $x < \frac{-5}{2}$
2) $x > -11$
3) $x \leq \frac{1}{3}$
4) $x \geq 2.3$
5) $x \geq -0.75$

Two/Multi Step Inequalities

1) $x \leq 0.4$
2) $x > -4.5$
3) $x \geq -\frac{3}{20}$
4) $x > 2$
5) $x \geq \frac{66}{95}$
6) $x > -\frac{38}{35}$
7) $x \geq \frac{-8}{19}$
8) $x > -2.2$
9) $x \leq 2$
10) $x > \frac{60}{167}$

Graphing Inequalities

1) $x \geq \frac{8}{3}$

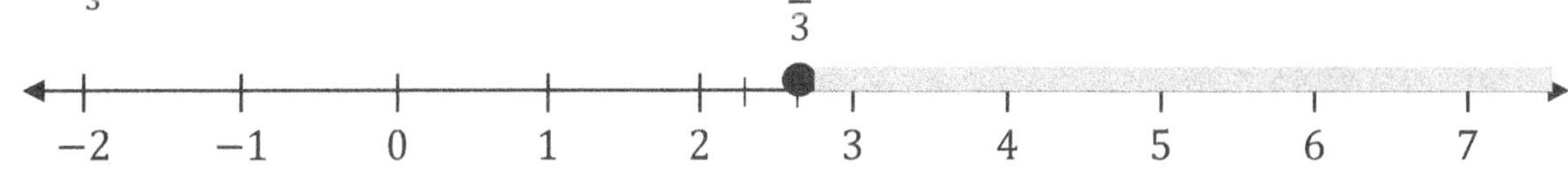

2) $x < \frac{13}{6}$

3) $x > 1$

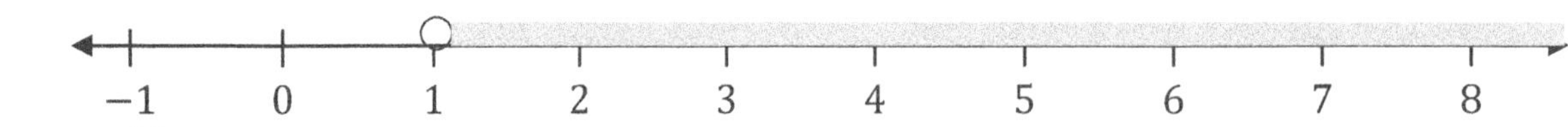

4) $x \leq \frac{16}{3}$

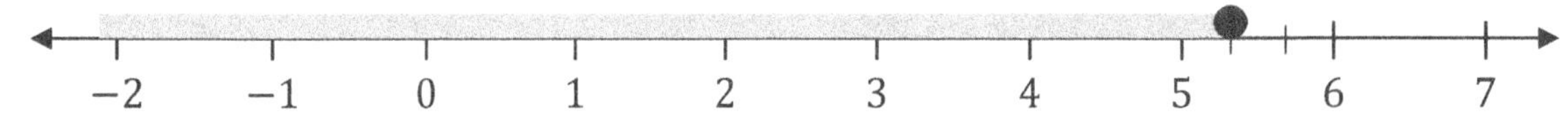

5) $x \leq 2$

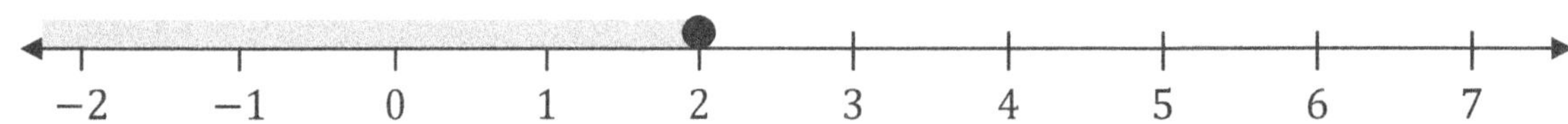

Classifying Systems of Equations

1) Inconsistent (no solution)
2) Consistent and independent (one solution)
3) Consistent and independent (one solution)
4) Consistent and dependent (infinitely many solution)
5) Inconsistent (no solution)

Systems of Equations Using Graphing

1) 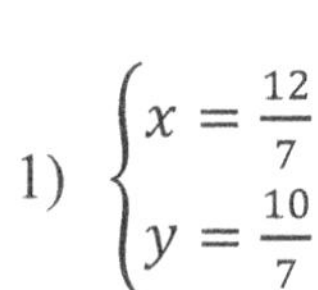

$$\begin{cases} x = \frac{12}{7} \\ y = \frac{10}{7} \end{cases}$$

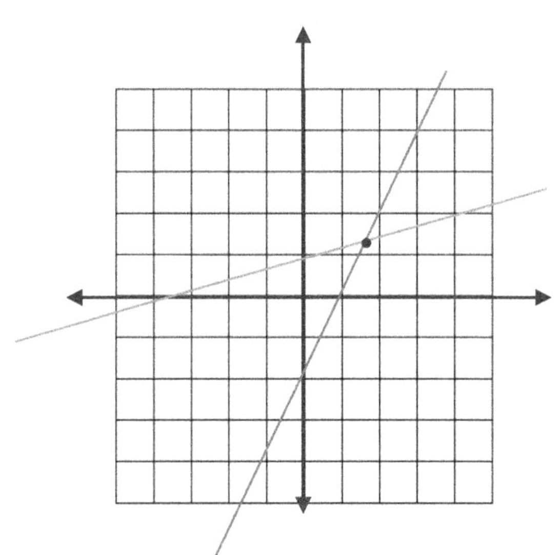

2) 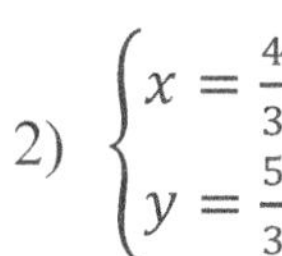
$\begin{cases} x = \frac{4}{3} \\ y = \frac{5}{3} \end{cases}$

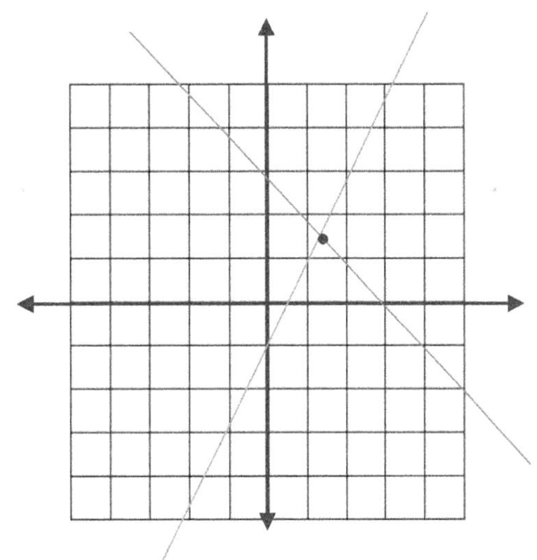

3) $\begin{cases} x = 1 \\ y = 2 \end{cases}$

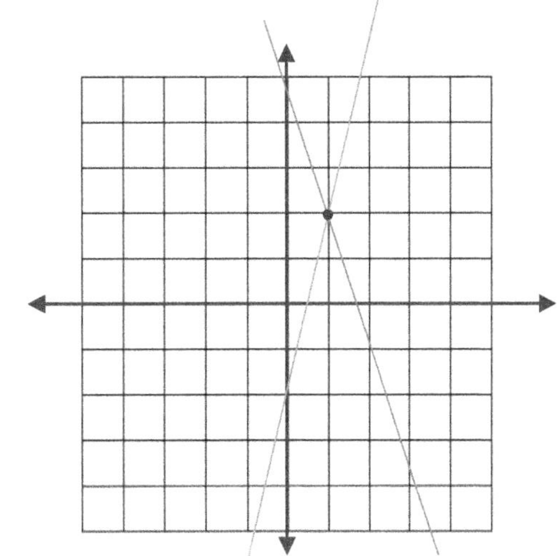

4) $\begin{cases} x = -1 \\ y = 1.5 \end{cases}$

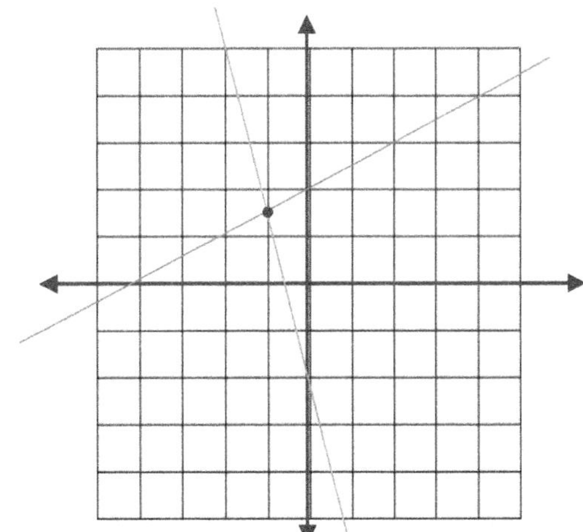

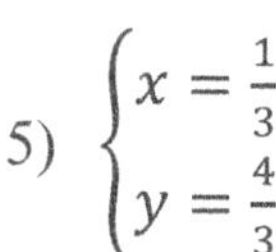

5) $\begin{cases} x = \frac{1}{3} \\ y = \frac{4}{3} \end{cases}$

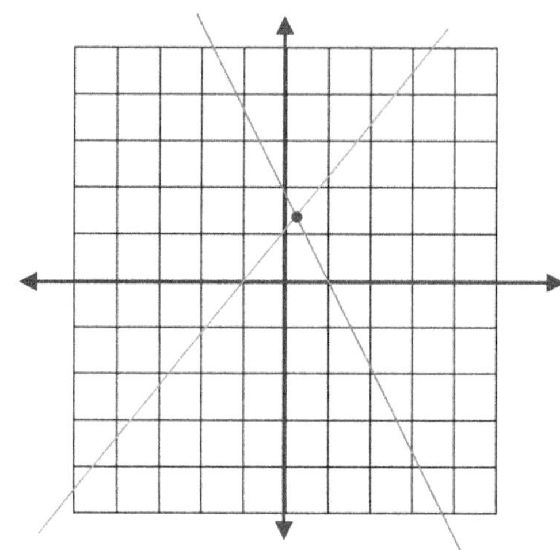

Systems of Equations Using Substitutions

1) $x = 4, y = 17$
2) $x = \frac{13}{5}, y = \frac{4}{5}$
3) $x = \frac{11}{3}, y = \frac{10}{3}$
4) $x = 2, y = 1$
5) $x = \frac{32}{9}, y = \frac{11}{9}$

Systems of Equations Using Elimination

1) $x = 3, y = 2$
2) $x = \frac{38}{11}, y = \frac{21}{11}$
3) $x = 1, y = -3$
4) $x = \frac{-12}{19}, y = \frac{-37}{38}$
5) $x = \frac{8}{19}, y = \frac{18}{19}$

Real World Applications

1) h: hours, $7 + 5h \geq 15$ and $h \geq 1.6$. Since John can't work a fraction of an hour, he must work at least 2 hours $\Rightarrow h \geq 2$
2) n: the number of nickels, d: the number of dimes, $0.05n + 0.10d = 3.25$ and $n = 13$ and $d = 26$.
3) S: the original amount of Chloe's savings, $\frac{S}{2} + 60 = S$, and $S = \$120$.
4) w: the number of weeks, $20 + 5w \geq 35$ and $w \geq 3$.
5) t: the number of toppings, $10 + 2t \leq 20$ and $t \leq 5$.
6) w: width and l: length, $\begin{cases} l = 2w + 5 \\ 2l + 2w = 34 \end{cases}$ and $w = 4$ and $l = 13$.
7) a: the number of adult tickets and c: the number of children tickets, $\begin{cases} a + c = 100 \\ 8a + 4c = 700 \end{cases}$ and $a = 75, c = 25$
8) n: the initial number of people, and t: the total amount of money, $\begin{cases} t = 1{,}000n \\ t = 1{,}200n - 2{,}400 \end{cases}$ and $n = 12, t = \$12{,}000$
9) x: the number of days first worker worked, $x = 9\ days$
   y: the number of days second worker worked, $y = 3\ days$
   r_1: the daily wage of first worker, $r_1 = \$300$
   r_2: the daily wage of second worker, $r_2 = \$600$
10) d: David's current age, l: Lucy's current age, $\begin{cases} d = \frac{3}{2}l \\ d + l = 30 \end{cases}$ and $d = 18$ and $l = 12$.

Chapter 7: Linear Equations

Topics that you'll learn in this chapter:

- ✓ Forms of Linear Equations
- ✓ Equations of Horizontal and Vertical Lines
- ✓ Slope From a Graph and Two Points
- ✓ Slopes of Horizontal and Vertical Lines
- ✓ Rate of Changes
- ✓ Graphing Line from an Equations
- ✓ Equation from a Graph
- ✓ Slopes of Parallel and Perpendicular Lines
- ✓ Writing Equations of Parallel and Perpendicular Lines
- ✓ Midpoint and Distance Formulas
- ✓ Real World Applications
- ✓ Worksheets
- ✓ Answer of Worksheets

Forms of Linear Equations

Linear equations can be expressed in several forms. Here are the main forms you'll encounter:

1. Slope-Intercept Form

This is the most common form of a linear equation (as we discussed on previous page) and is written as:

$$y = mx + b$$

2. Point-Slope Form

This form is useful when you know the slope of a line and a point on the line. It is written as:

$$y - y_1 = m(x - x_1)$$

where:

- (x_1, y_1) is a specific point on the line,
- m is the slope of the line.

3. Standard Form

This form is useful for certain types of problems, such as finding interception. It is written as:

$$Ax + By = C$$

where:

- A, B, and C are integers,
- x and y are variables.

Examples:

1) If the slope of a line is 2 and it crosses the y-axis at 3, write the equation of this line.

Solution:

Since the slope and the y-intercept are given, it is easy to use the first form of the linear equation:

$y = mx + b, m = 2$ and $b = 3$, so $y = 2m + 3$

2) Write a linear equation that passes through $(-1,2)$ and the slope is -5.

Solution:

Since we have one point and the slope, we can use the slope-point form of the linear equation:

$y - y_1 = m(x - x_1), m = -5$ and $(x_1, y_1) = (-1,2)$ so, $y - 2 = -5(x + 1) \rightarrow$

$y = -5x - 3$

Equations of Horizontal and Vertical Lines

Horizontal Lines:

A horizontal line is a straight line that goes from left to right. In other words, it has a constant y-value for all x-values. This means that no matter where you are on the line, the y-value does not change.

Equation of a Horizontal Line:

$$y = c$$

where:

- y is the dependent variable,
- c is a constant value.

For example, the equation $y = 3$ represents a horizontal line that crosses the y-axis at 3 and stays at $y = 3$ for all x-values.

Vertical Lines:

A vertical line is a straight line that goes up and down. This means it has a constant x-value for all y-values. The slope of a vertical line is undefined because there is no change in the x-values (the run is zero).

Equation of a Vertical Line:

$$x = k$$

where:

- x is the independent variable,
- k is a constant value.

For example, the equation $x = -2$ represents a vertical line that crosses the x-axis at -2 and stays at $x = -2$ for all y-values.

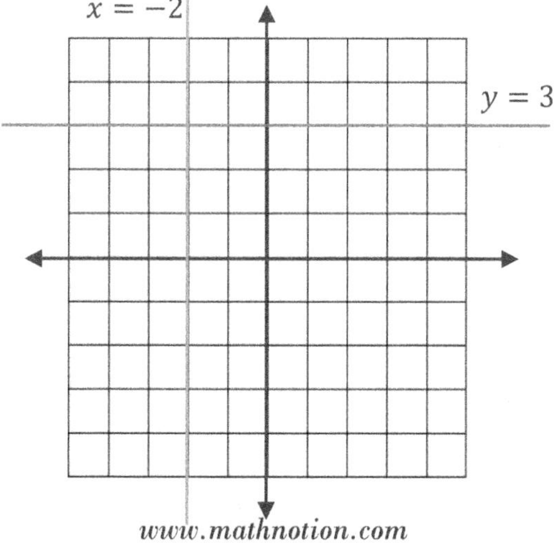

Slope From a Graph and Two Points

The slope m of a line can be calculated using the formula:

$$m = \frac{rise}{run}$$

where:

- **Rise** is the change in the y-coordinate (vertical change).
- **Run** is the change in the x-coordinate (horizontal change).

Alternatively, if you know two points (x_1, y_1) and (x_2, y_2) on the line, you can use the formula:

$$m = \frac{y_2 - y_1}{x_2 - x_1}$$

How to Find the Slope from a Graph:

1. **Identify Two Points on the Line:** Choose any two points on the line. It's helpful to pick points where the line crosses grid intersections for accuracy.
2. **Calculate the Rise and the Run:** Determine the vertical and the horizontal change between the two points.
3. **Divide the Rise by the Run:** Use the slope formula to calculate the slope.

Example:

Find the slope from the line.

Solution:

1. Identify two points on the line: We choose points $(2, 3)$ and $(-4, -2)$.
2. Calculate the rise and the run:

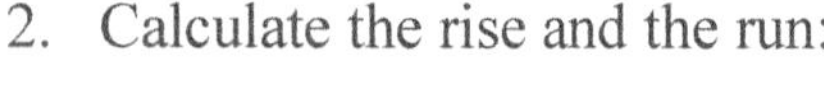

   Rise= $y_2 - y_1 = -2 - 3 = -5$

   Run= $x_2 - x_1 = -4 - 2 = -6$

3. Divide the rise by the run:

   $m = \frac{rise}{run} = \frac{y_2 - y_1}{x_2 - x_1} = \frac{-5}{-6} = \frac{5}{6}$

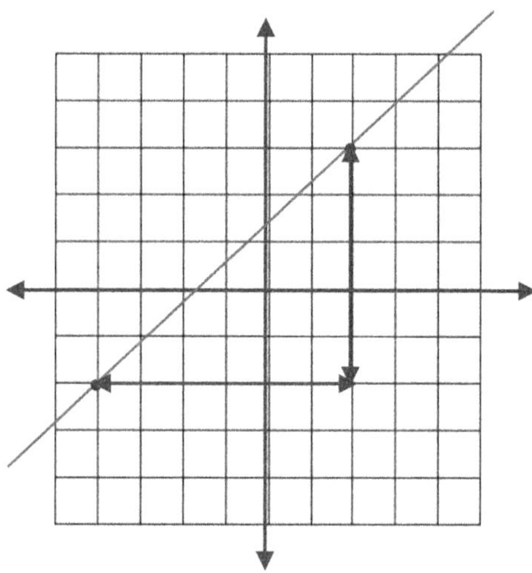

Slopes of Horizontal and Vertical Lines

Horizontal line: The main characteristic of a **horizontal line** is that it has the same y-value for all points on the line. This means there is no change in the y-value, regardless of how far you move along the x-axis.

Slope of a Horizontal Line:

The slope of a horizontal line is always 0 because there is no vertical change (rise) as you move horizontally (run).

$$m = \frac{Rise}{Run} = \frac{0}{Run} = 0$$

Vertical line: The main characteristic of a vertical line is that it has the same x −value for all points on the line. This means there is no change in the x-value, regardless of how far you move along the y-axis.

Slope of a Vertical Line:

The slope of a vertical line is undefined. This is because the run (horizontal change) is 0, and you cannot divide by zero.

$$m = \frac{Rise}{Run} = \frac{Rise}{0}$$

Since division by zero is undefined, the slope of a vertical line is also undefined.

Easy Memory Tip:

"Zero goes across, no slope goes up!" (This helps you match direction with the slope type).

"Horizontal = Zero slope" and "Vertical = No slope or undefined"

Examples:

1) Given the equation of a line $y = -3$, identify the slope.

   Solution:

   The line $y = -3$ is a horizontal line, so the slope is 0.

2) Find the slope of the line passing through the points $(-4, 5)$ and $(-4, 2)$.

   Solution:

   The equation of a line that passes through the points $(-4, 5)$ and $(-4, 2)$ is $x = -4$. This line is vertical, so the slope is undefined.

Rate of Changes

The rate of change tells us how one quantity changes in relation to another. In the context of linear equations, it describes how the y-value (dependent variable) changes as the x −value (independent variable) changes. Essentially, it's the measure of how steep a line is on a graph.

Formula for Rate of Change:

$$\text{Rate of Change} = \frac{\textit{Change in Dependent Variable}}{\textit{Change in Independent Variable}}$$

Average Rate of Changes of a Function:

The Average Rate of Change of a function is a measure of how the function's output value changes, on average, with respect to a change in the input value over a specific interval. In simpler terms, it's the "speed" at which the function is changing between two points.

Formula:

The Average Rate of Change of a function $f(x)$ between two points x_1 and x_2 is calculated using the following formula:

$$\text{Average Rate of Change} = \frac{f(x_2)-f(x_1)}{x_2-x_1}$$

Examples:

1) Imagine you have a plant that grows over time. You measure its height at different times to find out how fast it's growing.
   - At the beginning (Time 0 weeks), the plant is 10 cm tall.
   - After 3 weeks, the plant is 22 cm tall.

   Find the rate of change in height over these 3 weeks

   Solution:

   To find the rate of change in height over these 3 weeks, we'll use the formula for the rate of change:

   Rate of change $= \frac{\textit{Change in height}}{\textit{Change in time}} = \frac{22cm-10cm}{3weeks-0weeks} = \frac{12cm}{3weeks} = 4cm/week$

2) Find the average rate of change of the function: $f(x) = x^2 - 5$ from $x = -2$ to $x = 0$.

   Solution:

   Calculate the function values:
   - $f(-2) = (-2)^2 - 5 = 4 - 5 = -1$
   - $f(0) = 0^2 - 5 = 0 - 5 = -5$

   Use the formula: $\frac{f(0)-f(-2)}{0-(-2)} = \frac{-5-(-1)}{2} = \frac{-5+1}{2} = \frac{-4}{2} = -2$

Graphing Line from an Equations

To graph a line from an equation, you'll generally be working with a linear equation of the form: $y = mx + b$ where m is the slope and b is the y-intercept.

Steps to Graph the Line:

1. **Identify the slope and y-intercept:**
   - As it was discussed before, the slope m tells you how much the line goes up or down for each step to the right.
   - The y-intercept b is the point where the line crosses the y-axis. This means that when $x = 0$, the value of y is equal to b.
2. **Plot the y-intercept:** Start by plotting the point where the line crosses the y-axis, which is at $(0, b)$.
3. **Use the Slope:** The slope is written as a fraction $\frac{m}{1}$ (if it's a whole number, put it over 1). The numerator tells you how many units to go up (if positive) or down (if negative), and the denominator tells you how many units to move right.
4. **Plot another point using the slope**: From the y-intercept, use the slope to move to another point on the line. Repeat the process to find a few more points.
   ☑ To find different points on a line, we can also assign various values to x or y and obtain different results.
5. **Draw the line**: Once you have enough points (usually 2 or 3 is enough), draw a straight line through them. The line should extend in both directions.

Example:

Graph the line with the equation $y = -2x + 4$.

Solution:

The slope $m = -2$ means the line goes down 2 units for every 1 unit to the right.

The y-intercept $b = 4$ means the line crosses the y-axis at (0,4).

Start at the point (0,4).

From there, move down 2 units and right 1 unit to plot the next point at (1,2).

Draw the line through the points (0,4) and (1,2) and extend it in both directions.

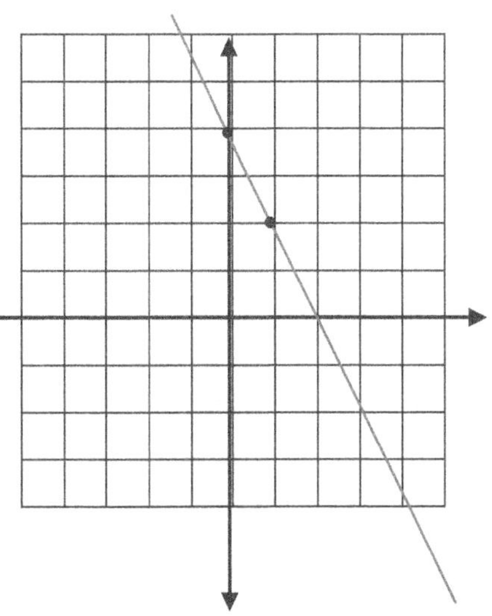

Equation from a Graph

To find the equation of a line from a graph, we need to figure out two things: the slope and the y-intercept.

Steps to Find the Equation from a Graph:

1. **Identify the y-intercept (b):** the y-intercept is the point where the line crosses the y-axis (when $x = 0$). When the graph clearly displays it, you can read it straight off. If not, utilize one of the points and the slope to determine it.
2. **Determine the slope (m):**
   - The slope is the ratio of the vertical change (rise) to the horizontal change (run) between two points on the line.
   - Find two points on the graph. Let's say you have points (x_1, y_1) and (x_2, y_2).
   - Calculate the slope using the formula: $m = \frac{y_2 - y_1}{x_2 - x_1}$
3. **Write the equation:** Write the equation of line in the form $y = mx + b$.

Example:

Write the equation of the graph.

Solution:

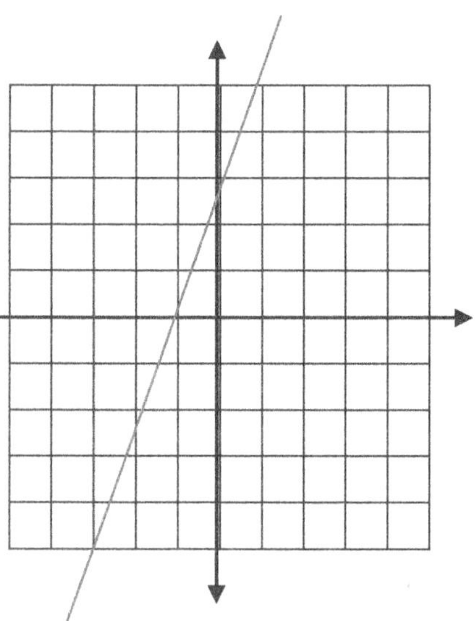

1. Identify two different points: The line passes through the points $(x_1, y_1) = (-1,0)$ and $(x_2, y_2) = (-3, -5)$.
2. Calculate the slope using formula:

   $m = \frac{y_2 - y_1}{x_2 - x_1} = \frac{-5-0}{-3-(-1)} = \frac{-5}{-2} = \frac{5}{2}$
3. Identify the y-intercept (b): Since it's not easy to determine the y-intercept from the graph, we can simply substitute a point (for example $(-1,0)$) into the equation and solve for the y-intercept.

   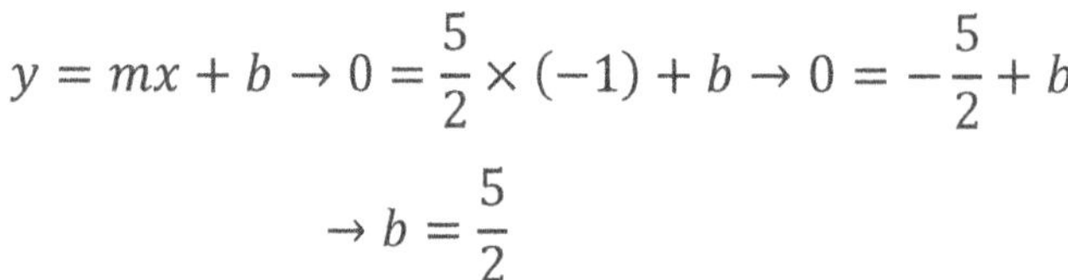

   $$y = mx + b \rightarrow 0 = \frac{5}{2} \times (-1) + b \rightarrow 0 = -\frac{5}{2} + b \rightarrow b = \frac{5}{2}$$
4. Write the equation: $y = \frac{5}{2}x + \frac{5}{2}$

Slopes of Parallel and Perpendicular Lines

Slopes of Parallel Lines: Parallel lines are lines that never intersect each other. They have the same steepness, so their slopes are equal.

Rule: If two lines are parallel, their slopes are the same.

$$m_1 = m_2$$

For example, two lines $y = -3x + 1$ and $y = -3x - 5$ are parallel because their slopes are equal.

Slopes of Perpendicular Lines: Perpendicular lines are lines that intersect at a right angle (90 degrees). Their slopes have a special relationship.

Rule: If two lines are perpendicular, their slopes are negative reciprocals of each other.

$$m_1 = -\frac{1}{m_2} \; or \; m_1 \times m_2 = -1$$

For example, two lines $y = 2x - 1$ and $y = -\frac{1}{2}x + 3$ are perpendicular because their slopes are negative reciprocals of each other $(2 \times -\frac{1}{2} = -1)$.

Examples:

1) The two lines below are parallel. Calculate the value of m.
   - $(2m - 1)x - y - 1 = 0$
   - $3x - 2y + 2 = 0$

   Solution:

   1. We need to find the slopes of both lines and set them equal.
   2. Rewrite both equations in slope-Intercept form ($y = mx + b$):

      $(2m - 1)x - y - 1 = 0 \rightarrow -y = -(2m - 1)x + 1 \rightarrow y = (2m - 1)x - 1 \rightarrow m_1 = 2m - 1$

      $3x - 2y + 2 = 0 \rightarrow -2y = -3x - 2 \rightarrow y = \frac{3}{2}x + 1 \rightarrow m_2 = \frac{3}{2}$
   3. Set the slopes equal ($m_1 = m_2$): $2m - 1 = \frac{3}{2} \rightarrow 4m - 2 = 3 \rightarrow 4m = 3 + 2 \rightarrow 4m = 5 \rightarrow m = \frac{5}{4}$

2) The two lines below are Perpendicular. Calculate the value of n.
   - $(1 - n)x + y - 3 = 0$
   - $3y - x + 1 = 0$

   Solution:

   1. Two lines are perpendicular if and only if the product of their slopes is -1.
   2. Rewrite both equations in slope-Intercept form ($y = mx + b$):
      - $(1 - n)x + y - 3 = 0 \rightarrow y = -(1 - n)x + 3 \rightarrow y = (n - 1)x + 3 \rightarrow m_1 = n - 1$
      - $3y - x + 1 = 0 \rightarrow 3y = x - 1 \rightarrow y = \frac{1}{3}x - \frac{1}{3} \rightarrow m_2 = \frac{1}{3}$
   3. Use the perpendicular condition:

      $m_1 \times m_2 = -1 \rightarrow (n - 1) \times \frac{1}{3} = -1 \rightarrow \frac{(n-1)}{3} = -1 \rightarrow n - 1 = -3 \rightarrow n = -2$

Writing Equations of Parallel and Perpendicular Lines

Steps to Write Equations of Parallel and Perpendicular Lines:

1. **Identify the Given Information:**
   - The slope of the original line.
   - A point that the new line passes through.
2. **Find the Slope of the New Line:**
   - For parallel lines, use the same slope as the original line.
   - For perpendicular lines, use the negative reciprocal of the original slope.
3. **Use the Point-Slope Form:** $y - y_1 = m(x - x_1)$
4. **Simplify to Slope-Intercept Form:** Convert the equation from point-slope form to slope-intercept form ($y = mx + b$) by solving for y.

Examples:

1) The equation of Line A is $y = \frac{3}{4}x - 2$. Line B is parallel to Line A and passes through the point $(8,5)$. Write the equation of Line B in slope-intercept form.

   Solution:

   1. Slope of line B: Since line B is parallel to line A, their slopes are the same. The slope of line A is $\frac{3}{4}$, so the slope of line B is also $\frac{3}{4}$.
   2. Equation of line B: Use the point-slope form to find the equation of line B. Here, $m = \frac{3}{4}$, and the point $(x_1, y_1) = (8,5)$. Substitute these values into the equation:

      $y - 5 = \frac{3}{4}(x - 8) \rightarrow y - 5 = \frac{3}{4}x - 6 \rightarrow y = \frac{3}{4}x - 6 + 5 \rightarrow y = \frac{3}{4}x - 1$

2) The equation of Line C is $y = -\frac{2}{5}x + 7$. Line D is perpendicular to line C and passes through the point $(-10,4)$. Write the equation of line D in slope-intercept form.

   Solution:

   1. Slope of line D: Since line D is perpendicular to line C, its slope is the negative reciprocal of the slope of line C. The slope of line C is $-\frac{2}{5}$, so the slope of line D is also $\frac{5}{2}$.
   2. Equation of line D: Use the point-slope form to find the equation of line D. Here, $m = \frac{5}{2}$, and the point $(x_1, y_1) = (-10,4)$. Substitute these values into the equation:

      $y - 4 = \frac{5}{2}\big(x - (-10)\big) \rightarrow y - 4 = \frac{5}{2}x + 25 \rightarrow y = \frac{5}{2}x + 25 + 4 \rightarrow y = \frac{5}{2}x + 29$

Midpoint and Distance Formulas

These formulas are super useful in geometry and algebra, and they help us find important information about points on a coordinate plane.

Midpoint Formula:

The midpoint formula is used to find the exact middle point between two points on a coordinate plane. If you have two points, (x_1, y_1) and (x_2, y_2), the midpoint M is calculated as:

$$M = (\frac{x_1 + x_2}{2}, \frac{y_1 + y_2}{2})$$

Distance Formula:

The distance formula is used to find the distance between two points on a coordinate plane. If you have two points, (x_1, y_1) and (x_2, y_2), the distance d between them is calculated as:

$$d = \sqrt{(x_2 - x_1)^2 + (y_2 - y_1)^2}$$

Tip:

"Make a right triangle and use the Pythagorean Theorem!"

It's just like finding the hypotenuse of a triangle.

Examples:

1) Find the midpoint between the points $A(2,4)$ and $B(6,8)$.

Solution:

1. Identify the coordinates:
   - $(x_1, y_1) = (2,4)$
   - $(x_2, y_2) = (6,8)$
2. Plug the values into the midpoint formula: $M = \left(\frac{2+6}{2}, \frac{4+8}{2}\right) = \left(\frac{8}{2}, \frac{12}{2}\right) = (4, 6)$

2) Find the distance between the points $C(1,3)$ and $D(4,7)$.

Solution:

1. Identify the coordinates:
   - $(x_1, y_1) = (1,3)$
   - $(x_2, y_2) = (4,7)$
2. Plug the values into the distance formula:

   $d = \sqrt{(4 - 1)^2 + (7 - 3)^2} = \sqrt{3^2 + 4^2} = \sqrt{9 + 16} = \sqrt{25} = 5$

Real World Applications

Linear equations are used in many areas of everyday life, business, and science. Here are some broader, real-world applications:

Economics and Business: In business, linear equations can model the relationship between costs, revenues, and profits.

Supply and Demand: The laws of supply and demand can be represented using linear equations. As the price of a product rises, the demand for it may decrease, which can be modeled with a linear equation to predict how price changes affect sales.

Construction and Engineering: Linear equations help in designing structures like buildings, bridges, or roads. For example, engineers may use linear equations to calculate the slope of a roof, or the strength of materials needed.

Environmental Science: In environmental science, linear equations can model population growth if the population grows at a constant rate. The equation helps predict future population sizes over time.

Technology and Computing: In computer science, linear equations are used to analyze the efficiency of algorithms, particularly in cases where the time complexity grows linearly with the size of the input.

Electricity and Energy: The relationship between power consumption and time is linear for many electrical devices. For instance, if a light bulb uses 100 watts per hour, the total energy consumption over time can be modeled with a linear equation.

Example:

You're helping to set up a new internet café and need to determine the total cost of purchasing computers. Each computer costs $500, and there is a one-time setup fee of $2,000 for the entire café. Write a linear equation to represent the total cost (C) based on the number of computers (n) you purchase.

Solution:

To represent this situation as a linear equation, you can use the following formula:

$$C = 500n + 2000$$

In this equation:

C represents the total cost in dollars

n represents the number of computers

500 is the cost per computer and 2,000 is the one-time setup free.

Worksheets

Forms of Linear Equations

Find answers.

1) Write the equation of a line that has a slope of 3 and passes through the point $(0, 2)$.
2) Convert the equation $y = 4x - 5$ to standard form $Ax + By = C$.
3) Find the x-intercept and y-intercept of the line with the equation $5x + 2y = 10$.
4) Determine the value of m such that the line $(m - 1)x - 2my = 0$ passes through the point $(2, -1)$.
5) Determine the value of a such that the point $(1 - a, 6)$ lies on the line $2x - \frac{5}{3}y = 0$.

Equations of Horizontal and Vertical Lines

Do the following problems:

1) Write the equation of a horizontal line that passes through the point $(3, 5)$.
2) Write the equation of a vertical line that passes through the point $(-2, 4)$.
3) Determine the x −intercept of the vertical line $x = 4$.
4) Find the equation of a horizontal line that passes through the point where the line $y = 2x + 3$ intersects the y −axis.
5) Write the equation of a vertical line that is 5 units to the right of the line $x = -3$.

Slope From a Graph and Two Points

Find the slope between the points:

1) $(-3, 5)$ *and* $(1,3)$
2) $(2, 5)$ *and* $(2, -9)$
3) $\left(\frac{-2}{3}, 1\right)$ *and* $\left(1, \frac{-2}{3}\right)$
4) $(0, 0)$ *and* $(1.25, 0.75)$
5) $\left(1\frac{2}{5}, -\frac{1}{4}\right)$ *and* $\left(\frac{5}{6}, -2\frac{1}{2}\right)$

Find the slope of the lines.

6)

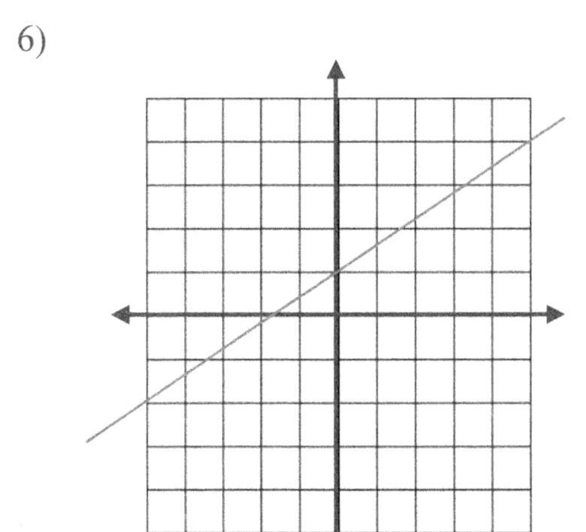

7)

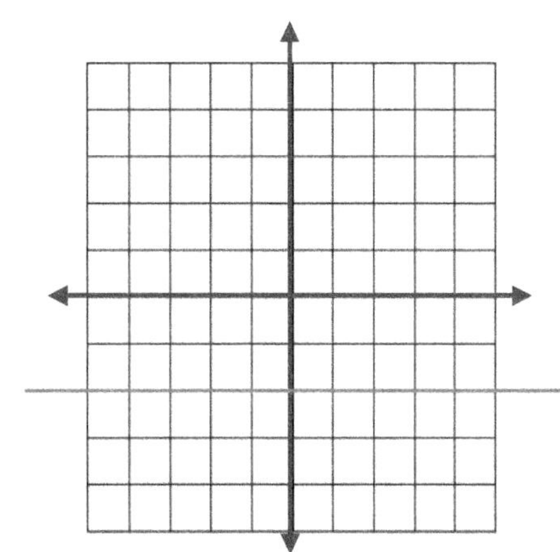

8)

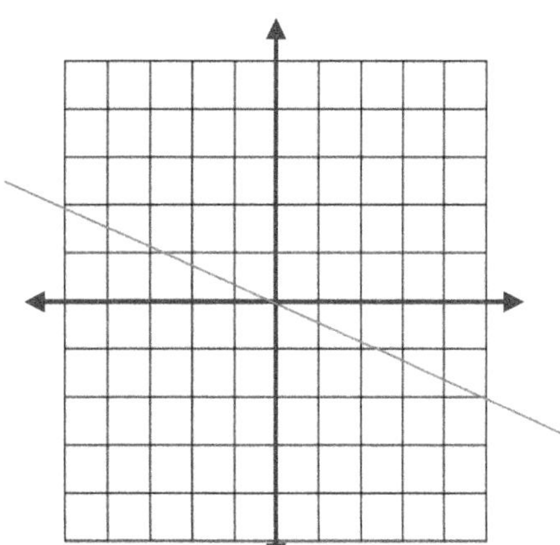

9)

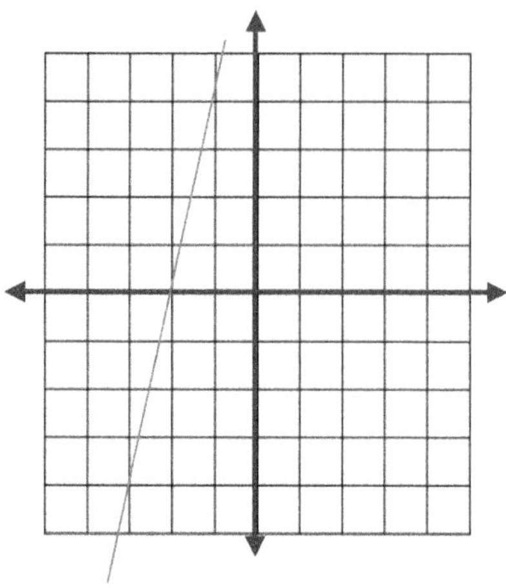

10)

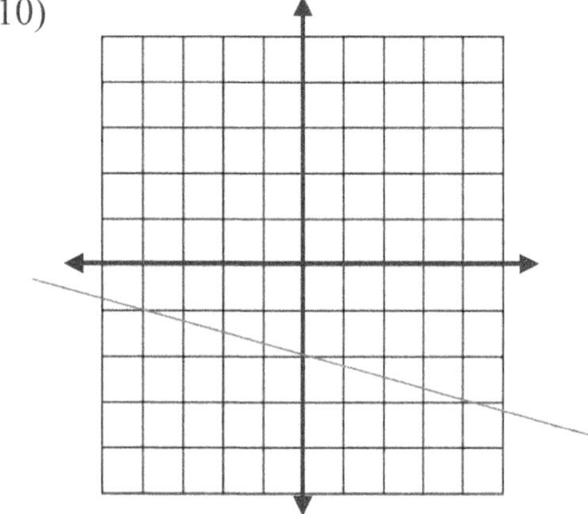

Slopes of Horizontal and Vertical Lines

Find the answers.

1) Find the slope of the horizontal line passing through the point $(4, 7)$.
2) Find the slope of a line that passes through the points $(2,3)$ and $(2, -4)$.
3) The slope of a horizontal line is $2m - 3$, find the m.
4) If the slope of a vertical line is calculated as $\frac{3ab}{2b-a}$, find the relationship between a and b.
5) Determine the value of m such that points $A: (m + 5, -3)$ and $B: (-1, 2m + 5)$ lie on a line that is perpendicular to the line $x = -5$.

Rate of Changes

Do the following problems:

1) Find the rate of change of the distance traveled by car if it covers 100 kilometers in 2 hours.
2) Given the function $f(x) = 3x + 2$, find the average rate of change from $x = 1$ to $x = 4$.
3) The temperature in a city rises from $15°C$ at 8 AM to $30°C$ at 2 PM. Find the average rate of change of temperature per hour.
4) The population of a town increased from 50,000 to 80,000 over a period of 10 years. What is the average rate of change in population per year?
5) What is the average rate of change of the function: $\sqrt{3x^3 - 2x^2}$ from $x = 2$ to $x = 3$?

Graphing Line from an Equations

Graph following linear equations:

1) $y = -3x$
2) $y = 4x + 3$
3) $2y - x = -5$
4) $\frac{3}{4}y - \frac{1}{2}x + \frac{1}{4} = 0$
5) $\frac{y-x}{3} = \frac{5}{6}$

Equation from a Graph

Write the equation of the following lines:

1)

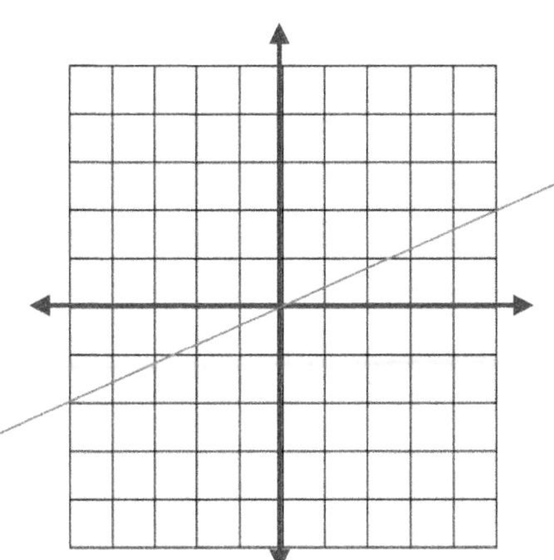

2)

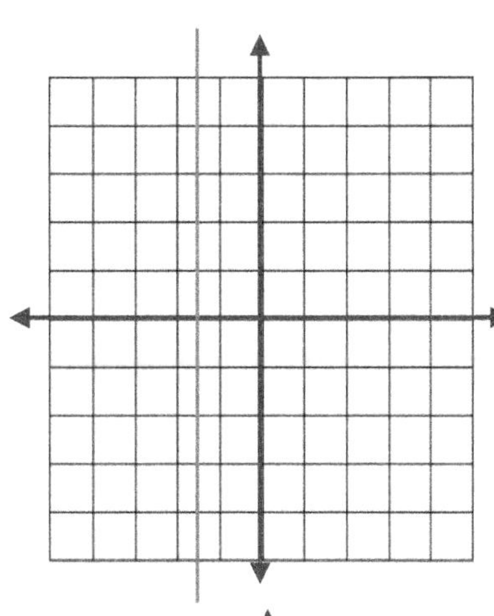

3)

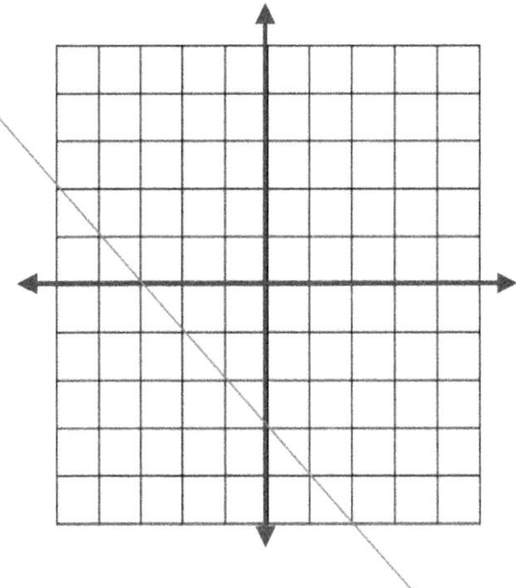

4)

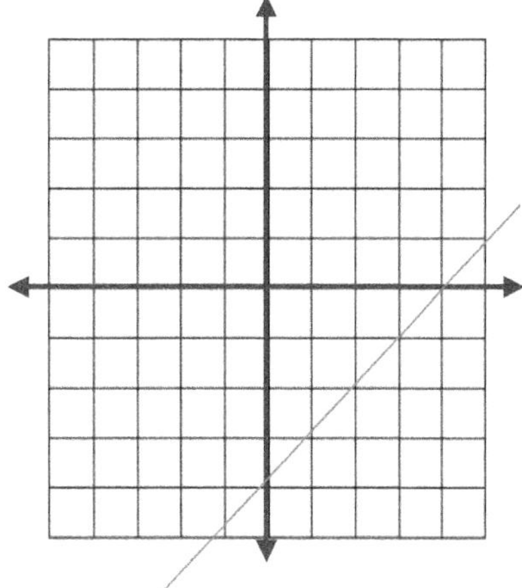

5)

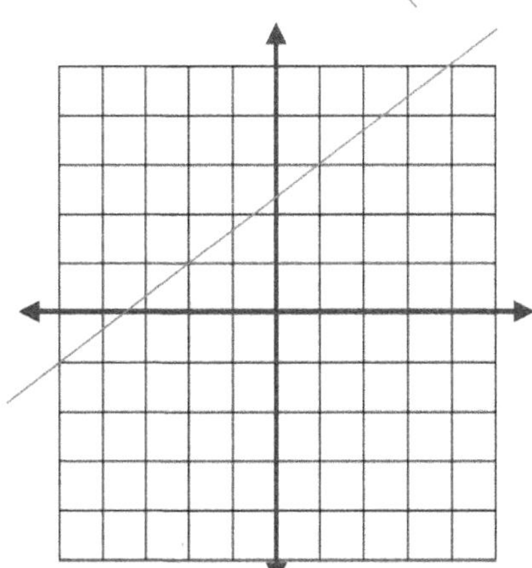

Slopes of Parallel and Perpendicular Lines

Do the problems.

1) Determine if the lines given by the equations $y = -2x + 5$ and $2y - 4x - 1 = 0$ are parallel.
2) Determine if the lines given by the equations $3y + 5x = 0$ and $y = \frac{3x-2}{5}$ are perpendicular.
3) If the equation of a line is $-y = -0.2x + 3$, what is the slope of a line that is parallel to it?
4) If the two lines $y = 4x + 1$ and $y = (-2a + 1)x + 7$ are perpendicular, find the value of a.
5) If the two lines $y = (2a)x + 1$ and $2y + bx = 5$ are parallel, find the relationship between a and b.

Writing Equations of Parallel and Perpendicular Lines

Do the following problems:

1) Write the equation of a line that is parallel to $y = 2.5x + 4$ and passes through the point $(-1, 4)$.
2) Write the equation of a line that is perpendicular to $y = -\frac{1}{4}x + 2$ and passes through the point $(-3, 0)$.
3) Write the equation of a line that is perpendicular to the line passing through the points $(6, -2)$ and $(8, 7)$, and passes through the point $(-1, -5)$.
4) Write the equation of a line that intersects the line $x - 2y = 3$ at a point with $x = -5$ and is parallel to the line $\frac{x}{2} = \frac{y}{4} + 1$.
5) Write the equation of a line that is perpendicular to the line $\frac{x+2y}{3} = \frac{y-1}{2}$ at the point where it intersects the y-axis.

Midpoint and Distance Formulas

Solve.

1) Find the midpoint of the line segment connecting the points $(-2, 1.5)$ and $(5, -2.5)$.
2) Find the distance between the points $(-1, 0)$ and $(-4, -6)$.
3) The midpoint of the line segment connecting the points $(3, 1)$ and $(-x, 9)$ is $(5, 5)$ Find the value of x.
4) If the three vertices of a triangle are as $A: (2,5), B: (-6, -3)$ and $C: (4, 1)$ find the perimeter of the triangle.
5) Write the equation of a line that passes through the midpoint of points $A: (-3, 0)$ and $B: (5, 2)$ and is perpendicular to the line $y = -2x + 5$

Real World Applications

Do the word problems.

1) You buy 3 notebooks for \$2 each and a pack of pens for \$5. Write a linear equation to represent the total cost (C) of buying x packs of pens.
2) The standards require that the ramp has a slope of $1: 12$, meaning for every inch of vertical rise, the ramp should extend 12 inches horizontally. Write the linear equation that represents the relationship between the vertical rise (y) and the horizontal run (x) of the ramp.

3) A company sells widgets for $50 each and has fixed costs of $2,000. The cost to produce each widget is $30. Write a linear equation to represent the profit (P) based on the number of widgets (w) sold.
4) You need to mix a 10% salt solution with a 20% salt solution to make 50 liters of a 15% salt solution. Write a linear equation to represent the amounts of the two solutions needed.
5) A family uses 200 gallons of water in the first month. After that, they use 150 gallons each month. Write a linear equation to represent the total amount of water used w after m months. How much water will they use in 6 months?
6) A company produces gadgets at a cost of $8 per gadget and sells them for $12 each. Write a linear equation to represent the company's profit (P) based on the number of gadgets (g) sold.
7) An economist models the relationship between the supply (S) of a product and its price (P) as $S = 2P - 10$. Meanwhile, the demand (D) is modeled as $D = 30 - 3P$. Determine the price (P) at which supply equals demand.
8) A chemist needs to mix a solution that is 30% acid with a solution that is 60% acid to get 10 liters of a 50% acid solution. Write a system of linear equations to represent this problem and solve the amounts of each solution needed.
9) Three friends, Alice, Bob, and Charlie, are discussing their ages. They know the following:
   - The sum of their ages is 91 years.
   - Alice is 5 years older than Bob.
   - Charlie is twice as old as Alice.

   The problem is to find their individual age.
10) A construction company is planning to build a rectangular garden with a walkway surrounding it. The garden's length is 3 times its width. The walkway is 2 meters wide all around the garden. The total area of the garden and the walkway combined is 80 square meters more than just the garden's area. Find the dimensions of the garden.

Answer of Worksheets

Forms of Linear Equations

1) $y = 3x + 2$
2) $4x - y = 5$
3) $(2,0)$ and $(0,5)$
4) $m = \frac{1}{2}$
5) $a = -4$

Equations of Horizontal and Vertical Lines

1) $y = 5$
2) $x = -2$
3) $(4,0)$
4) $y = 3$
5) $x = 2$

Slope From a Graph and Two Points

1) $-\frac{1}{2}$
2) Undefined
3) -1
4) $\frac{3}{5}$
5) $\frac{135}{34}$
6) $\frac{3}{5}$
7) 0
8) $\frac{-2}{5}$
9) 4
10) $-\frac{1}{4}$

Slopes of Horizontal and Vertical Lines

1) 0
2) Undefined
3) $\frac{3}{2}$
4) $a = 2b$
5) $m = -4$

Rate of Changes

1) 50 km/hour
2) 3
3) $2.5°C$
4) 3,000 people
5) $3\sqrt{7} - 4$

Graphing Line from an Equations

1) $y = -3x$

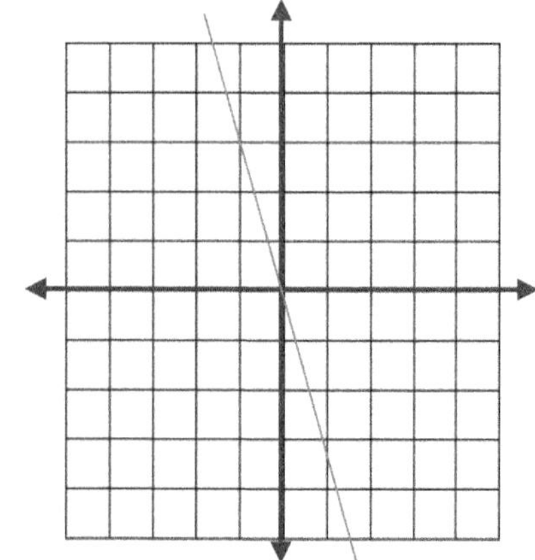

2) $y = 4x + 3$

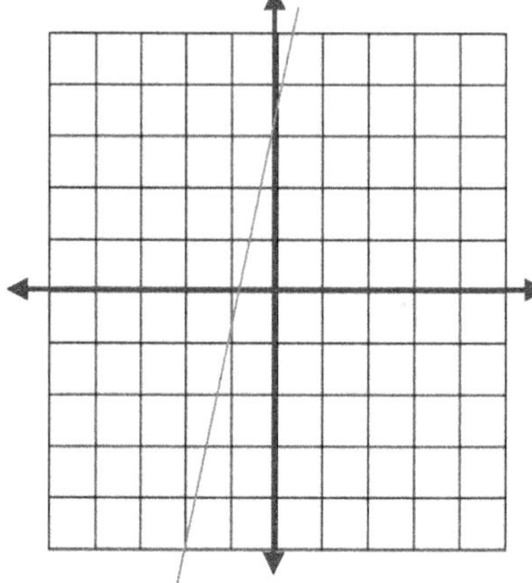

3) $2y - x = -5$

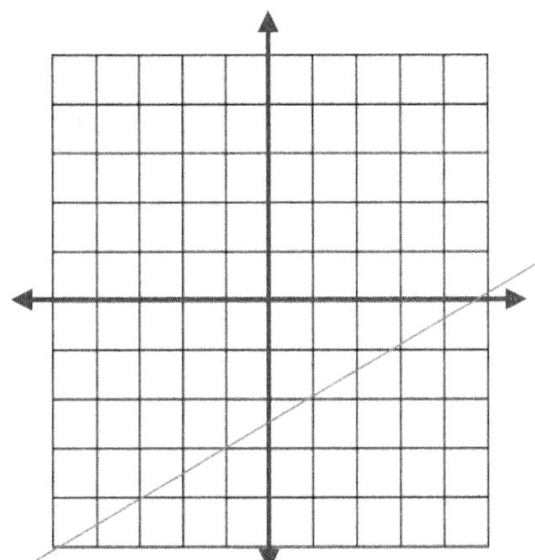

4) $\frac{3}{4}y - \frac{1}{2}x + \frac{1}{4} = 0$

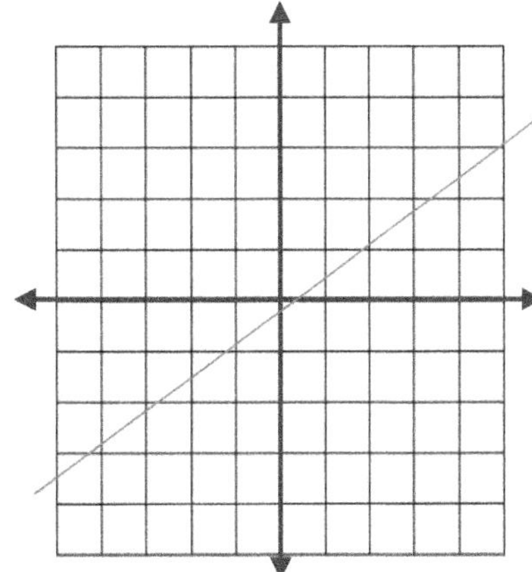

5) $\frac{y-x}{3} = \frac{5}{6}$

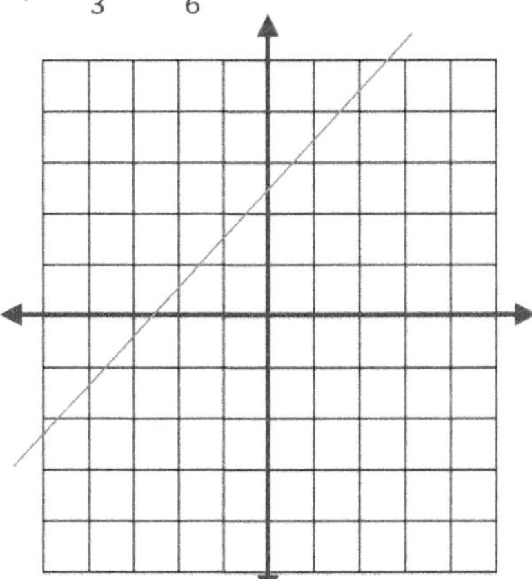

Equation from a Graph

1) $y = \frac{2}{5}x$
2) $x = -1.5$
3) $y = -x - 3$
4) $y = x - 4$
5) $y = \frac{2}{3}x + \frac{7}{3}$

Slopes of Parallel and Perpendicular Lines

1) Not parallel.
2) Perpendicular
3) 0.2
4) $\frac{5}{8}$
5) $b = -4a$

Writing Equations of Parallel and Perpendicular Lines

1) $y = 2.5x + 6.5$
2) $y = 4x + 12$
3) $y = -\frac{2}{9}x - \frac{47}{9}$
4) $y = 2x + 6$
5) $y = \frac{1}{2}x - 3$

Midpoint and Distance Formulas

1) $(1.5, -0.5)$
2) $3\sqrt{5}$
3) -7
4) $8\sqrt{2} + 2\sqrt{29} + 2\sqrt{5}$
5) $y = \frac{1}{2}x + \frac{1}{2}$

Real World Applications

1) $C = 6 + 5x$
2) $y = \frac{1}{12}x$
3) $P = 20w - 2000$
4) $\begin{cases} x + y = 50 \\ 0.10x + 0.20y = 7.5 \end{cases}$
5) $w = 200 + 150(m - 1), m = 6 \rightarrow w = 950$
6) $P = 4g$
7) $P = 8$
8) $\begin{cases} x + y = 10 \\ 0.30x + 0.60y = 5 \end{cases}$, $x = \frac{10}{3}$ liters and $y = \frac{20}{3}$ liters
9) Bob: 19 years, Alice: 24 years and Charlie: 48 years
10) Width: 4 meters and Length:12 meters

Chapter 8: Polynomials

Topics that you'll learn in this chapter:

- ✓ Polynomials and Algebraic Expressions
- ✓ Writing Polynomials in Standard Form
- ✓ Simplifying Polynomials
- ✓ Adding and Subtracting Polynomials
- ✓ Multiplying and Dividing Monomials
- ✓ Multiplying Polynomials and Monomials
- ✓ Multiplying Binomials
- ✓ Division of Polynomials
- ✓ Factoring Trinomials
- ✓ Real World Applications
- ✓ Worksheets
- ✓ Answer of Worksheets

Polynomials and Algebraic Expressions

Polynomials:

A polynomial is a specific type of algebraic expression that consists of variables (also known as indeterminates), coefficients, and non-negative integer exponents. The terms in a polynomial are combined using addition, subtraction, and multiplication.

Examples: $2x^3 - x^2 + 9, y^3 - 5$ and 7 (a constant polynomial)

Algebraic Expressions:

As we discussed in the fifth chapter, an algebraic expression is a broader category that includes polynomials but also encompasses other types of expressions involving variables, coefficients, and arithmetic operations (addition, subtraction, multiplication, division, and exponentiation). Algebraic expressions can include terms with negative exponents, fractional exponents, or even variables in the denominator.

Examples: $2x + \frac{3}{x}, y^2 + \sqrt{y} - 5$ and $\frac{1}{a+1}$

Key Differences:

1. **Exponents:**
   - **Polynomials:** Only non-negative integer exponents.
   - **Algebraic Expressions**: Can have negative, fractional, or radical exponents.
2. **Operations:**
   - **Polynomials**: Only addition, subtraction, and multiplication of terms.
   - **Algebraic Expressions**: Can include division, roots, and more complex operations.
3. **Types of Terms:**
   - **Polynomials**: Terms are combined linearly.
   - **Algebraic Expressions:** Terms can involve more complex relationships, including rational and radical terms.

Example:

Determine which of the following expressions are polynomials and which are algebraic expressions.

a) $10x^2 - x^3 + 3x^4$ b) $2\sqrt{x^2 - 4} + 5x$ c) $\frac{1}{3}x^{-2} + 2x + 3$

Solution:

a) Polynomial b) Algebraic expression c) Algebraic expression

Writing Polynomials in Standard Form

Here are some key points about polynomials:

- **Terms:** Each individual part of the polynomial separated by addition or subtraction is called a term. For example, $3x^3, 2x^2, -5x,$ and 7 are terms of the polynomial $3x^3 + 2x^2 - 5x + 7$.
- **Coefficients:** The numerical factors multiplied by the variable(s) are called coefficients. In $3x^3$, the coefficient is 3.
- **Constant Term:** The term in the polynomial without any variables is called the constant term. In $3x^3 + 2x^2 - 5x + 7$, the constant term is 7.
- **Degree:** The degree of the polynomial is the highest sum of exponents. To determine the degree of a polynomial, follow these steps:
    - For each term, add up the exponents of all variables present in that term.
    - If a term has no variable, its degree is 0.
    - If a term has only one variable, the sum of exponents is just the exponent of that variable.

  For example, the highest degree of $3x^3y + 2yx^2 - 5x + 7$ is 4 (term $x^3y: 3 + 1 = 4$).

Steps to Write Polynomials in Standard Form:

1. **Identify the Terms**: Break down the polynomial into its individual terms.
2. **Order the Terms:** Arrange the terms so that the exponents of the variable decrease from left to right.
3. **Combine Like Terms:** If there are any like terms (terms with the same variable and exponent), combine them by adding or subtracting their coefficients.
4. **Write the Final Expression:** Write the ordered terms as a single polynomial expression.

Example:

Convert the polynomial $-2x^3 + 7x^2 + 12x + 6 - 4x^5$to standard form.

Solution:

1. Identify the terms: $-2x^3, 7x^2, 12x, 6$ and $-4x^5$
2. Order the terms: Arrange the terms in descending order of the exponents:
    - $-4x^5$
    - $-2x^3$
    - $7x^2$
    - $12x$
    - 6
3. Combine like terms: There are no like terms to combine in this case.
4. Write the final expression: The polynomial is: $-4x^5 - 2x^3 + 7x^2 + 12x + 6$

Simplifying Polynomials

Steps to Simplify Polynomials:

1. **Identify Like Terms:** Look for terms that have the same variable raised to the same exponent.
2. **Combine Like Terms:** Add or subtract the coefficients of like terms.
3. **Rewrite the Polynomial:** Write the simplified expression.

Example: Simplify the polynomial: $7a^2b - 3ab + 2a^2b + 4ab - 5$

Solution:

1. Identify like terms: $7a^2b$ and $2a^2b$, $-3ab$ and $4ab, -5$ (constant term)
2. Combine like terms: $7a^2b + 2a^2b = 9a^2b, -3ab + 4ab = ab$
3. Rewrite the polynomial: $9a^2b + \ ab - 5$

Adding and Subtracting Polynomials

To add or subtract polynomials, follow these steps:

1. **Align Like Terms**: Write the polynomials so that like terms (terms with the same variable and exponent) are aligned vertically.
2. **Distribute the Minus Sign (For subtracting Polynomials)**: Distribute the minus sign to each term of the polynomial being subtracted.
3. **Combine Like Terms**: Add or subtract the coefficients of the like terms.
4. **Rewrite the Result**: Write the resulting polynomial.

Examples:

1) Add $2xy^2 - 3x^2y + 5x^2$ and $9y^2x - y^2 + 5yx^2 - x^2$

Solution:

Align like terms: $2xy^2$ and $9y^2x, -3x^2y$ and $5yx^2, 5x^2$ and $-x^2$

Combine like terms:

$(2xy^2 + 9y^2x) + (-3x^2y + 5yx^2) + (5x^2 - x^2) - y^2 = 11xy^2 + 2yx^2 + 4x^2 - y^2$

2) Subtract $-x^3 - 2x^2y^2 + 4y^2 - x$ from $3x^2y^2 - x^3 + 2y^2 + 6x$

Solution:

Distribute the minus sign: $-(-x^3 - 2x^2y^2 + 4y^2 - x) = x^3 + 2x^2y^2 - 4y^2 + x$

Align like terms: $-x^3$ and $x^3, 2x^2y^2$and $3x^2y^2, -4y^2$and $2y^2, x$ and $+6x$

Combine like terms:

$(-x^3 + x^3) + (2x^2y^2 + 3x^2y^2) + (-4y^2 + 2y^2) + (x + 6x) = 5x^2y^2 - 2y^2 + 7x$

Multiplying and Dividing Monomials

A monomial is a type of polynomial that consists of a single term. It can be a constant, a variable, or a product of constants and variables raised to non-negative integer powers.

Here are the key characteristics of monomials:

1. **Single Term**: A monomial contains just one term.
2. **No Addition or Subtraction**: Monomials do not have addition or subtraction. If you have multiple terms, it's not monomial.
3. **Can Include Constants**: Constants (like $5, -3,$ or π) are considered monomial.

Multiplying Monomials:

1. **Multiply the coefficients (numerical parts)**: If there are numbers in front of the variables, multiply those first.
2. **Multiply the variables**: For variables with the same base, add their exponents.

Dividing Monomials:

1. **Divide the coefficients (numerical parts).**
2. **Subtract the exponents** of the same base variables.

Examples:

1) Multiply $2x^3$ and $-5x^2$

   Solution:

   1. Multiply the coefficients: $2 \times (-5) = -10$
   2. Multiply the variables: $x^3 \times x^2 = x^5$ (For variables with the same base (in this case, x), add the exponents)
   3. Putting it all together: $2x^3 \times -5x^2 = -10x^5$

2) Divide $\frac{12x^5y^3}{3x^2y}$

   Solution:

   1. Divide the coefficients: $\frac{12}{3} = 4$
   2. Subtract the exponents of the same base variables:
      - $\frac{x^5}{x^2} = x^3$
      - $\frac{y^3}{y} = y^2$
   3. Putting it all together: $\frac{12x^5y^3}{3x^2y} = 4x^3y^2$

Multiplying Polynomials and Monomials

Multiplying a polynomial by a monomial involves distributing the monomial to each term of the polynomial and then simplifying the result.

Steps to Multiply a Polynomial by a Monomial:

1. **Distribute the monomial to each term** in the polynomial.
2. **Multiply the coefficients** (numerical parts).
3. **Multiply the variables:** For variables with the same base, add their exponents.
4. **Combine the results** to get the final expression.

Example: Multiply $3x^2$ by the polynomial $4x^2 + 2x - 5$.

Solution:

1. Distribute $3x^2$ to each term in the polynomial: $3x^2.(4x^2 + 2x - 5)$
2. Multiply $3x^2$ by each term individually:
   - $3x^2 \times 4x^2 = 12x^4$
   - $3x^2 \times 2x = 6x^3$
   - $3x^2 \times -5 = -15x^2$
3. Combine the results: $12x^4 + 6x^3 - 15x^2$

Multiplying Binomials

A binomial is a type of polynomial that contains exactly two terms separated by either addition or subtraction. Each term can involve constants, variables, and/or powers of variables, but there are no other operations like division or multiplication between terms (except for the multiplication of the constants and variables within each term). For example, $5a^2 - 3a$ is a binomial.

Steps to Multiply Binomials using FOIL:

1. **First:** Multiply the first terms in each binomial.
2. **Outer:** Multiply the outer terms in each binomial.
3. **Inner:** Multiply the inner terms in each binomial.
4. **Last:** Multiply the last terms in each binomial.
5. **Combine all these products and simplify** if needed.

Example: Multiply $3x + 2$ and $x - 4$

Solution: According to the steps mentioned above:

$(3x + 2)(x - 4) = 3x \cdot x + 3x \cdot (-4) + 2 \cdot x + 2 \cdot (-4) = 3x^2 - 12x + 2x - 8 = 3x^2 - 10x - 8$

Division of Polynomials

Polynomial division is similar to long division of numbers. It involves dividing a polynomial (the dividend) by another polynomial (the divisor) to get a quotient and sometimes a remainder. Here's a step-by-step guide to the process:

Step-by-Step Guide to Polynomial Division:

Example: Let's divide $2x^3 + 3x^2 + 4x + 5$ by $x + 1$.

1. Set up the Division:
   - Write the dividend $(2x^3 + 3x^2 + 4x + 5)$ inside the division symbol.
   - Write the divisor $(x + 1)$ outside the division symbol.
2. Divide the Leading Terms:
   - Divide the leading term of the dividend by the leading term of the divisor.
   - $\frac{2x^3}{x} = 2x^2$
   - Write $2x^2$ as the first term of the quotient.
3. Multiply and Subtract:
   - Multiply the entire divisor by the term just found $(2x^2)$ and subtract this from the dividend.
   - $(2x^2) \cdot (x + 1) = 2x^3 + 2x^2$
   - Subtract: $(2x^3 + 3x^2 + 4x + 5) - (2x^3 + 2x^2) = x^2 + 4x + 5$
4. Repeat the Process:
   - Use the new dividend $(x^2 + 4x + 5)$ and repeat steps 2 and 3.
   - Divide the leading term: $\frac{x^2}{x} = x$
   - Multiply and subtract: $(x) \cdot (x + 1) = x^2 + x$
   - Subtract: $(x^2 + 4x + 5) - (x^2 + x) = 3x + 5$
5. Continue:
   - Repeat until the degree of the new dividend is less than the degree of the divisor.
   - Divide the leading term: $\frac{3x}{x} = 3$
   - Multiply and subtract: $(3) \cdot (x + 1) = 3x + 3$
   - Subtract: $(3x + 5) - (3x + 3) = 2$
6. Final Result:
   - The quotient is $2x^2 + x + 3$ and the remainder is 2.

Factoring Trinomials

A trinomial is a type of polynomial that contains exactly three terms separated by either addition or subtraction. Like binomials, the terms in a trinomial can include constants, variables, and/or powers of variables. For example, $x^2 + 5x + 6$ is a trinomial.

Factoring trinomials involve expressing the trinomial as the product of two binomials. Here's a step-by-step guide to help you through the process:

Steps to Factor Trinomials of the Form $ax^2 + bx + c$:

1. **Identify** a, b, **and** c in the trinomial $ax^2 + bx + c$.
2. **Find two numbers that multiply to** $a \cdot c$ **and add up to** b.
3. **Rewrite the middle term (bx) using the two numbers found in step 2.**
4. **Group the terms into two pairs.**
5. **Factor out the common factor from each pair.**
6. **Factor out the common binomial factor.**

Easy Memory Phrases:

These phrases guide you step by step to break down and factor any trinomial!

- **"Multiply, add, split, group!" (for a ≠ 1)**
- **"Multiply to last, add to middle!" (for a = 1)**

Example:

Factor $2x^2 + 7x + 3$.

Solution:

1. Identify $a, b,$ and c: $a = 2, b = 7, c = 3$
2. Find two numbers that multiply to: $a \cdot c = 2 \cdot 3 = 6$ and add up to $b = 7$: The numbers are 6 and 1 because $6 \cdot 1 = 6$ and $6 + 1 = 7$.
3. Rewrite the middle term using the two numbers found: $2x^2 + 6x + 1x + 3$
4. Group the terms into two pairs: $(2x^2 + 6x) + (1x + 3)$
5. Factor out the common factor from each pair: $2x(x + 3) + 1(x + 3)$
6. Factor out the common binomial factor $(x + 3)$: $(2x + 1)(x + 3)$

So, the factored form of $2x^2 + 7x + 3$ is $(2x + 1)(x + 3)$.

Real World Applications

Polynomials are more than just theoretical constructs; they have a wide range of practical applications in various fields. Here are a few examples:

Engineering:

- **Control Systems:** Polynomials are used in designing control systems for machinery and robotics to determine system stability.
- **Signal Processing:** Polynomials help with filtering and analyzing signals in telecommunications and audio processing.

Physics:

- **Kinematics:** Polynomials describe the motion of objects, such as the position, velocity, and acceleration of a moving body.
- **Optics:** Polynomial equations model the behavior of light and lens systems.

Medicine:

- **Medical Imaging:** Polynomial equations are used in reconstructing images in techniques like MRI and CT scans.
- **Drug Dosage:** Polynomials model the concentration of drugs in the bloodstream over time.

Environmental Science:

- **Modeling Pollution Levels:** Polynomial functions help in predicting and analyzing pollution levels in the environment.
- **Climate Change Models:** Polynomials are used in climate models to simulate changes in temperature, precipitation, and other factors.

Example:

When you throw a ball, its path can be described using a quadratic polynomial. The equation of its height as a function of time t is given by: $h(t) = -4.9t^2 + vt + h_0$where:

- $h(t)$ is the height of the ball at time t.
- v is the initial velocity (speed) of the ball in meters per second (m/s)
- h_0 is the initial height from which the ball is thrown in meters (m)
- The constant -4.9 comes from the effect of gravity on the ball (in meters per second squared)

Suppose you throw a ball with an initial velocity of 10 m/s from a height of 1.5 meters. Let's find the height of the ball after 1 second.

Solution: Substitute the values into the polynomial equation:

$h(t) = -4.9t^2 + vt + h_0 \rightarrow h(1) = -4.9(1)^2 + 10(1) + 1.5 = 6.6$ meters.

So, after 1 second, the ball will be at a height of 6.6 meters.

Worksheets

Difference Between Polynomials and Algebraic Expressions

Determine whether each of the following expressions is a polynomial or an algebraic expression:

1) $2x^3 - \frac{2}{5}x^2 + 4$
2) $\frac{1}{y-5}$
3) $\frac{12x^3+x^2-4}{3}$
4) $\sqrt{x^4+3} - 2x^3$
5) $2x^{-2} - 3x^4 + x^{-3}$

Writing Polynomials in Standard Form

Rewrite the given polynomials in their standard form:

1) $3x^2 - 5x + 4x^5 - x^3$
2) $-y + 6y^2 + 5y + 2y^3$
3) $-2x(4 - 5x^2) + 6x^2 - 12$
4) $4a^2 - 5ba^3 + 2b - b^2a^3 + 6b^3$
5) $12y(4z^2 - 5y^3z^3 + y^3) - z^2(y - 5)$

Simplifying Polynomials

Simplify.

1) $-6x^2 + 5x - 3x^2 - 7x + x^3$
2) $-3(2y^2 - 4x^2 + x) + 6y^2 - 5x$
3) $4x^3 + 2x^2y + 3y^2x - yx(-2x + y)$
4) $\frac{-x^2+2x^4}{2} - \frac{10x^3+15x^2}{5} + x^3 + 5x^4$
5) $-3a(ab + 2ba^2) + b(-5ab - a^2)$

Adding and Subtracting Polynomials

Add and subtract following polynomials:

1) $(4xy^2 - x^2 + 2) + (-y^2 + 5xy^2 + 2x^2 - 4)$
2) $(2x^4 - x^3 - 3x^2) - (x^5 + 2x^4 - 3x^4 - x^3)$
3) $(4x^3 - 2x^2 + 7x - 5) + (3x^3 + 6x^2 - 4x + 2)$
4) $(-3ab^3 + b^2 - 2a) - (-4ab^3 - 6ab + 5b^2)$
5) $(6x^4 - 5x^3 + x^2 - 4x + 7) - (3x^4 + 2x^3 - 8x^2 + x - 5) + (x^4 - 4x^3 + 6x^2 - 3x + 2)$

Multiplying and Dividing Monomials

Multiply and divide the following monomials:

1) $3x^2 \cdot 4x^3$
2) $5a^3b^2 \cdot 0.2ab^4$
3) $-\frac{1}{10}x^4y^3 \cdot 0.6x^2y^2z^6$
4) $\frac{-(xy)^3y^4z^2}{3} \cdot 15x^2(zy)^3$
5) $1.5x^2y^4z^3 \cdot 0.2xy^2(zy)^2$
6) $\frac{12x^6}{-6x^4}$
7) $\frac{48x^5y^2z^5}{-12x^3z^3}$
8) $\frac{-25x^4y^2z^3}{10(xy)^2z^2}$
9) $\frac{6(ab^2)^3c^3}{-2(ac^2)^2} \cdot \frac{c^2}{3a}$
10) $\frac{25xy^4}{z^3} \cdot \frac{14(zx)^3x^2}{-35y^2}$

Multiplying Polynomials and Monomials

Multiply:

1) $-3y^2 \cdot (-5xy + 2x^2 + 7y)$
2) $2x^2y \cdot (3x^3y + 5y^2 - xy + 3)$
3) $(ab^2)^3 \cdot (-12(ab)^2 + 3a^2b - ab + 1)$
4) $(-2x^3y + y^2x + x^2 - 5x) \cdot (-3x^3)$
5) $\frac{-a^2b^3+2b^2a-3ab}{5} \cdot 20a^3$

Multiplying Binomials

Multiply:

1) $(a - b)(a + b)$
2) $(3x - 2)(5x + 4)$
3) $(x^2 + 3y)(-y^2 - 2x)$
4) $(-2(xy)^2 - 1)(x^3 + 1)$
5) $(3ab + 2a)^2$

Division of Polynomials

Determine the quotient and remainder.

1) $(2x^3 - 6x^2 + 8x) \div x^2$
2) $(6x^2 + 5x - 1) \div (3x - 1)$
3) $(-4x^4 - 3x^2 + 5x + 2) \div (2x^2 - 3)$
4) $(x^4 - 3x^3 + 5x^2 - 7x + 2) \div (x^2 - x + 1)$
5) $(3x^5 - 2x^4 + 6x^3 - 8x^2 + 5x - 1) \div (x^3 - 2x + 1)$

Factoring Trinomials

Factor following polynomials:

1) $x^2 - 7x + 10$
2) $2x^2 + 3x - 2$
3) $3x^2 - 8x + 4$
4) $4x^2 - 4x - 3$
5) $7x^4 + 2x^2 - 3$

Real World Applications

Do the following problems.

1) A rectangular garden has a length of $2x + 5$ meters and a width of $x + 3$ meters. Write a polynomial expression for the area of the garden.
2) The height of a ball thrown in the air is given by the polynomial $h(t) = -5t^2 + 20t + 3$, where h is the height in meters and t is the time in seconds. Find the height of the ball after 2 seconds.
3) The volume of a cylindrical tank is modeled by the polynomial $V(r) = \pi r^2h$, where r is the radius and h is the height. If the height of the tank is $h = 10$ and the radius is $r = 5$, calculate the volume of the tank.
4) The area of a rectangular garden is given by the polynomial expression $A = 10x^2 - 3x - 1$, where $(5x + 1)$ is the length of one side of the garden in meters. find an expression for the perimeter of the garden.
5) The profit $P(x)$ for a company selling x units of a product is given by the polynomial expression $P(x) = -4x^2 + 200x - 1000$. Find the number of units that must be sold to maximize the profit.

Answer of Worksheets

Polynomials and Algebraic Expressions

1) Polynomial
2) Algebraic expression
3) Polynomial
4) Algebraic expression
5) Algebraic expression

Writing Polynomials in Standard Form

1) $4x^5 - x^3 + 3x^2 - 5x$
2) $2y^3 + 6y^2 + 4y$
3) $10x^3 + 6x^2 - 8x - 12$
4) $-b^2a^3 - 5ba^3 + 6b^3 + 4a^2 + 2b$
5) $-60y^4z^3 + 12y^4 + 47yz^2 + 5z^2$

Simplifying Polynomials

1) $x^3 - 9x^2 - 2x$
2) $12x^2 - 8x$
3) $4x^3 + 4x^2y + 2y^2x$
4) $6x^4 - x^3 - \frac{7}{2}x^2$
5) $-6a^3b - 4a^2b - 5ab^2$

Adding and Subtracting Polynomials

1) $9xy^2 + x^2 - y^2 - 2$
2) $-x^5 + 3x^4 - 3x^2$
3) $7x^3 + 4x^2 + 3x - 3$
4) $ab^3 - 4b^2 + 6ab - 2a$
5) $4x^4 - 11x^3 + 15x^2 - 8x + 14$

Multiplying and Dividing Monomials

1) $12x^5$
2) a^4b^6
3) $-0.06x^6y^5z^6$
4) $-5x^5y^{10}z^5$
5) $0.3x^3y^8z^5$
6) $-2x^2$
7) $-4x^2y^2z^2$
8) $-2.5x^2z$
9) $-b^6c$
10) $-10x^6y^2$

Multiplying Polynomials and Monomials

1) $15xy^3 - 6x^2y^2 - 21y^3$
2) $6x^5y^2 + 10x^2y^3 - 2x^3y^2 + 6x^2y$
3) $-12a^5b^8 + 3a^5b^7 - a^4b^7 + a^3b^6$
4) $6x^6y - 3x^4y^2 - 3x^5 + 15x^4$
5) $-4a^5b^3 + 8a^4b^2 - 12a^4b$

Multiplying Binomials

1) $a^2 - b^2$
2) $15x^2 + 2x - 8$
3) $-x^2y^2 - 2x^3 - 3y^3 - 6xy$
4) $-2x^5y^2 - 2(xy)^2 - x^3 - 1$
5) $9a^2b^2 + 12a^2b + 4a^2$

Division of Polynomials

1) Quotient:$2x - 6$ Remainder: $8x$.
2) Quotient: $2x + \frac{7}{3}$ Remainder: $\frac{4}{3}$
3) Quotient: $-2x^2 - \frac{9}{2}$ Remainder: $5x - \frac{23}{2}$

4) Quotient: $x^2 - 2x + 2$ Remainder: $-3x$
5) Quotient: $3x^2 - 2x + 12$ Remainder: $-15x^2 + 31x - 13$

Factoring Trinomials

1) $(x-5)(x-2)$
2) $(2x-1)(x+2)$
3) $(3x-2)(x-2)$
4) $(2x+1)(2x-3)$
5) $(7x^2-3)(x^2+1)$

Real World Applications

1) $2x^2 + 11x + 15$
2) 23 meters
3) 250π
4) $14x$
5) 25 units

Chapter 9: Functions and Quadratics

Topics that you'll learn in this chapter:

- ✓ Relations or Functions
- ✓ Function Notation
- ✓ Identify Linear and Nonlinear Functions
- ✓ Domain and Range by Ordered Pairs
- ✓ Domain and Range by Mapping Diagrams
- ✓ Domain and Range by Graphs
- ✓ Domain and Range by Equations
- ✓ Evaluating Function
- ✓ Adding and Subtracting Functions
- ✓ Multiplying and Dividing Functions
- ✓ Composition of Functions
- ✓ Quadratic Equation
- ✓ Solving Quadratic Equations
- ✓ Quadratic Formula and the Discriminant
- ✓ Graphing Quadratic Functions
- ✓ Real World Applications
- ✓ Worksheets
- ✓ Answer of Worksheets

Relations or Functions

Relations

A relation is simply a set of ordered pairs. Each ordered pair consists of two elements: one from the first set (called the domain) and one from the second set (called the range). For example, let's say we have two sets:

- Set A: $\{1, 2, 3\}$
- Set B: $\{4, 5, 6\}$

A relation between these two sets could be: $\{(1,4), (2,5), (3,6)\}$.

This means that the number 1 from Set A is related to the number 4 from Set B, the number 2 is related to the number 5, and so on.

Functions

A function is a special type of relation where every element in the domain (the first set) is related to exactly one element in the range (the second set). In other words, for each input, there is only one output.

Using the previous example, if we define a function f such that:

- $f(1) = 4$
- $f(2) = 5$
- $f(3) = 6$

Here, each element in Set A (the domain) is related to one and only one element in Set B (the range). If any element in Set A were related to more than one element in Set B, it would no longer be considered a function.

Example

Determine if the following sets of ordered pairs are relations or functions:

a) $\{(1, 5), (2, 7), (-3, 6), (5, 9)\}$

b) $\{(-2, 3), (8, 2), (-2, 1), (3, 5)\}$

Solution:

a. This is a relation and a function because each element in the domain is related to exactly one element in the range.

b. This is a relation but not a function because the number -2 from the domain is related to both 3 and 1 in the range.

Function Notation

Function notation uses the symbol $f(x)$ to represent a function. Here's how it works:

- f: This is the name of the function. It could be any letter, but f is most commonly used.
- x: This represents the input value or the variable.
- $f(x)$: This represents the output value or the function's value at x.

How It Works:

A **function** is like a machine that takes an input, does something to it, and gives an output. When you see $f(x)$, you can think of it as a rule that assigns a specific output to each input x. For example, if we have a function f such that: $f(x) = x + 3$

This means for any input x, the function will add 3 to it to get the output.

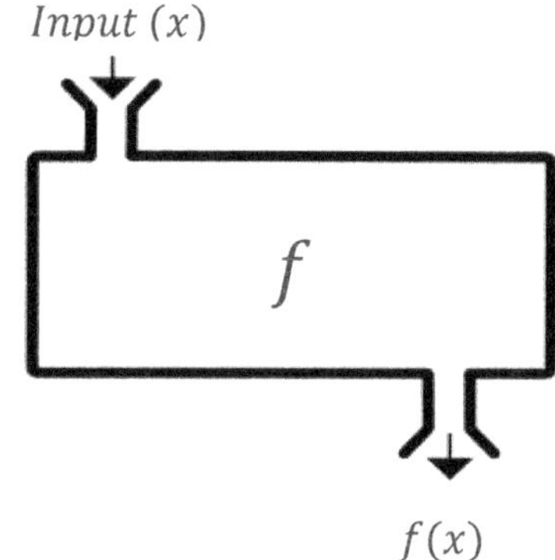

Examples:

1) Given $h(x) = 3x - 2$, find $h(4)$.

   Solution:

   To find $h(4)$, substitute x with 4: $h(4) = 3(4) - 2 = 12 - 2 = 10$

2) Given $g(x) = 2x^2 - 5x + 2$, find $g(-3)$.

   Solution:

   To find $g(-3)$, substitute x with -3:

   $g(-3) = 2(-3)^2 - 5(-3) + 2 = 2 \times 9 + 15 + 2 = 18 + 15 + 2 = 35$

3) Given $f(x) = -4x + 3$, find the value of x such that $f(x) = -1$.

   Solution:

   For $f(x) = -4x + 3$: $-4x + 3 = -1 \rightarrow -4x = -1 - 3 \rightarrow -4x = -4 \rightarrow x = \frac{-4}{-4} = 1$

Identify Linear and Nonlinear Functions

Linear Functions:

As we learned in the previous chapters, a linear function is a function that creates a straight line when graphed. It can be written in the form: $y = mx + b$

Characteristics:

- The graph is a straight line.
- The rate of change (slope) is constant.
- The function has a constant rate of increase or decreases.

Nonlinear Functions:

A nonlinear function is a function that does not create a straight line when graphed. Nonlinear functions can have various forms, such as quadratic ($y = ax^2 + bx + c$), exponential ($y = a \cdot b^x$), or absolute value ($y = |x|$) functions.

Characteristics:

- The graph is not a straight line.
- The rate of change is not constant.
- The function can curve, bend, or change direction.

How to Identify Linear and Nonlinear Functions:

By Equation:

- Linear functions have the form $y = mx + b$.
- If the equation has variables raised to powers other than 1 (e.g., x^2, x^3), or involves absolute value or exponential terms, it is nonlinear.

By Graph:

- A graph that is a straight line represents a linear function.
- A graph that curves, bends, or changes direction represents a nonlinear function.

By Table of Values:

- For a linear function, the difference in y values divided by the difference in x values (slope) will be constant.
- For a nonlinear function, the rate of change will not be constant.

Examples of Identifying Linear and Nonlinear Functions:

1) Determine if the following functions are linear or nonlinear:

a) $y = -0.2x + 4$

b) $y = x^2 - 3$

c) $y = 2^x + 1$

d) $y = |x - 4|$

Solution:

a) $y = -0.2x + 4$ →Linear

b) $y = x^2 - 3$ → Nonlinear (Quadratic)

c) $y = 2^x + 1$ → Nonlinear (Exponential)

d) $y = |x - 4|$ → Nonlinear (Absolute Value)

2) Look at the graph and determine if the function is linear or nonlinear:

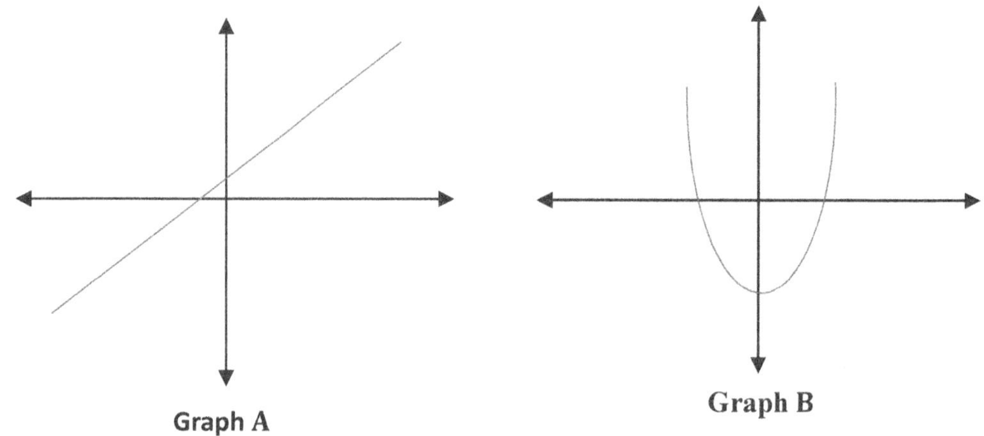

Solution:

- Graph A is Linear (the graph is a straight line)
- Graph B is Nonlinear (Quadratic)

3) Given the table of values, determine if the function is linear or nonlinear:

x	−2	−1	0	1	2	3
y	5	2	1	2	5	7

Solution:

The function is Nonlinear since the rate of change is not constant.

Domain and Range by Ordered Pairs

Domain and Range:

1. **Domain:** The domain of a function is the set of all possible input values (independent variable) that the function can accept. It is usually represented by the variable x.
2. **Range:** The range of a function is the set of all possible output values (dependent variable) that the function can produce. It is usually represented by the variable y.

Sets of Ordered Pairs:

A set of ordered pairs is a collection of pairs, where each pair consists of an input value (from the domain) and a corresponding output value (from the range). The format for each pair is (x, y).

Steps to Determine Domain and Range from Ordered Pairs:

1. **List the x-Values (Domain)**: Go through each ordered pair and list the first element of each pair, which represents the domain.
2. **List the y-Values (Range)**: Go through each ordered pair and list the second element of each pair, which represents the range.

Tips:

- **Domain = all the x-values (Inputs)**
- **Range = all the y-values (Outputs)**
- **No repeats!**

Examples:

1) Given the set of ordered pairs $\{(4,1), (1,-3), (-1,2), (7,-2)\}$, determine the domain and range.

Solution:

- Domain: $\{4, 1, -1, 7\}$
- Range: $\{1, -3, 2, -2\}$

2) Given the set of ordered pairs $\{(2a+1, -2), (5, -1), (3, b-1)\}$, the domain and the range, find the value of a and b.

- Domain: $\{2, 5, 3\}$
- Range: $\{-2, -1, 7\}$

Solution: Given the provided domain and range, we must set $2a + 1 = 2$ and $b - 1 = 7$:

- $2a + 1 = 2 \rightarrow 2a = 2 - 1 \rightarrow 2a = 1 \rightarrow a = \frac{1}{2}$
- $b - 1 = 7 \rightarrow b = 7 + 1 \rightarrow b = 8$

Domain and Range by Mapping Diagrams

Mapping diagrams are a visual way to understand the concepts of domain and range in functions. Let's break down these terms and how they are represented using mapping diagrams.

Mapping Diagrams:

A mapping diagram visually shows the relationship between the elements of the domain and the elements of the range. It helps to see how each input (domain) is mapped to an output (range).

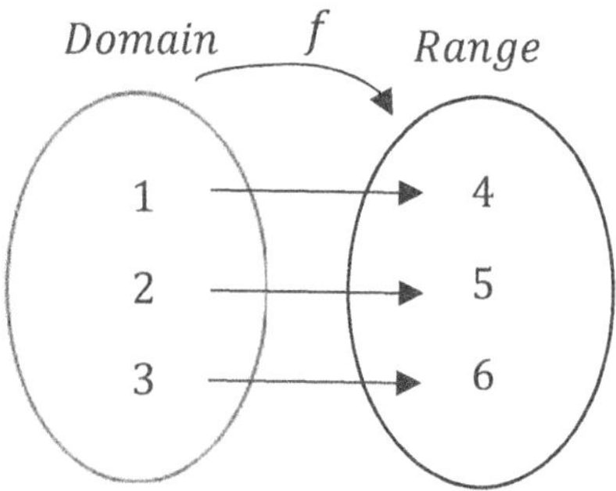

How to Create a Mapping Diagram:

List the Elements: Write the elements of the domain in a column on the left. Write the elements of the range in a column on the right.

Draw Arrows: Draw arrows from each element in the domain to its corresponding element in the range, based on the function's definition.

Example:

Create a mapping diagram for the function g defined by the set of ordered pairs:

$g: \{(3,5), (-1,2), (0,5), (2,-3)\}$

Solution:

Domain:$\{3, -1,0,2\}$ and Range: $\{5, 2, -3\}$

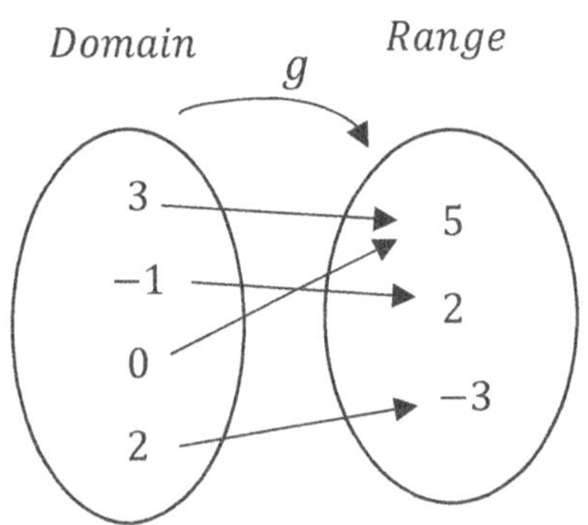

Domain and Range by Graphs

1. **Domain (from a graph):** On a graph, the domain corresponds to the horizontal extent of the graph (how far the graph extends left and right).

How to Find the Domain from a Graph:

- Look at the x-axis and identify the smallest and largest x-values where the graph exists.
- If the graph continues infinitely in either direction, the domain is all real numbers (or $-\infty$ $to + \infty$)

2. **Range (from a graph):** On a graph, the range corresponds to the vertical extent of the graph (how far the graph extends up and down).

How to Find the Range from a Graph:

- Look at the y-axis and identify the smallest and largest y-values that the graph reaches.
- If the graph continues infinitely upward or downward, the range is all real numbers (or $-\infty$ $to + \infty$)

Key Points to Remember:

- **Domain:** Focus on the x-axis (horizontal).
- **Range:** Focus on the y-axis (vertical).
- If there are gaps or holes in the graph, exclude those values from the domain or range.

Example:

Given the provided graph, determine its domain and range.

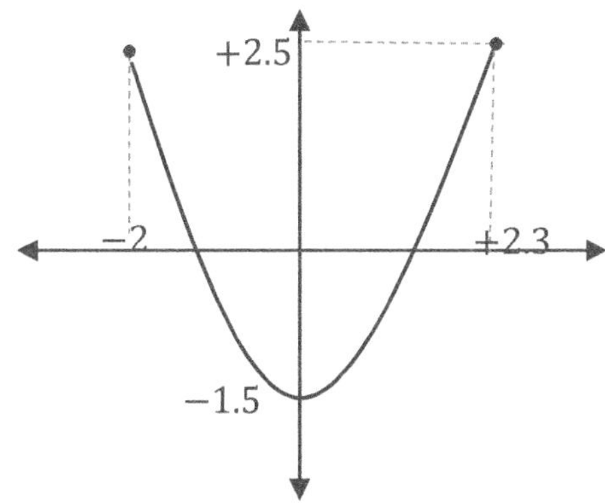

Solution:

Domain: The lowest x-value is -2 and the highest x-value is $+2.3$, so the domain is $[-2, 2.3]$

Range: The lowest y-value is -1.5 and the highest y-value is $+2.5$, so the range is $[-1.5, 2.5]$

Domain and Range by Equations

1. **Domain (from an equation):**
   - Look for values of x that make the equation undefined or impossible.
   - For most equations, the domain is all real numbers unless there's a restriction.

Common Restrictions:

- **Fractions (Rational Expressions):** The denominator cannot be zero.
  - Example: $y = \frac{1}{x}$,The denominator x cannot be zero, so the domain is all real numbers except $x = 0$.
- **Square Roots:** The expression inside the square root must be greater than or equal to zero.
  - Example: $y = \sqrt{x}$ The expression inside the square root (x) must be ≥ 0, so the domain is $x \geq 0$.

2. **Range (from an equation):**
   - Look at the possible output values (y-values) the equation can produce.
   - For most equations, the range is all real numbers unless there's a restriction.

Common Restrictions:

- Quadratic Equations (Parabolas): The range depends on whether the parabola opens upward or downward.
  - Example: $y = x^2$, the smallest y-value is 0, and the parabola goes infinitely upward, so the range is $y \geq 0$.
- Absolute Value Equations: The range starts at the vertex and goes infinitely upward or downward.
  - Example: $y = |x|$, The smallest y-value is 0, so the range is $y \geq 0$.

Example: Determine the domain and range of the following equations:

a. $y = 2x + 5$

b. $y = x^2 - 4$

Solution:

a. Domain: All real numbers (no restrictions).
   Range: All real numbers (no restrictions).

b. Domain: All real numbers (no restrictions).
   Range: The smallest y-value is -4, and the parabola goes infinitely upward, so the range is $y \geq -4$.

Evaluating Function

Evaluating a function means finding the output (y-value) when you plug in a specific input (x-value) into the function.

Steps to Evaluate a Function:

1. **Write down the function.**
2. **Replace x with the given input value.**
3. **Simplify the equation** to find the output**.**

Example:

For $f(x) = \frac{x^3-2}{x}$, evaluate $f(-1)$.

Solution:

$f(-1) = \frac{(-1)^3-2}{-1} = \frac{-1-2}{-1} = 3$

Adding and Subtracting Functions

When you add or subtract two functions, you combine their outputs (y-values) for the same input (x-value).

Key Points to Remember:

1. **Add or subtract the corresponding terms** in the functions.
2. **Combine like terms** (terms with the same variable and exponent).
3. **Distribute negative signs** when subtracting functions.
4. **Simplify the result** as much as possible.

Examples:

1) Add functions:

$f(x) = -2x^2 + 3x - 4$ $g(x) = x^3 - 3$

Solution:

$(f+g)(x) = f(x) + g(x) = -2x^2 + 3x - 4 + x^3 - 3 = x^3 - 2x^2 + 3x - 7$

2) Subtract functions:

$f(x) = \frac{1}{x}$ $g(x) = \frac{2}{x}$

Solution:

$(f-g)(x) = f(x) - g(x) = \frac{1}{x} - \frac{2}{x} = \frac{-1}{x}$

Multiplying and Dividing Functions

When you multiply or divide two functions, you combine their outputs (y-values) for the same input (x-value).

- $(f \cdot g)(x) = f(x) \cdot g(x)$
- $\left(\frac{f}{g}\right)(x) = \frac{f(x)}{g(x)}$

Steps to Multiply or Divide Functions:

1. Multiplying Functions

- Multiply the two functions term by term.
- Simplify the result by combining like terms (if possible).

2. Dividing Functions

- Divide the first function by the second function.
- Simplify the result (if possible).
- Note: The denominator cannot be zero, so exclude any x-values that make the denominator zero.

Examples:

1) Multiply functions:

   $f(x) = x^2 + 3x - 7$ $\quad\quad$ $g(x) = 2x$

   Solution:

   $(f \cdot g)(x) = f(x) \cdot g(x) = (x^2 + 3x - 7)(2x)$

   Distribute $2x$: $2x \cdot x^2 + 2x \cdot 3x - 2x \cdot 7 = 2x^3 + 6x^2 - 14x$

   So, $(f \cdot g)(x) = 2x^3 + 6x^2 - 14x$

2) Divided functions:

   $f(x) = x^2 - 4$ $\quad\quad$ $g(x) = x - 2$

   Solution:

   $\left(\frac{f}{g}\right)(x) = \frac{f(x)}{g(x)} = \frac{x^2-4}{x-2}$, factor the numerator: $\frac{x^2-4}{x-2} = \frac{(x-2)(x+2)}{(x-2)}$

   cancel the common term $(x - 2)$: $\frac{(x-2)(x+2)}{(x-2)} = x + 2\ (for\ x \neq 2)$

   So, $\left(\frac{f}{g}\right)(x) = x + 2$ but $x \neq 2$ (because the denominator cannot be zero).

Composition of Functions

Composition of functions is like putting two functions together, where the output of one function becomes the input of another. It's like a "function inside a function."

If you have two functions, $f(x)$ and $g(x)$, the composition $f(g(x))$ means:

- First, plug x into $g(x)$ to get $g(x)$.
- Then, take the result of $g(x)$ and plug it into $f(x)$.

How to Find the Composition of Functions:

1. **Write down the two functions. For example:**
   - $f(x) = 2x + 1$
   - $g(x) = x - 3$
2. **Plug $g(x)$ into $f(x)$:**
   - $f(g(x)) = f(x - 3)$
3. **Replace x in $f(x)$ with $g(x)$:**
   - $f(x - 3) = 2(x - 3) + 1$
4. **Simplify the result:**
   - $f(x - 3) = 2x - 6 + 1 = 2x - 5$

So, $f(g(x)) = 2x - 5$.

Key Points to Remember:

1. **Composition** $f(g(x))$ means "plug $g(x)$ into $f(x)$."
2. **Order matters!** $f(g(x))$ is not the same as $g(f(x))$.
3. **Simplify the result** by combining like terms or expanding.

Easy Trick:

"Put the output of one into the input of the next!"

(Think of it like a math machine: one function feeds into another.)

Example:

Find the $g(f(x))$:

$f(x) = x^2 + 1$ $\qquad$ $g(x) = \frac{1}{x-1}$

Solution:

$g(f(x)) = g(x^2 + 1) = \frac{1}{(x^2+1)-1} = \frac{1}{x^2}$; So, $g(f(x)) = \frac{1}{x^2}$

Quadratic Equation

A quadratic equation is an equation where the highest power of the variable (usually x) is 2. It has the general form:

$$ax^2 + bx + c = 0$$

Where:

- a, b, and c are constant (numbers), and $a \neq 0$.
- x is the variable.

Key Features of Quadratic Equations:

1. **Degree:** The highest power of x is 2.
2. **Graph:** The graph of a quadratic equation is a parabola (a U-shaped curve).
3. **Direction of the Parabola:**
   - If $a > 0$, the parabola opens upward.
   - If $a < 0$, the parabola opens downward.

Here is the graph of $y = x^2$:

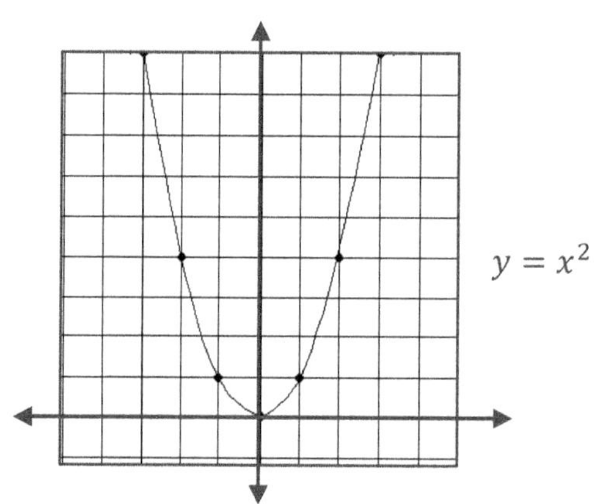

Example:

Determine if the following parabolas open upwards or downwards.

- $y = -3x^2 + x - 4$
- $y = 2x^2 + 5x - 9$

Solution

Considering the sign of 'a' in the above equations, in the first equation ($y = -3x^2 + x - 4$) a (-3) is negative, so the parabola opens downwards, and in the second equation ($y = 2x^2 + 5x - 9$), a (2) is positive, so the parabola opens upwards.

Solving Quadratic Equations

Methods to Solve Quadratic Equations:

1. **Factoring:**
   - Express the quadratic equation in standard form.
   - Factor the quadratic expression.
   - Set each factor equal to zero and solve for x.
2. **Completing the Square:**
   - Move the constant term to the other side of the equation.
   - Make the coefficient of x^2equal to 1 (if it isn't already)
   - Add the square of half the coefficient of x to both sides.
   - Factor the left side and solve.

Key Points to Remember:

- Factoring works best when the equation can be easily factored.
- Completing Square is a method you might learn later on, but it's important to know that it also helps solve quadratics.

Examples:

1) Solve $x^2 + 5x + 6 = 0$ using factoring method.

   Solution:

   1. Factor the quadratic expression: Find two numbers that multiply to 6 (the constant term) and add up to 5 (the coefficient of x). The numbers are 2 and 3.
   2. Write the factored form: $(x + 2)(x + 3) = 0$
   3. Set each factor equal to zero and solve: $x + 2 = 0\ or\ x + 3 = 0 \rightarrow x = -2$ or $x = -3$

   Thus, the solutions are $x = -2$ and $x = -3$.

2) Solve $2x^2 - 7x - 4 = 0$ using completing the square method.

   Solution:

   1. Move the constant term to the other side of the equation: $2x^2 - 7x = 4$
   2. Make the coefficient of x^2equal to 1: $\frac{2x^2}{2} - \frac{7x}{2} = \frac{4}{2} \rightarrow x^2 - \frac{7}{2}x = 2$
   3. Add the square of half the coefficient of x to both sides. $x^2 - \frac{7}{2}x + \frac{49}{16} = 2 + \frac{49}{16}$
   4. Factor the left side and solve. $(x - \frac{7}{4})^2 = \frac{81}{16} \rightarrow x - \frac{7}{4} = \frac{9}{4}\ or\ x - \frac{7}{4} = -\frac{9}{4}$

   $x = \frac{9}{4} + \frac{7}{4} = \frac{16}{4} = 4$ or $x = -\frac{9}{4} + \frac{7}{4} = -\frac{2}{4} = -\frac{1}{2}$

Quadratic Formula and the Discriminant

The quadratic formula is a method for solving quadratic equations of the form $ax^2 + bx + c = 0$. The formula is:

$$x = \frac{-b \pm \sqrt{b^2 - 4ac}}{2a}$$

Here's what each part of the formula represents:

- a, b, and c are the coefficients of the quadratic equation.
- The symbol $\pm$ indicates that there are generally two solutions.
- The expression under the square root sign, $b^2 - 4ac$, is called the discriminant.

Discriminant:

The discriminant of a quadratic equation is a key part of the quadratic formula. It is given by:

$$\Delta = b^2 - 4ac$$

The value of the discriminant determines the nature and number of the roots of the quadratic equation:

- **If** $\Delta > 0$: The equation has two distinct real roots.
- **If** $\Delta = 0$: The equation has exactly one real root (also called a repeated or double root).
- **If** $\Delta < 0$: The equation has no real roots.

$\Delta > 0$

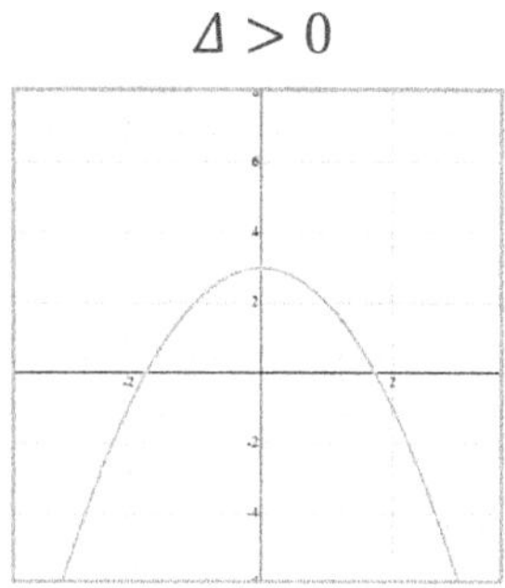

$\Delta = 0$

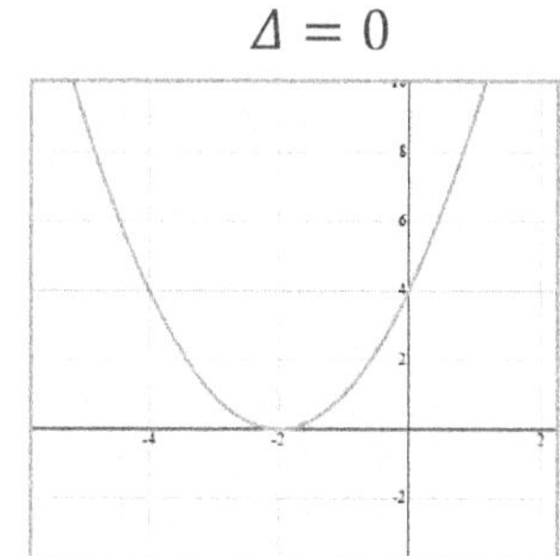

$\Delta < 0$

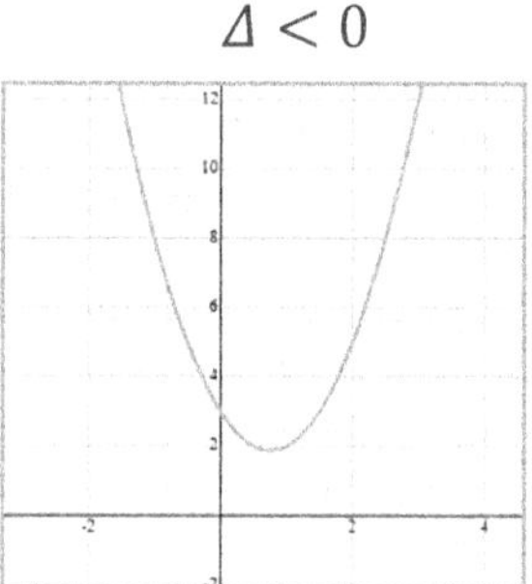

Example: Solve $2x^2 + 3x - 2 = 0$ using quadratic formula.

Solution:

- Identify the coefficients: $a = 2, b = 3, c = -2$.
- Plug these values into the quadratic formula: $x = \frac{-b\pm\sqrt{b^2-4ac}}{2a} = \frac{-3\pm\sqrt{3^2-4(2)(-2)}}{2(2)}$

$x = \frac{-3\pm\sqrt{9+16}}{4} \rightarrow x = \frac{-3\pm\sqrt{25}}{4} \rightarrow x = \frac{-3\pm5}{4}$

- Calculate the solutions: For $x = \frac{-3-5}{4} = -2$ and for $x = \frac{-3+5}{4} = 0.5$

Graphing Quadratic Functions

Steps to Graph a Quadratic Function:

1. **Identify the Vertex**: The vertex is the highest or lowest point of the parabola. For the quadratic equation $y = ax^2 + bx + c$, the vertex can be found using the formula:
$$x = -\frac{b}{2a}$$
Plug this x value back into the original equation to find the corresponding y value.
2. **Find the Axis of Symmetry:** The axis of symmetry is a vertical line that passes through the vertex. It can be represented as:
$$x = -\frac{b}{2a}$$
3. **Calculate the *Y*-Intercept:** The y-intercept is where the graph crosses the y-axis. This happens when x is 0. For the quadratic equation $y = ax^2 + bx + c$, the y-intercept is c.
4. **Determine Additional Points:** Choose several values of x (both positive and negative) and plug them into the quadratic equation to find corresponding y values. This will give you more points to plot and create a more accurate graph.
5. **Plot the Points and Draw the Parabola:** Plot the vertex, axis of symmetry, y-intercept, and the additional points on a graph. Connect the points with a smooth, curved line to form the parabola.

Example: Graph the quadratic function $y = x^2 - 4x + 3$.

Solution:

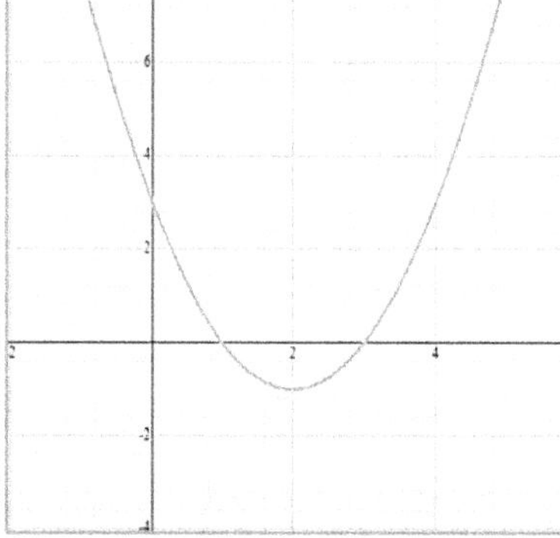

1. Identify the vertex: $x = -\frac{-4}{2(1)} = 2$, Plugging $x = 2$ into the equation:
$y = (2)^2 - 4(2) + 3 = -1$. So, the vertex is $(2, -1)$.
2. Axis of symmetry: The axis of symmetry is $x = 2$.
3. y-intercept: When $x = 0$: $y = (0)^2 - 4(0) + 3 = 3$, so, the y-intercept is $(0,3)$.
4. Additional points:
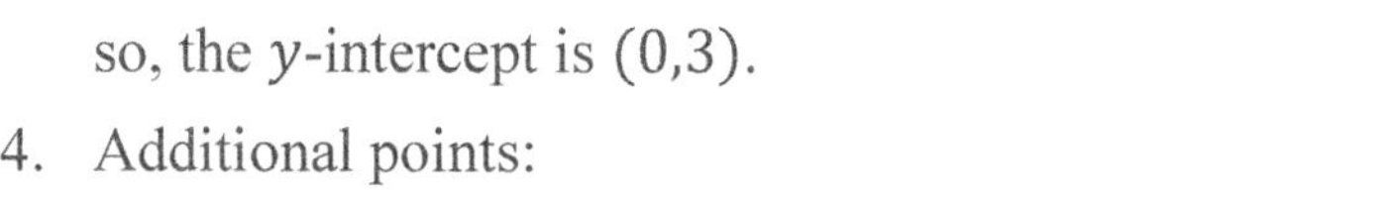
For $x = 1$: $y = (1)^2 - 4(1) + 3 = 0$, for $x = 3$:
$y = (3)^2 - 4(3) + 3 = 0$
So, we have additional points $(1, 0)$ and $(3, 0)$.
5. Plot and draw: Plot the vertex $(2, -1)$, axis of symmetry $x = 2$, y-intercept $(0, 3)$, and additional points $(1, 0)$ and $(3, 0)$. Draw the parabola through these points.

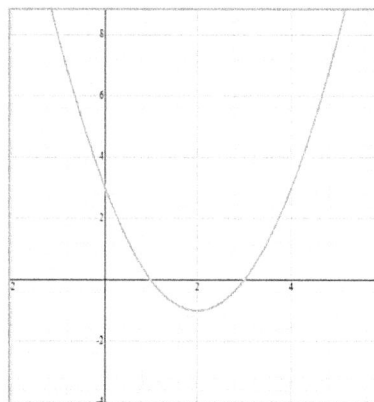

Real World Applications

Functions and quadratic equations have a wide range of real-world applications across various fields. Here are some examples:

Economics and Finance:

- **Supply and Demand:** Functions are used to model the relationship between the price of goods and the quantity supplied or demanded.
- **Interest Calculations:** Functions are used to calculate compound interest, annuities, and amortization schedules.

Engineering:

- **Electrical Circuits:** Functions describe the behavior of electrical components like resistors, capacitors, and inductors.
- **Control Systems:** Functions model the behavior of dynamic systems and are used in designing control systems for machinery and robots.
- **Design:** Quadratic equations are used in designing structures, bridges, and roller coasters to ensure safety and stability.

Projectile Motion: Quadratic equations model the trajectory of objects in projectile motion, such as the path of a ball thrown into the air.

Optimization Problems: Quadratic equations help in finding maximum or minimum values in various optimization problems, such as maximizing profit or minimizing cost.

Example:

Imagine you run a small business that produces handmade candles. You want to find out how many candles you need to produce and sell to maximize your profit. The profit Pin dollars from selling x candles can be modeled by the quadratic equation:

$$P(x) = -2x^2 + 60x - 200$$

Find the number of candles that maximizes your profit.

Solution:

To find the number of candles that maximize your profit, we need to find the vertex of the quadratic equation, as the vertex represents the maximum (or minimum) value of the parabola.

The vertex formula for a quadratic equation $x = -\frac{b}{2a}$

Here, $a = -2, b = 60, and\ c = -200$. Plugging in the values: $x = -\frac{60}{2(-2)} = -\frac{60}{-4} = 15$

So, producing and selling 15 candles will maximize your profit.

Worksheets

Relations or Functions

Determine if the following sets of ordered pairs are relations or functions.

1) $\{(5,1),(-3,2),(-1,1),(4,2)\}$
2) $\{(1,0),(0,1),(0,-2.5),(4,0),(-5,0)\}$
3) $\{(3,-5),(9,-5),(1,-5),(11,-5)\}$
4) $\{(4,6),(-2,3),(4,1),(-5,-9),(3,7)\}$
5) $\{(-2,7),(7,3),(3,10),(10,0)\}$

If the following relations are functions, find the missing values.

6) $\{(-2,1),(3,a-2),(5,2),(3,8)\}$
7) $\{(5,b+2),(-3,7),(5,-4),(9,2)\}$
8) $\{(7,6),(1,3),(5,-a^2),(5,-9),(3,1)\}$
9) $\{(1,a+b),(-3,b-2),(1,9),(-3,0)\}$
10) $\{(0,7),(-3,2a-3b),(0,a^3-1),(-3,-3)\}$

Function Notation

Evaluate the function at a specific point:

1) $f(x)=4x+10$, find $f(2.5)$
2) $f(x)=3x^3-x^2$, find $f(-2)$
3) $f(x)=\frac{3}{5x-1}$, find $f(0)$
4) $f(x)=-7x+3$, find $f(-x-3)$
5) $f(x)=\sqrt{-5x+5}$, find $f(-4)$

Solve for x using function notation:

6) $f(x)=4x-7$, solve for x when $f(x)=9$
7) $f(x)=-2x^2+3x$, solve for x when $f(x)=0$
8) $f(x)=\frac{-1}{x+4}$, solve for x when $f(x)=-3$
9) $f(x)=\sqrt{2x+1}$, solve for x when $f(x)=3$
10) $f(x)=-\sqrt[3]{6x^2}$, solve for x when $f(x)=-6$

Identify Linear and Nonlinear Functions

Determine is the following functions are linear or nonlinear.

1) $g(x)=x^2-4x+4$
2) $h(x)=\sqrt{x}+5$
3) $k(x)=x-\frac{3}{x}$
4) $t(x)=\sqrt{(-x^2+3)^2}$
5) $f(x)=\frac{-8x+2}{-7}$

Given the table of values, determine if the function is linear or nonlinear:

6)

x	1	2	3	4	5
y	1	4	9	16	25

7)

x	−5	−4	−3	−2	−1
y	7	10	13	16	19

8)

x	0	1	2	3	4
y	−1	0	7	26	65

9)

x	1	2	3	4	5
y	$\frac{1}{5}$	$\frac{2}{7}$	$\frac{3}{9}$	$\frac{4}{11}$	$\frac{5}{13}$

10)

x	-1	0	1	2	3
y	-6	-1	4	9	14

Domain and Range by Ordered Pairs

Given the following relations, determine the domain and range.

1) $A = \{(1,-2),(3,-4),(4,2)\}$
2) $B = \{(-5,10),(-4,9),(-3,8),(-2,7)\}$
3) $R = \{(-1,3),(0,5),(2,7),(3,9)\}$
4) $Q = \{(4,-2),(0,-2),(3,-2),(1,-2)\}$
5) $K = \{(1.5,0),(2,4),(7,0),(-3,4)\}$

Domain and Range by Mapping Diagrams

Create a mapping diagram for the following functions defined by the set of ordered pairs:

6) $A = \{(3,-1),(1,7),(2,5),(-9,0)\}$
7) $B = \{(-2,0),(1,-3),(4,7),(3,0)\}$
8) $C = \{12,-1),(13,-2),(14,-3),(15,-4)\}$
9) $D = \{(1,5),(7,5),(-3,5),(-1.3,5)\}$
10) $E = \{(3,6),(1,-2.5),(10,6),(4,-2.5),(6,-2.5)\}$

Domain and Range by Graphs

Determine the domain and the range of given functions:(In all the figures below, the graph extends infinitely to the left and right, and up and down).

1)

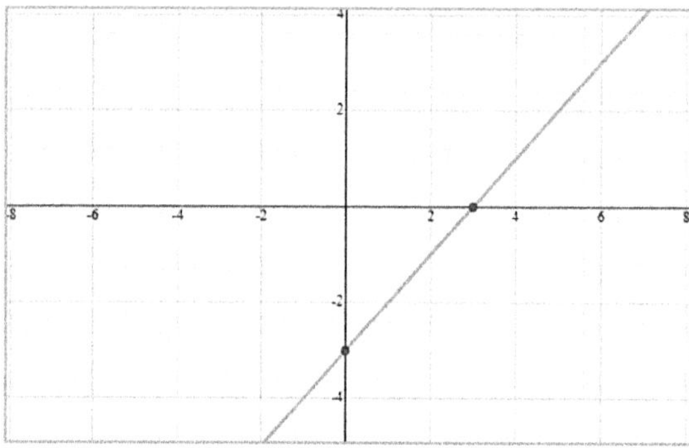

2)

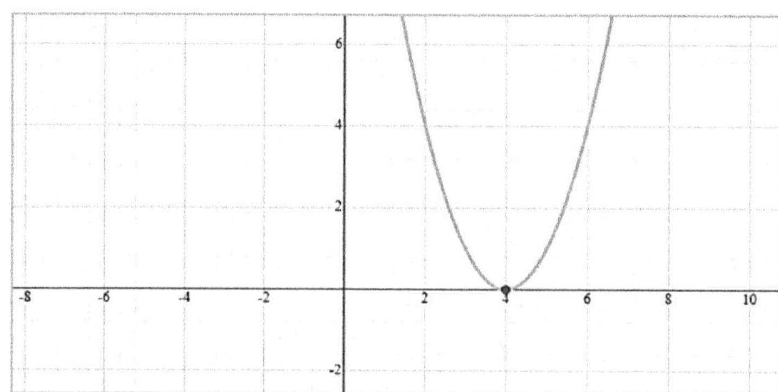

3)

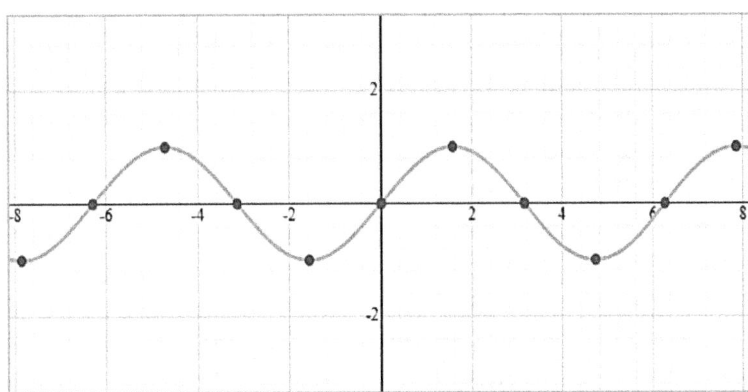

4)

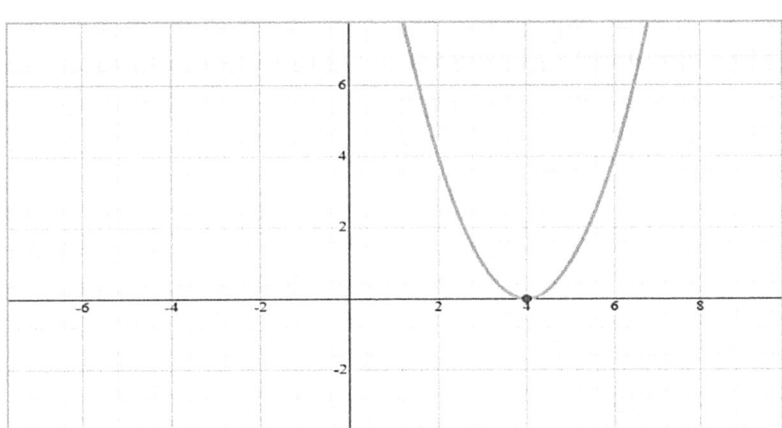

5)

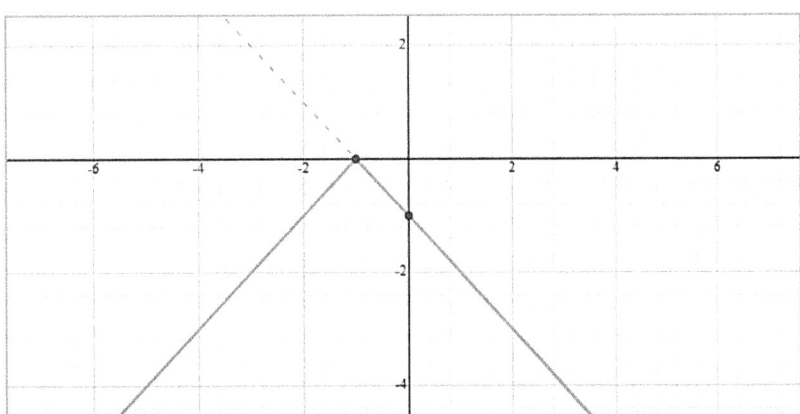

Domain and Range by Equations

Determine the domain and the range.

1) $f(x) = 2x + 10$
2) $f(x) = -|3x + 1|$
3) $f(x) = \sqrt{x - 3}$
4) $f(x) = \frac{1}{x-2}$
5) $f(x) = x^2 + 1$

Evaluating Function

Find the value of the following functions based on the given value of x:

1) $f(x) = 2x^4 - 5, x = -1$
2) $f(x) = \frac{3x-5}{2x}, x = 2$
3) $f(x) = \sqrt{x^2 + 1}, x = \sqrt{5}$
4) $f(x) = 2^x - 2, x = 0$
5) $f(x) = \sqrt[3]{5x + 3}, x = -2$

Adding and Subtracting Functions

Add and subtract functions:

1) $f(x) = -6x + 1, g(x) = 3x, (f - g)(x)$
2) $f(x) = -2x^3 + x^2, g(x) = -x^2 + 3, (f + g)(x)$
3) $f(x) = \sqrt{x + 5}, g(x) = \sqrt{x - 3}, (f - g)(x)$
4) $f(x) = \frac{x-3}{x}, g(x) = \frac{2x+3}{2x^2}, (f + g)(x)$
5) $f(x) = 2^x - 5, g(x) = 2^x + 1, (f - g)(x)$

Multiplying and Dividing Functions

Multiply and divide functions:

1) $f(x) = 7x + 3, g(x) = 2x - 2, (f \cdot g)(x)$
2) $f(x) = \sqrt{x}, g(x) = \sqrt{x}, (\frac{f}{g})(x)$
3) $f(x) = x^3 - x^2, g(x) = \frac{1}{x}, (f.g)(x)$
4) $f(x) = x^2 + 2x - 15, g(x) = x - 3, (\frac{f}{g})(x)$
5) $f(x) = \frac{4x-1}{5x+2}, g(x) = \frac{5x^3+2x^2}{x}, (f \cdot g)(x)$

Composition of Functions

Find the composition of functions.

1) $f(x) = 4x + 1, g(x) = \frac{x}{2}, f(g(x))$
2) $f(x) = x^2, g(x) = 1 - x, f(g(x))$
3) $f(x) = \frac{1}{x}, g(x) = \frac{1}{x}, g(f(x))$
4) $f(x) = 2x^2 - 3x, g(x) = \sqrt{x}, f(g(x))$
5) $f(x) = \frac{1}{x+2}, g(x) = \frac{x}{x-1}, g(f(x))$

Quadratic Equation

Determine if the following parabolas open upwards or downwards.

1) $y = 3x^2 - 5x + 9$
2) $y = -x^2 + x - 2$
3) $y = 4x^2 + 5x$
4) $y = x - 6x^2 + 10$

Identify the signs of a and c on the parabola below.

5)

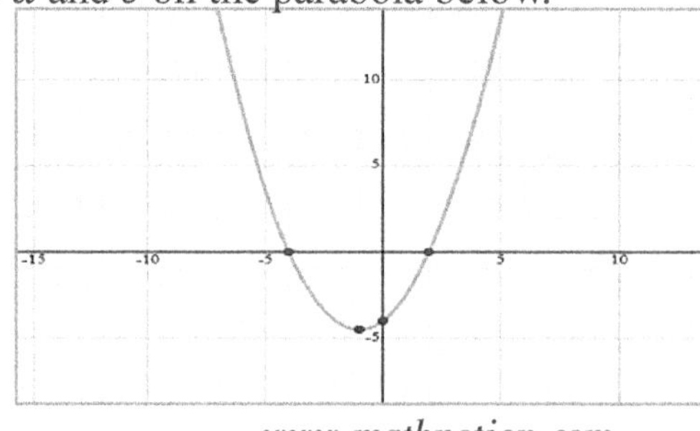

Solving Quadratic Equations

Solve the quadratic equations:

1) $x^2 - 4x - 5 = 0$
2) $2x^2 + 3x - 2 = 0$
3) $x^2 + x - 12 = 0$
4) $4x^2 - 2x - 6 = 0$
5) $3x^2 - x - 4 = 0$

Quadratic Formula and the Discriminant

Solve the following quadratic equations using formula:

1) $4x^2 - 4x - 8 = 0$
2) $x^2 - 9x + 4 = 0$
3) $-2x^2 - 11x + 3 = 0$
4) $7x^2 + 2x - 3 = 0$
5) $-x^2 - 4x + 1 = 0$

Graphing Quadratic Functions

Graph the following quadratic functions:

1) $y = x^2 + x + 1$
2) $y = -x^2 + 4x - 3$
3) $y = \frac{1}{2}x^2 - 2x + 4$
4) $y = -\frac{1}{3}x^2 + 2x - 6$
5) $y = -2x^2 + x - 1$

Real World Applications

Do the following problems.

1) The population $P(t)$ of a city is projected to grow according to a function over a 50-year period, starting from 1 million people. The function is only valid for the next 50 years, and the population grows between 1 million and 5 million. What is the domain and range of this population function?
2) A car's speed, in kilometers per hour, is recorded over a 5-hour road trip. The speeds are modeled as $v(t) = -2t^2 + 20t + 40$, where t is the time in hours. What is the maximum speed of this car during its trip?
3) You are designing a rectangular garden with an area of 54 square meters. The length is 3 meters more than twice its width. Find the dimensions of the garden.
4) The function $C(F) = \frac{5}{9}(F - 32)$ converts Fahrenheit (F) to Celsius (C).
   a) Find $C(77)$
   b) If $C = 20$, find F
5) The daily temperature of a city is approximated by $T(h) = -3(h - 12)^2 + 36$, where h represents hours after midnight (from 0 to 24). What is the maximum temperature of this city during 12 hours?
6) A rock is thrown upward from the top of a 100-meter cliff. Its height h(t) at time t seconds is given by $h(t) = -5t^2 + 20t + 100$. After how many seconds will the rock hit the ground?
7) A person consumes $C(x)$ calories per hour during a workout, and the function $x(t)$ represents the number of hours they work out per day based on the day of the week t. If $C(x) = 500x$ and $x(t) = 2 + 0.5t$, find the total calories burned on the 3rd day of the week.

8) A university calculates the tuition $T(c)$ based on the number of credits c a student takes, and the number of credits depends on the number of classes $f(n)$, where n is the number of semesters. If $T(c) = 300c$ and $f(n) = 5n$, find the tuition for a student in their 4th semester.
9) The value of an investment $V(t)$ after t years depends on the annual growth g, and g depends on the initial investment i, represented by $g(i)$. If $V(t) = i \cdot (1 + g)^t$ and $g(i) = 0.05i$, find the value of an investment of $1,000 after 2 years.
10) A parabolic bridge arch is represented by the equation $y = -x^2 + 6x + 8$, where x is the horizontal distance (in meters) from one end of the arch, and y is the height of the arch (in meters). What is the maximum height of the arch, and how far apart are the two ends of the arch?

Answer of Worksheets

Relations or Functions

1) Relation and function
2) Relation but not a function
3) Relation and function
4) Relation but not a function
5) Relation and function
6) $a = 10$
7) $b = -6$
8) $a = 3$ or $a = -3$
9) $a = 7$ and $b = 2$
10) $a = 2$ and $b = \frac{7}{3}$

Function Notation

1) 20
2) -28
3) -3
4) $7x + 24$
5) 5
6) $x = 4$
7) $x = 0$ or $x = \frac{3}{2}$
8) $x = -\frac{11}{3}$
9) $x = 4$
10) $x = 6$ or $x = -6$

Identify Linear and Nonlinear Functions

1) Nonlinear
2) Nonlinear
3) Nonlinear
4) Nonlinear
5) Linear
6) Nonlinear
7) Linear
8) Nonlinear
9) Nonlinear
10) Linear

Domain and Range by Ordered Pairs

1) Domain= $\{1,3,4\}$ and Range= $\{-2,-4,2\}$
2) Domain= $\{-5,-4,-3,-2\}$ and Range= $\{10,9,8,7\}$
3) Domain= $\{-1,0,2,3\}$ and Range= $\{3,5,7,9\}$
4) Domain= $\{4,0,3,1\}$ and Range= $\{-2\}$
5) Domain= $\{1.5,2,7,-3\}$ and Range= $\{0,4\}$

Domain and Range by Mapping Diagrams

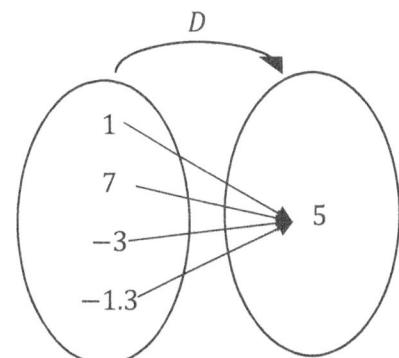

1)

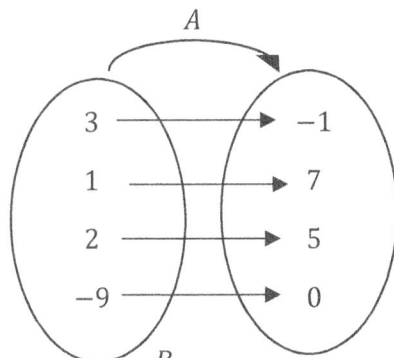

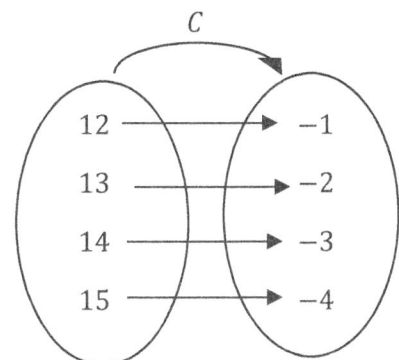

A

3 → −1
1 → 7
2 → 5
−9 → 0

2)

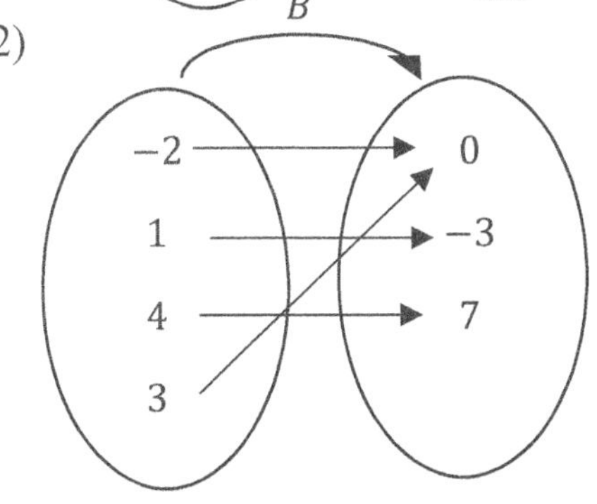

3)

C

12 → −1
13 → −2
14 → −3
15 → −4

4)

D

1, 7, −3, −1.3 → 5

www.mathnotion.com

5)

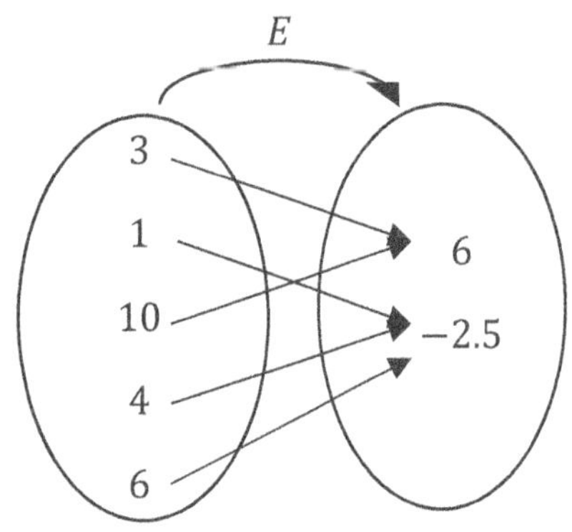

Domain and Range by Graphs

1) Domain: $(-\infty, +\infty)$ and Range: $(-\infty, +\infty)$
2) Domain: $(-\infty, +\infty)$ and Range: $[0, +\infty)$
3) Domain: $(-\infty, +\infty)$ and Range: $[-1,1]$
4) Domain: $(-\infty, +\infty)$ and Range: $[0, +\infty)$
5) Domain: $(-\infty, +\infty)$ and Range: $(-\infty, 0]$

Domain and Range by Equations

1) Domain: $(-\infty, +\infty)$ and Range: $(-\infty, +\infty)$
2) Domain: $(-\infty, +\infty)$ and Range: $(-\infty, 0]$
3) Domain: $[3, +\infty)$ and Range: $[0, +\infty)$
4) Domain: All real numbers except $x = 2$ and Range: All real numbers except $y = 0$
5) Domain: $(-\infty, +\infty)$ and Range: $[1, +\infty)$

Evaluating Function

1) -3
2) $\frac{1}{4}$
3) $\sqrt{6}$
4) -1
5) $\sqrt[3]{-7}$

Adding and Subtracting Functions

1) $(f-g)(x) = -9x + 1$
2) $(f+g)(x) = -2x^3 + 3$
3) $(f-g)(x) = \sqrt{x+5} - \sqrt{x-3}$
4) $(f+g)(x) = \frac{2x^2-4x+3}{2x^2}$
5) $(f-g)(x) = -6$

Multiplying and Dividing Functions

1) $(f \cdot g)(x) = 14x^2 - 8x - 6$
2) $\left(\frac{f}{g}\right)(x) = 1$ for $x > 0$
3) $(f.g)(x) = x^2 - x$ but $x \neq 0$
4) $\left(\frac{f}{g}\right)(x) = x + 5$ but $x \neq 3$
5) $(f \cdot g)(x) = 4x^2 - x$ but $x \neq 0$ and $x \neq -\frac{2}{5}$

Composition of Functions

1) $f(g(x)) = 2x + 1$
2) $f(g(x)) = x^2 - 2x + 1$
3) $g(f(x)) = x$ but $x \neq 0$
4) $f(g(x)) = 2x - 3\sqrt{x}$
5) $g(f(x)) = \frac{1}{-x-1}$ but $x \neq -1$ and $x \neq -2$

Quadratic Equation

1) upwards
2) downwards
3) upwards
4) downwards
5) $a > 0$ and $c < 0$

Solving Quadratic Equations

1) $x = 5, -1$
2) $x = -2, \frac{1}{2}$
3) $x = -4, 3$
4) $x = -1, \frac{3}{2}$
5) $x = -1, \frac{4}{3}$

Quadratic Formula and the Discriminant

1) $x = 2, -1$
2) $x = \frac{9+\sqrt{65}}{2}, \frac{9-\sqrt{65}}{2}$
3) $x = \frac{11+\sqrt{145}}{-4}, \frac{11-\sqrt{145}}{-4}$
4) $x = \frac{-1+\sqrt{22}}{7}, \frac{-1-\sqrt{22}}{7}$
5) $x = -2 + \sqrt{5}, -2 - \sqrt{5}$

Graphing Quadratic Functions

1) $y = x^2 + x + 1$

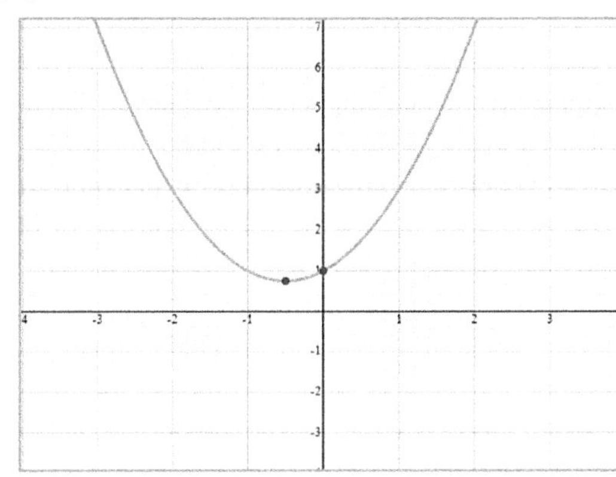

2) $y = -x^2 + 4x - 3$

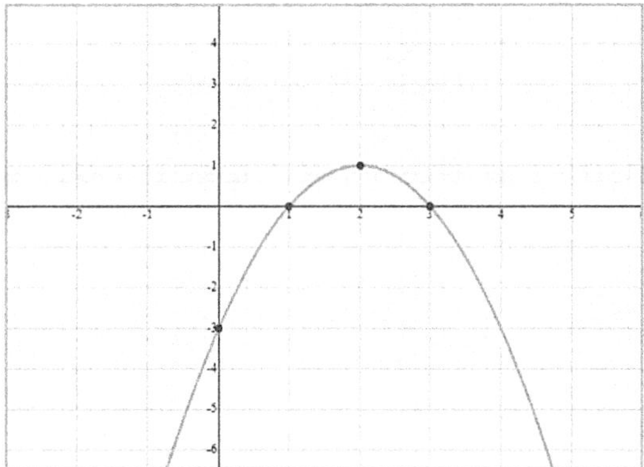

3) $y = \frac{1}{2}x^2 - 2x + 4$

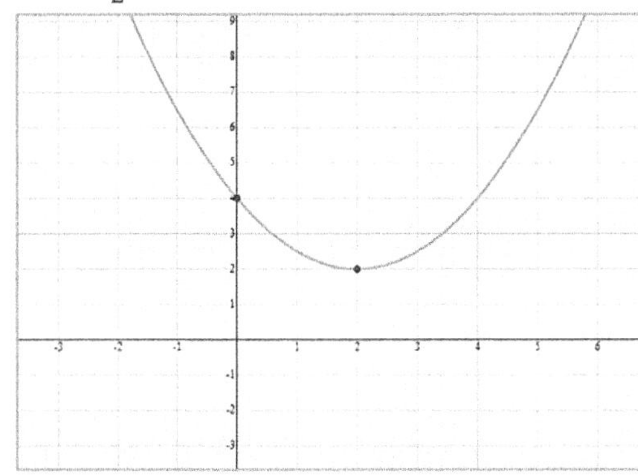

4) $y = -\frac{1}{3}x^2 + 2x - 6$

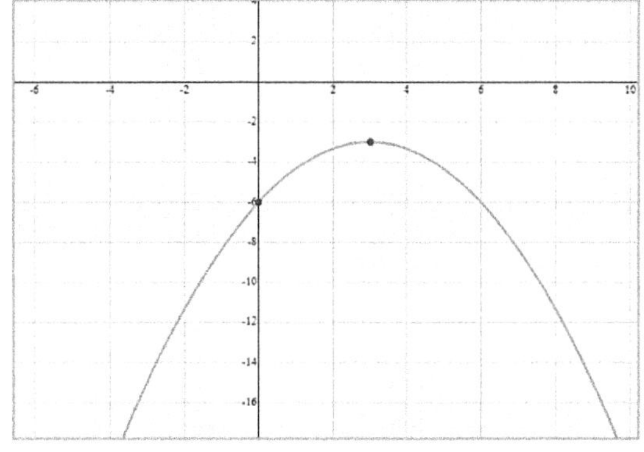

5) $y = -2x^2 + x - 1$

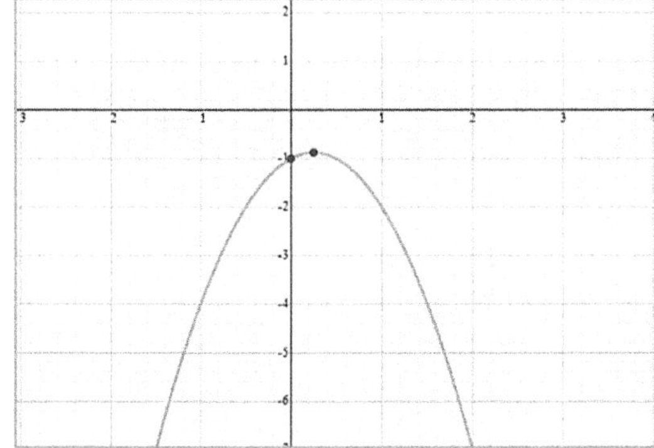

Real World Applications

1) Domain: [0,50] and range: [1,5] million people
2) 90 km/h
3) Width: 4.5m, Length:12m
4) a) 25°C and b) 68°F
5) 36 degrees
6) Approximately 6.9 seconds
7) 1750 calories
8) 6,000 dollars
9) 1102.5 dollars
10) Approximately 8.24 meters

Chapter 10: Sequences

Topics that you'll learn in this chapter:

- ✓ Identify Arithmetic Sequences
- ✓ Identify Geometric Sequences
- ✓ Sum of Arithmetic and Geometric Sequences
- ✓ Real World Applications
- ✓ Worksheets
- ✓ Answer of Worksheets

Identify Arithmetic Sequences

An arithmetic sequence is a sequence of numbers where the difference between any two consecutive numbers is always the same. This difference is called the common difference.

For example, in the sequence: $2,5,8,11,14,17, \ldots$

The common difference is $+3$, because each number is 3 more than the number before it.

Key Features of an Arithmetic Sequence:

1. **Common Difference (d):** This is the number you add or subtract to get from one term to the next.
   - In the sequence above, the common difference is 3.
   - If the numbers are decreasing, the common difference will be negative. For example, in the sequence $10, 7, 4, 1, -2, \ldots$, the common difference is -3.
2. **First Term (a_1):** This is the first number in the sequence. For the sequence $2,5,8,11, \ldots$ the first term a_1 is 2.
3. **General Formula:** You can find any term in the sequence using the formula:

$$a_n = a_1 + (n-1) \times d$$

where:

- a_n is the n-th term (the term you want to find),
- a_1 is the first term,
- n is the position of the term in the sequence,
- d is the common difference.

Examples:

1) Find the $20th$ term of the sequence $2,5,8,11,14, \ldots$

   Solution

   - The first term $a_1 = 2$
   - Common difference $d = 3$
   - Use the formula: $a_n = a_1 + (n-1) \times d = 2 + (20-1) \times 3 = 2 + 57 = 59$

   So, the $20th$ term is 59.

2) An arithmetic sequence has the first term $a_1 = 4$ and a common difference $d = 6$. The last term is 100. How many terms are there in this sequence?

   Solution: Substitute the values $a_1 = 4, d = 6$ and $a_n = 100$ into the formula:

   $100 = 4 + (n-1) \times 6 \rightarrow 100 = 4 + 6n - 6 \rightarrow 100 - 4 + 6 = 6n$

   $102 = 6n \rightarrow n = \frac{102}{6} = 17$. So, the sequence has 17 terms.

Identify Geometric Sequences

A geometric sequence is a sequence of numbers where each term is found by multiplying the previous term by a constant factor. This factor is called the common ratio.

For example, in the sequence: $3, 6, 12, 24, 48, \dots$

The common ratio is 2, because each number is obtained by multiplying the previous number by 2.

Key Features of a Geometric Sequence:

1. **Common Ratio (r)**: This is the number you multiply by to get from one term to the next. It can be positive, negative, or even a fraction.
   - In the example above, the common ratio is $r = 2$.
2. **First Term (a_1)**: This is the first number in the sequence.
   - In the sequence 3,6,12,24, ... the first term a_1 is 3.
3. **General Formula**: You can find any term in a geometric sequence using the formula:

$$a_n = a_1 \times r^{(n-1)}$$

Examples:

1) Find the $7th$ term of the sequence $1, 5, 25, 125, 625, \dots$

   Solution

   1. Identify the common ratio (r): The ratio between consecutive term is: $r = \frac{5}{1} = 5$
   2. Use the formula for the $n\text{-}th$ term of a geometric sequence: $a_1 = 1, r = 5$ and $n = 7$

      $a_n = a_1 \times r^{(n-1)} = 1 \times 5^{7-1} = 5^6 = 15{,}625$

2) A geometric sequence starts with $a_1 = 1$ and has a common ratio $r = 3$. The last term is 729. How many terms are in the sequence?

   Solution:

   1. Substitute the values $a_1 = 1, r = 3$ and $a_n = 729$ into the formula:

      $a_n = a_1 \times r^{(n-1)} \rightarrow 729 = 1 \times 3^{n-1} \rightarrow 729 = 3^{n-1}$
   2. Express 729 as a power of 3: $729 = 3^6$

      Thus: $3^6 = 3^{n-1}$ Since the bases are equal, the exponents must be equal: $n - 1 = 6 \rightarrow n = 6 + 1 = 7$

Sum of Arithmetic and Geometric Sequences

1. Sum of an Arithmetic Sequence

The formula for the sum of the first n terms of an arithmetic sequence is:

$$S_n = \frac{n}{2} \cdot (a_1 + l)$$

Where:

- S_n= sum of the first n terms,
- a_1 = first term,
- l = last term,
- n = number of terms

Alternatively, if the common difference d is known, use: $S_n = \frac{n}{2} \cdot [2a_1 + (n-1)d]$

2. Sum of a Geometric Sequence

The formula for the sum of the first n terms of a geometric sequence is:

$$S_n = \frac{a_1 \times (1 - r^n)}{1 - r}$$

where:

- S_n is the sum of the first n terms,
- a_1 is the first term,
- r is the common ratio,
- n is the number of terms.

Examples:

1) Find the sum of the first 20 terms of the arithmetic sequence where the first term is 3 and the common difference is 5.

   Solution:

   Substitute the values into the formula: $a_1 = 3, d = 5$ and $n = 20$:

   $S_{20} = \frac{20}{2} \cdot [2(3) + (20 - 1)(5)] = 10 \cdot [6 + 19 \cdot 5] = 10 \cdot 101 = 1010$

2) Calculate the sum of the first 8 terms of a geometric sequence where the first term is 2 and the common ratio is 3.

   Solution: Substituting the values into the formula: $a_1 = 2, r = 3$ and $n = 8$:

$$S_8 = \frac{2 \times (1 - 3^8)}{1 - 3} = \frac{2 \times (1 - 6561)}{-2} = 6560$$

Real World Applications

Sequences play an important role in a variety of real-world applications across different fields. Here are some examples:

Daily Life

- **Staircases:** Arithmetic sequences can determine the arrangement of steps in a staircase.
- **Savings Plans:** They help calculate how much one needs to save overtime to reach financial goals.

Architecture and Art

- **Design Patterns:** Arithmetic and geometric sequences appear in architectural designs and artwork to create aesthetically pleasing proportions (e.g., the Fibonacci sequence).
- **Music Composition:** Sequences are often used to create rhythms and melodies in music.

Physics and Engineering

- **Waveforms:** Sequences model periodic phenomena such as sound waves, light waves, and vibrations.
- **Signal Processing:** Geometric sequences are used in analyzing and compressing signals for communication systems.

Computer Science

- **Algorithms:** Sequences are crucial in designing and analyzing algorithms, such as sorting and searching techniques.
- **Data Storage:** Memory allocation and data indexing often rely on sequences for efficiency.

Example:

Emma is saving money to go on a trip to Disney World. She decides to save a specific amount of money every month. She starts by saving $20 in the first month, $30 in the second month, $40 in the third month, and so on, increasing the amount she saves by $10 each month. What is the total amount Emma will have saved by the end of 12 months?

Solution:

The formula for the sum of the first n terms of an arithmetic sequence is:

$S_n = \frac{n}{2} \cdot [2a_1 + (n-1)d] \rightarrow S_{12} = \frac{12}{2} \cdot (2 \cdot 20 + (12-1) \cdot 10)$

$S_{12} = 6 \cdot (40 + 110) = 6 \cdot 150 = 900$

By the end of 12 months, Emma saves a total of $900.

Worksheets

Identify Arithmetic Sequences

Do the following problems.

1) Find the $10th$ term of the arithmetic sequence where the first term is $a_1 = 5$ and the common difference is $d = 3$.
2) An arithmetic sequence starts with $a = 4$ and $d = 6$. If the last term is 70, how many terms are in the sequence?
3) Given the arithmetic sequence: $-7, -4, -1, 2, 5, \ldots$ what $30th$ term of this sequence?
4) Find the general formula for the nth term of the arithmetic sequence: $4, 9, 14, 19, \ldots$?
5) In an arithmetic sequence, the $3rd$ term is 12, and the $7th$ term is 28. Find the first term and the common difference.

Identify Geometric Sequences

Find the answers.

1) Find the $7th$ term of a geometric sequence where the first term is $a = 2$ and the common ratio is $r = 3$.
2) The $5th$ term of a geometric sequence is 32, and the common ratio is 2. Find the first term and the $8th$ term.
3) The $3rd$ term of a geometric sequence is 18, and the $7th$ term is 486. Write the general formula for the nth term.
4) Find the general formula for the nth term of geometry sequence: $-2, 8, -32, 128, \ldots$
5) The $3rd$ term of a geometric sequence is 16, and the $6th$ term is 128. What is the $8th$ term of the sequence?

Sum of Arithmetic and Geometric Sequences

Do the following problems.

1) In a geometric sequence, the $3rd$ term is 128, and the $6th$ term is 16. Find the $10th$ term and the sum of the first 10 terms.
2) The sum of the first n terms of an arithmetic sequence is given by $S_n = n(5n + 3)$. Find the first term and the common difference.
3) Find the sum of the first 12 terms of the arithmetic sequence: $2, 5, 8, 11, \ldots$
4) Find the sum of the first 10 terms of the geometric sequence: $50, 25, 12.5, 6.25, \ldots$
5) The first term of an arithmetic sequence is 4, and the $30th$ term is 88. Find the sum of the first 30 terms of the sequence.
6) The sum of the first 5 terms of a geometric series is 189. The first term is 3. Find the common ratio.
7) If the $6th$ term of an arithmetic sequence is 25, and the $15th$ term is 70, calculate the sum of the first 15 terms.
8) Find the sum of the first 100 terms of this sequence: $\frac{1}{2}, \frac{1}{4}, \frac{1}{8}, \frac{1}{16}, \ldots$
9) What is the sum of all two-digit numbers whose remainder when divided by 7 is 5?
10) In an arithmetic sequence with the first term a, if 1 is added to the common difference, how much will be added to the sum of the first 20 terms?

Real World Applications

Find the answers.

1) A bus route has stops at distances of 2 km, 5 km, 8 km, 11 km, and so on. If there are 12 stops in total, what is the total distance covered by the bus?
2) A ball is dropped from a height of 100 meters and bounces back to 70% of its previous height after each bounce. What is the total vertical distance traveled by the ball after 3 bounces?
3) A bacteria colony doubles in size every hour. If the colony starts with 1 bacteria, how many bacteria will there be after 10 hours? What is the total number of bacteria that have existed after 10 hours?
4) Liam deposits $100 in a savings account. The account grows by 5% each month due to interest. What is the total amount of money the account has accumulated over the 6 months?
5) A stadium has 50 rows of seats. The first row has 20 seats, the second row has 24 seats, the third row has 28 seats, and so on. How many total seats are there in the stadium?
6) A staircase has 20 steps. The height of the first step is 5 cm, and the height of the $10th$ step is 9 cm. Assuming the step heights increase in an arithmetic sequence, find the height of the last step and the total height of the staircase.
7) A car rental company charges $200 as a base fee and $15 for each additional day rented. If someone rents the car for $1, 2, 3, 4$, etc. days, find the total cost for 30 days.
8) A savings account balance grows geometrically. The balance after 1 year is $1,200, and after 3 years, it is $4,800. Find the initial balance and the growth factor.
9) A swimming pool has rows of tiles, where the first row has 30 tiles, the second row has 28 tiles, the third row has 26 tiles, and so on. If the pool has 20 rows, calculate the total number of tiles.
10) Each time a text is edited, half of its errors are corrected. After, how many edits will at least 99% of the errors in the text be corrected?

Answer of Worksheets

Identify Arithmetic Sequences

1) 32
2) 12 terms
3) 80
4) $a_n = 5n - 1$
5) $a_1 = 4$ and $d = 4$

Identify Geometric Sequences

1) 1458
2) $a_1 = 2$ and the $8th = 256$
3) $a_n = 2 \cdot 3^{n-1}$
4) $a_n = -2 \cdot (-4)^{n-1}$
5) 512

Sum of Arithmetic and Geometric Sequences

1) The $10th = 1$ and $S_n = 1023$
2) $a_1 = 8$ and $d = 10$
3) 222
4) Approximately 100
5) 1380
6) 2
7) 525
8) Approximately 1
9) 702
10) 190

Real World Applications

1) 35 km
2) 406.6 meters
3) After 10 hours, there are 512 bacteria, and the total number of bacteria that have existed is 1023.
4) ≈ \$134.01
5) 5900 seats
6) The height of the last step is approximately 13.44 cm, and the total height is approximately 184.44 cm.
7) \$650
8) The initial balance is \$1,200, and the growth factor is 2.
9) 220
10) At least 7 edits

Chapter 11: Transformation

Topics that you'll learn in this chapter:

- ✓ Transformations
- ✓ Translations
- ✓ Rotations
- ✓ Reflections
- ✓ Reflections Over Common Axes
- ✓ Dilations
- ✓ Scale Factor
- ✓ Sequences of Congruence Transformations
- ✓ Real World Applications
- ✓ Worksheets
- ✓ Answer of Worksheets

Transformations

In mathematics, a transformation is a way of changing the position, size, or shape of a figure on a coordinate plane while following certain rules. Transformations can be divided into two main groups based on whether they preserve size or allow changes in size:

1. **Congruence Transformations (Rigid Transformations)**:
   - Congruence transformations, also known as rigid transformations or isometries, are transformations in which the shape and size of a figure remain unchanged. In other words, the original figure and the transformed figure are congruent, and they have the same size, shape, angles, and side lengths. These transformations simply reposition the figure without altering its dimensions.
   - **Types:**
     - **Translation**
     - **Rotation**
     - **Reflection**
2. **Similarity Transformations**:
   - Similarity transformations are transformations in which a figure changes its size but maintains the same shape. These transformations preserve the angles of the figure and the proportionality of its side lengths, so the original figure (pre-image) and the resulting figure (image) are similar rather than congruent. This means they have the same shape but might differ in size.
   - Types:
     - **Dilation**
     - **Combination** of congruence transformations and dilation.

Example: Determine if following transformation is congruence or similarity.

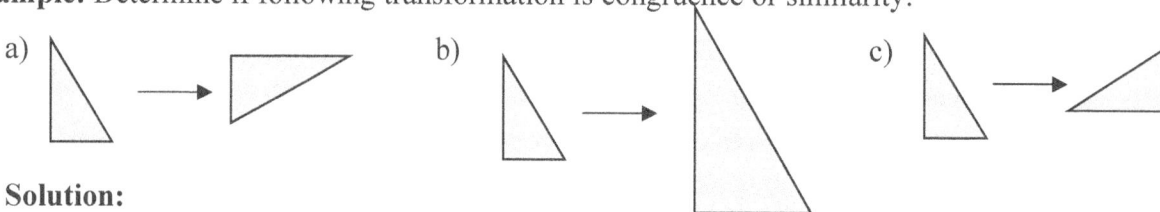

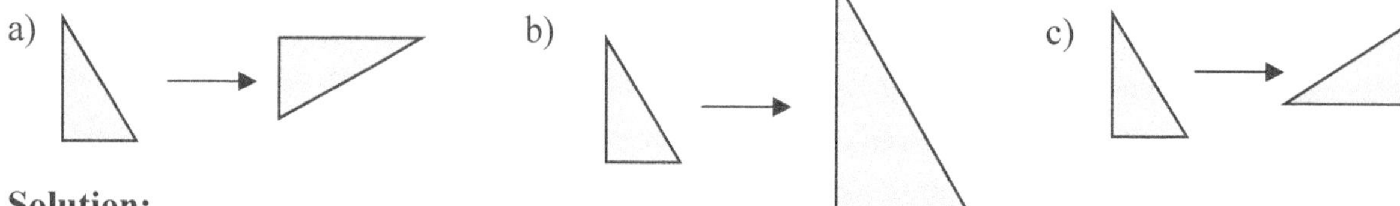

Solution:

a) Congruence transformation (the shape and size has remained unchanged)

b) Similarity transformation (the shape has remained unchanged, but the size has changed)

c) Congruence transformation (the shape and size has remained unchanged)

Translation

Translation is a type of transformation where a figure is moved from one location to another on a coordinate plane without changing its size, shape, or orientation. It's like sliding the figure in a straight line.

Key Features of Translation:

1. **Slide, Don't Twist**: A translation only moves the shape in a straight line. It doesn't rotate (turn) or reflect (flip) the shape. Thus, the shape looks exactly the same before and after.
2. **Same Shape, Same Size**: The shape doesn't stretch, shrink, or change in any way. It's like a perfect copy but in a new spot.
3. **Using Coordinates**: In translation, we describe how far to slide the shape using numbers. For example:
    - **Move right/left**: This changes the x-coordinate.
    - **Move up/down**: This changes the y-coordinate.
    - If a point starts at (x, y), and we move it a units right or left, and b units up or down, the new position is:

$$(x', y') = (x + a, y + b)$$

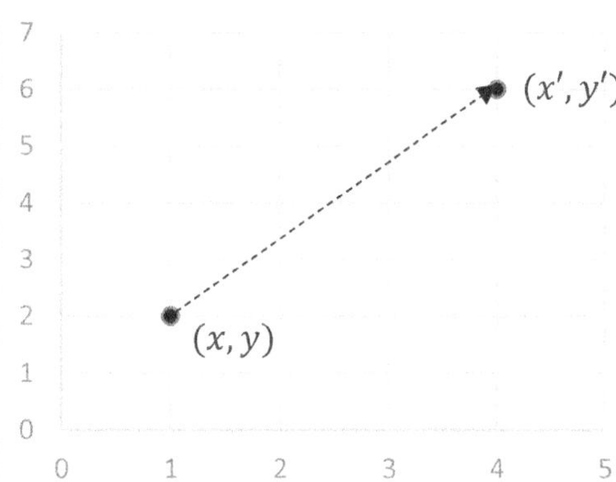

Example:

If you have a point at $(-2, 6)$, and you move it 5 steps to the left and 3 steps up, what is new point?

Solution:

- Subtract 5 from x-coordinate: $-2 - 5 = -7$
- Add 3 to the y-coordiante: $6 + 3 = 9$
- The new point is: $(-7, 9)$

Rotation

In mathematics, **rotation** is a type of transformation that turns a shape or object around a fixed point, called the center of rotation.

Key Features of Rotation:

1. **Angle of Rotation:** This tells you how far the shape is rotated. It's usually measured in degrees (like $90°$, $180°$, $270°$, etc.).
2. **Direction:** The rotation can be clockwise (turning to the right) or counterclockwise (turning to the left).
3. **Center of Rotation:** This is the point around which the object rotates. It can be a point on the shape, inside the shape, or even outside the shape.
4. **Properties:**
   - The size and shape of the object don't change during rotation.
   - The distance from any point on the shape to the center of rotation stays the same.

Rotation Rules Around the Origin:

- **90° Clockwise:** $(x, y) \rightarrow (y, -x)$
- **90° Counterclockwise:** $(x, y) \rightarrow (-y, x)$
- **180° (Clockwise or Counterclockwise):** $(x, y) \rightarrow (-x, -y)$
- **270° Clockwise (or 90° Counterclockwise):** $(x, y) \rightarrow (-y, x)$
- **270° Counterclockwise (or 90° Clockwise):** $(x, y) \rightarrow (y, -x)$

Example:

Rotate a triangle with vertices $(-3, 1), (-1, 1)$ and $(-1, 4)$,$90°$ clockwise around the origin (0,0).

Solution

This triangle has three vertices $(-3, 1), (-1, 1)$ and $(-1, 4)$ that, using the rules of rotation around the origin, will be transformed to these points: $(1, 3), (1, 1), (4, 1)$

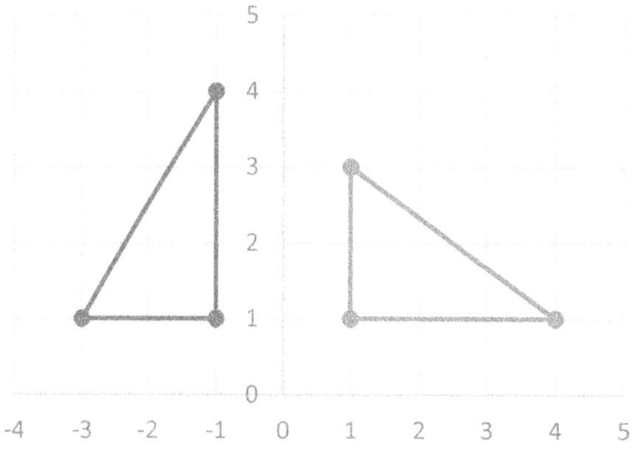

Reflections

In mathematics, reflection is a type of transformation that "flips" a shape over a line, called the line of reflection. It's like looking at a mirror image of the shape.

Key Features of Reflections:

1. **Line of Reflection:** This is the "mirror line" over which the shape is reflected. It could be the x-axis, y-axis, or any other line.
2. **Flipped Image:**
   - After the reflection, the original shape (pre-image) and its reflected shape (image) are symmetrical with respect to the line of reflection.
   - Each point on the shape is the same distance from the line of reflection, but on the opposite side.
3. **No Changes to Size or Shape:** The reflected shape has the same size and dimensions as the original shape. It's just reversed.

How to Reflect a Shape Over a Line:

1. **Take One Point at a Time:** Start with one vertex (corner) of the shape and reflect it over the line.
2. **Measure the Distance:** Use a ruler or measure the distance between the point and the line of reflection. This distance will be the same on the opposite side of the line after reflection.
3. **Find the Reflected Point:** Place the reflected point on the opposite side of the line of reflection, keeping it the same distance away.
4. **Repeat for All Points:** Do the same for all the other vertices or points of shape.
5. **Connect the Reflected Points:** After reflecting all the points, connect them in the same order to recreate the shape.

Example:

Reflect the shape with following coordinates over the given line: $(1,4)$, $(3,4)$, $(3,7)$ and $(1,5)$

Solution:

By measuring the distance of all points from the given line, reflect them on the opposite side of the line and then connect them in the same order:

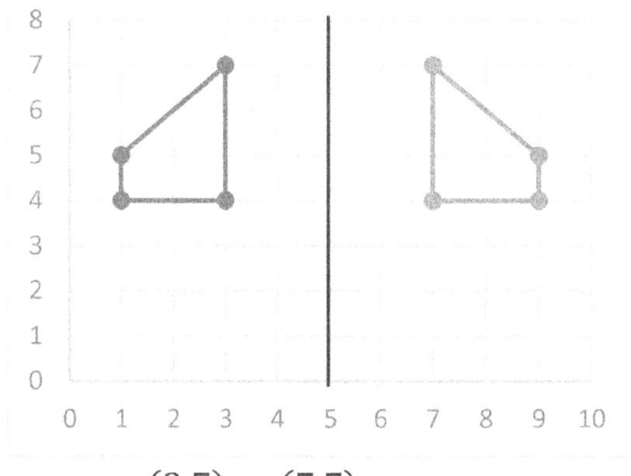

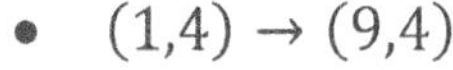

- $(1,4) \rightarrow (9,4)$
- $(3,4) \rightarrow (7,4)$
- $(3,7) \rightarrow (7,7)$
- $(1,5) \rightarrow (9,5)$

Reflections Over Common Axes

Reflections over common axes, like the x-axis and y-axis, are a type of transformation in mathematics where a shape or a point is flipped across these axes.

Common Reflection Rules in a Coordinate Plane:

- Reflection over the x-axis: $(x, y) \rightarrow (x, -y)$
- Reflection over the y-axis: $(x, y) \rightarrow (-x, y)$
- Reflection over the line $y = x$: $(x, y) \rightarrow (y, x)$
- Reflection over the line $y = -x$: $(x, y) \rightarrow (-y, -x)$

Examples:

1) Reflect the point $A: (3, 5)$ over the x-axis and y-axis.

   Solution:
   - Using the reflection rules, the reflected point over the x-axis is point $A': (3, -5)$
   - Using the reflection rules, the reflected point over the y-axis is point $A'': (-3, 5)$

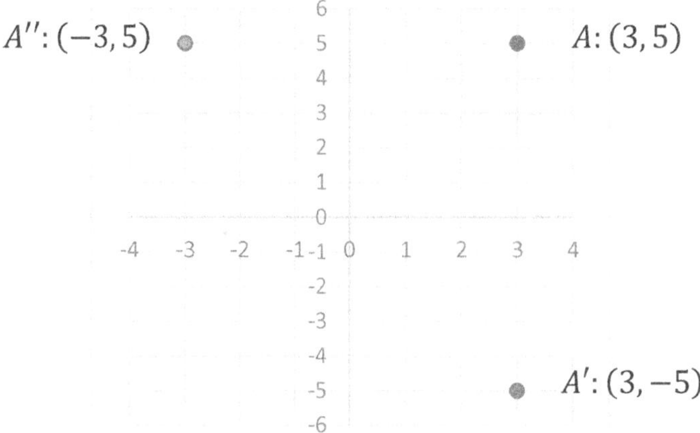

2) Reflect the shape with the coordinates $(2,4), (3,4), (3,5), (4,6)$ and $(2,6)$ over the line $y = x$.

   Solution:
   Using the reflection rules, the reflected shape over the $y = x$ will be $(4,2), (4,3), (5,3), (6,4)$ and $(6,2)$:

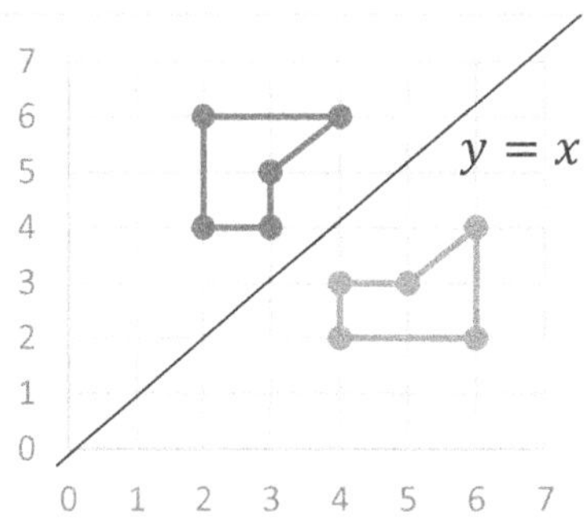

Dilation

Dilation is a transformation that changes the size of a shape while keeping its proportions and overall structure the same. It's like zooming in or out on the shape without distorting it.

Key Concepts of Dilation:

1. **Center of Dilation**:
    - This is the fixed point from which the shape is enlarged or reduced. All points on the shape move away from or toward this center during dilation.
2. **Scale Factor**:
    - A number that tells you how much to enlarge or reduce the shape.
    - **Greater than 1**: The shape becomes larger (enlargement).
    - **Between 0 and 1**: The shape becomes smaller (reduction).
    - **Equal to 1**: The shape stays the same size (no dilation occurs).
3. **Proportionality**:
    - The sides of the shape change in length according to the scale factor, but angles remain the same.
4. **Coordinate change:** When the center of dilation is the origin $(0,0)$, Multiply both the x- and y-coordinates of each point by the scale factor.

$$(x, y) \rightarrow (kx, ky)$$

Where k is the scale factor.

Example:

Take a rectangle with points $(1,1)$, $(3,1)$, $(3,2)$ and $(1,2)$, Dilate it with a scale factor of 2 from the origin.

Solution:

Using the coordinate change, new points will be: $(2,2)$, $(6,2)$, $(6,4)$ and $(2,4)$

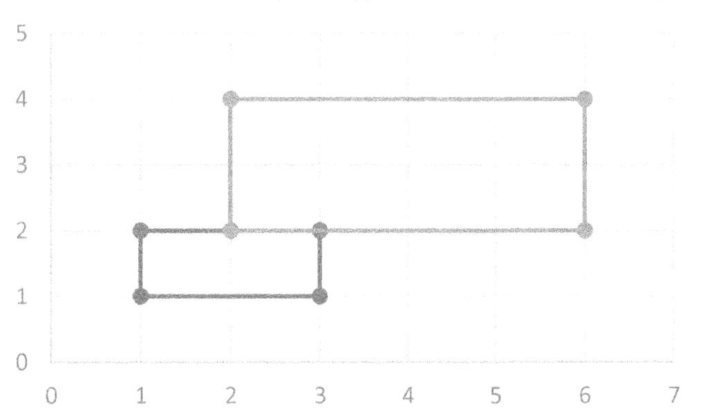

Scale Factor

As we learned in thc prcvious scction, The scale factor in a transformation refers to the ratio by which a figure is enlarged or reduced during a dilation. It determines how much the size of a figure changes while preserving its shape.

The scale factor is the multiplier applied to the dimensions of a figure (like lengths, widths, or heights) to create a similar figure that is either larger or smaller. It is often denoted as k.

Types of Scale Factor:

- **Enlargement:** If $k > 1$, the figure becomes larger.
- **Reduction:** If $0 < k < 1$, the figure becomes smaller.
- **No Change:** If $k = 1$, the figure remains the same size.

How to Calculate Scale Factor:

You can calculate the scale factor using the formula:

$$\text{Scale Factor}= \frac{New\ Dimension}{Original\ Dimension}$$

Impact of Scale Factor on Area and Volume:

- **For 2D shapes (like rectangles or circles):** The area changes by the square of the scale factor (k^2).
- **For 3D shapes (like cubes or spheres):** The volume changes by the cube of the scale factor (k^3).

Examples:

1) The original dimension of a rectangle is 4 cm by 6 cm and the scale factor is 1.5. What are the new dimensions?

   Solution:
   Using the scale formula: New width= $4 \times 1.5 = 6\ cm$ and new length= $6 \times 1.5 = 9\ cm$

2) The area of the square is initially $16\ cm^2$. After a dilation, the new area is $64\ cm^2$. What is the scale factor of the dilation?

   Solution:

   1. When a figure is dilated, the area changes by the square of the scale factor (k^2). So: New Area= k^2 ×Original Area
   2. Substitute the values given: The original area is $16\ cm^2$, and the new area is $64\ cm^2$. Substitute these into the equation: $64 = k^2 \times 64 \rightarrow k^2 = \frac{64}{16} = 4 \rightarrow k = \sqrt{4} = 2$
   So, the scale factor of the dilation is 2.

Sequences of Congruence Transformations

A **sequence of congruence transformations** is just a series of transformations (translation, reflection, or rotation) done one after the other to change the position or orientation of a shape. The final result will be a new position or orientation of the shape, but it will still be congruent (the same shape and size) to the original.

Example:

Imagine you have a triangle with vertices at points $A(1,1), B(4,1)$, and $C(3,3)$ on a coordinate plane. First move the triangle 3 units up and 2 units right, then rotate the triangle 180° counterclockwise around the origin and finally reflect the triangle over the y-axis. What are the final vertices of the triangle?

Solution:

1. The new vertices after translation: Move each vertex 3 units up and 2 units right:
   - $A(1,1) \to A'(3,4)$
   - $B(4,1) \to B'(6,4)$
   - $C(3,3) \to C'(5,6)$
2. Rotate the triangle 180° counterclockwise: Use the rule for a 180° rotation: $(x,y) \to (-x,-y)$:
   - $A'(3,4) \to A''(-3,-4)$
   - $B'(6,4) \to B''(-6,-4)$
   - $C'(5,6) \to C''(-5,-6)$
3. Reflect the triangle over the y-axis: Use the rule for reflection over the y-axis: $(x,y) \to (-x,y)$:
   - $A''(-3,-4) \to A'''(3,-4)$
   - $B''(-6,-4) \to B'''(6,-4)$
   - $C''(-5,-6) \to C'''(5,-6)$

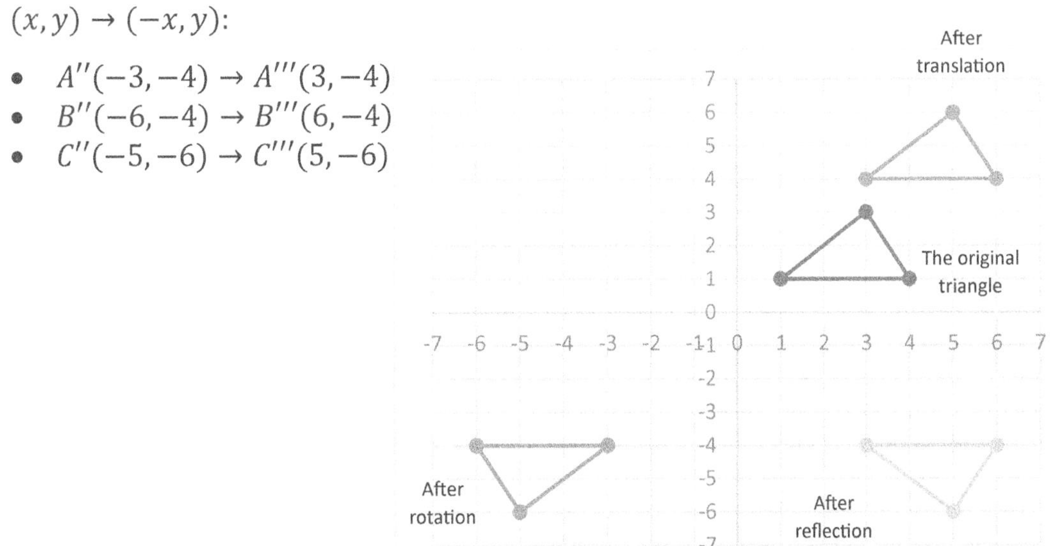

Real World Applications

Transformations in algebra and geometry have numerous real-world applications across various fields. Here are some examples:

Computer Graphics and Animation: Transformations like translation, rotation, reflection, and scaling are fundamental to creating animations, video games, and 3D modeling. For example:

- Characters in video games move using translation.
- Objects are rotated and scaled to create realistic scenes in animations and movies.

Robotics: Robots use transformations to navigate and interact with the environment. For example: A robotic arm uses rotations and translations to perform precise tasks like assembling products or performing surgeries.

Image Processing: Transformations play a crucial role in editing and analyzing images. For example:

- Scaling an image to adjust its size.
- Rotating or flipping an image for better alignment or perspective.

Satellite Navigation and Mapping:

- GPS and mapping systems use transformations to overlay maps, calculate distances, and determine directions accurately.
- Rotations and translations are used to align geographic data from different sources.

Physics and Astronomy: Transformations are used to study motion and forces. For example:

- Rotational transformations help simulate planetary orbits and the motion of celestial bodies.
- Translations and reflections help in understanding trajectories of particles.

Example:

Imagine you're designing a rectangular poster for an event, and the original poster size is 24 cm by 36 cm. You need to resize it proportionally to fit a smaller space while keeping the same shape. The smaller version should have a width of 10 cm. What will the height of the resized poster be?

Solution:

The scale factor is the ratio of the new dimension to the original dimension. Since the new width is 10 cm and the original width is 24 cm, the scale factor (k) is: $k = \frac{new\ width}{original\ width} = \frac{10}{24} = \frac{5}{12}$

To maintain proportionality, multiply the original height by the scale factor:

New height=Original height$\times\ k = 36 \times \frac{5}{12} = \frac{36\times5}{12} = 15cm$

Worksheets

Transformations

Determine if following transformation is congruence, similarity or none of them.

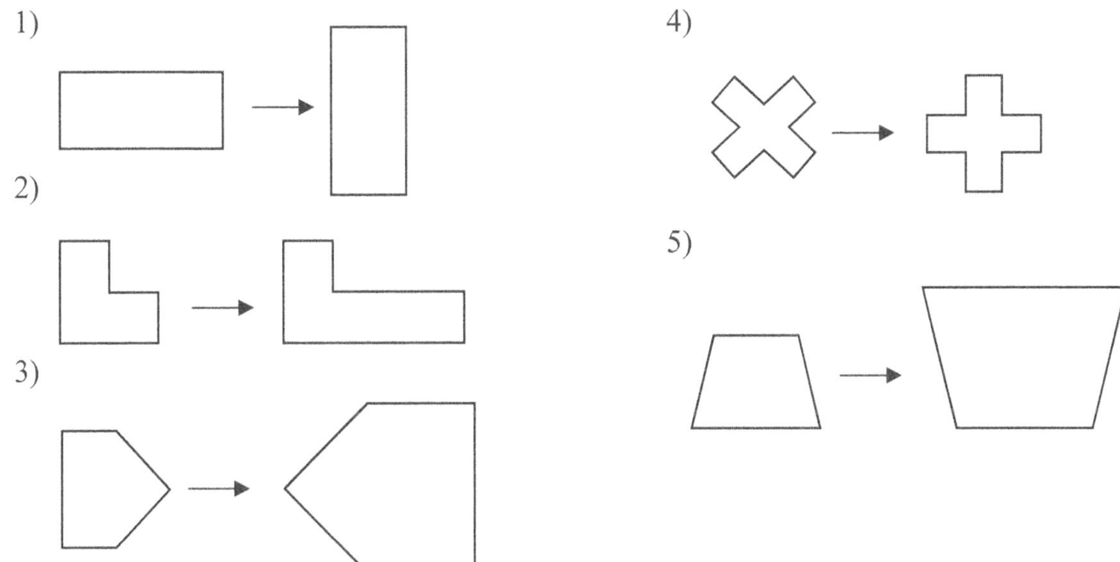

Translations

Do the following problems.

1) Translate the triangle with vertices $(0,0), (2,0), (1,3)$ 3 units to the right and 2 units up. What are the new vertices?
2) A square has vertices $(-1,-1), (-1,1), (1,1), (1,-1)$. Translate it 4 units left and 5 units up. Find the new coordinates of the vertices.
3) If the vertices of a rectangle after translation by $\vec{a} = (-2,5)$ match these vertices: $A(3,1), B(1,3), C(-2,0)$and $D(0,-2)$, find the original vertices of rectangle?
4) If point $A(-2,1)$ is translated three times by three vectors: $\vec{a} = (3,0), \vec{b} = (-4,2)$ and $\vec{c} = (-1,5)$, find the final translated point.
5) The equation of a line is $y = 2x + 3$. After translating the line 4 units to the right and 5 units down, what is the new equation of the line?

Rotations

Find the answers.

1) Rotate the point $B(-4,5)$ 90° clockwise about the origin. What are the new coordinates?
2) A square has vertices $(1,1), (1,4), (4,4), (4,1)$. Rotate the square 180° about the origin. What are the coordinates of the vertices after the rotation?
3) Rotate the point $D(5,-3)$ 90° clockwise, then 180° counterclockwise about the origin. What are the final coordinates?
4) A triangle has vertices $A(-2,3), B(0,-1), C(4,2)$. Rotate the triangle 270° counterclockwise around the origin. What are the coordinates of the new vertices?
5) Rotate the point $C(4,2)$ 90° counterclockwise about the point $(2,2)$. What are the new coordinates?

Reflections

Reflect following shapes over the given line:

1)

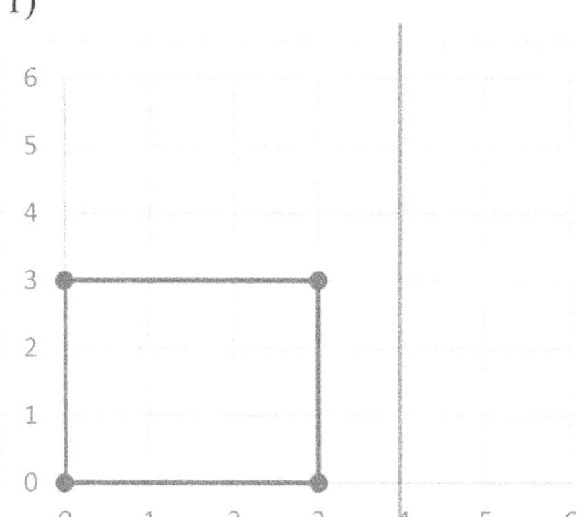

2)

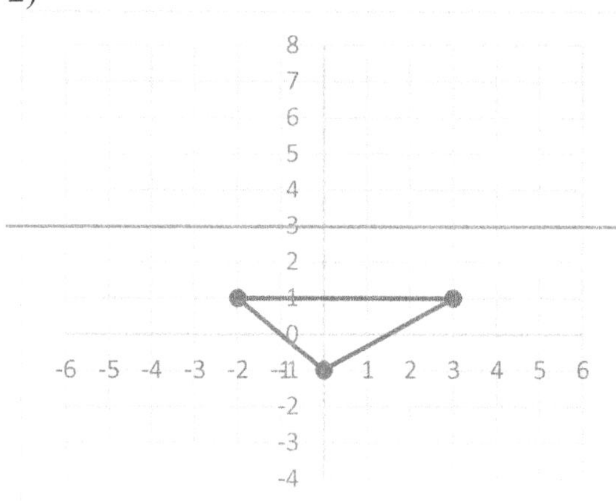

3)

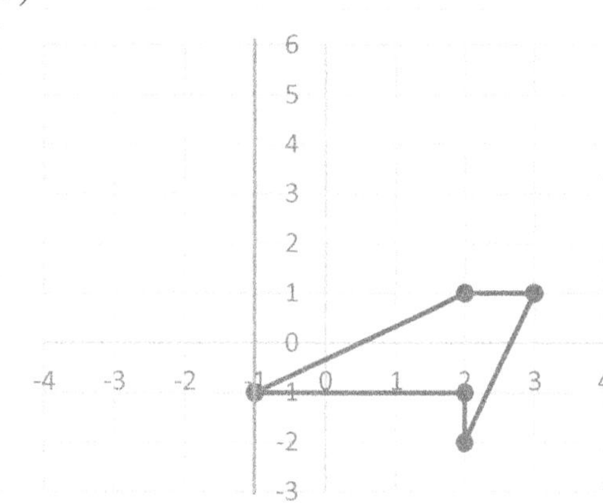

4)

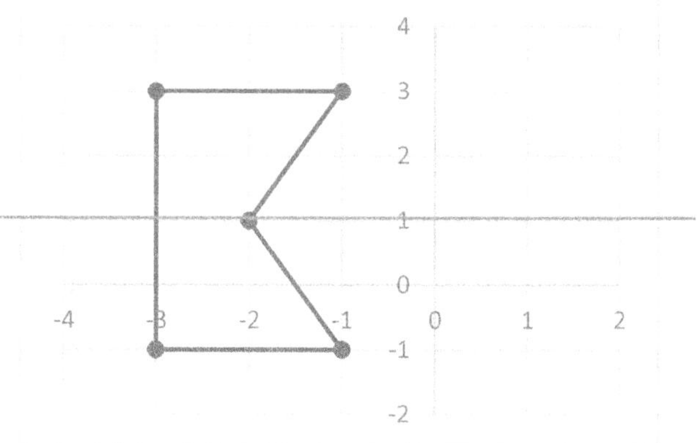

5)

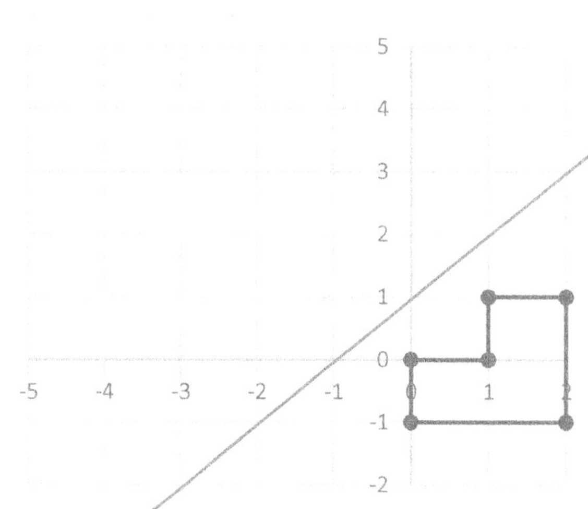

Reflections Over Common Axes

Do the following problems.

1) Reflect the point $B(-2,5)$ over the y-axis. What is the coordination of the reflected point?
2) Reflect the point $C(1,6)$ over the line $y = x$. What are the coordinates of the reflected point?
3) A triangle has vertices at $D(1,2)$, $E(3,5)$, and $F(4,1)$. Reflect the triangle over the x-axis. What are the coordinates of the reflected vertices?
4) A quadrilateral has vertices at $K(2,3)$, $L(4,6)$, $M(5,2)$, and $N(3,1)$. Reflect the quadrilateral over the line $y = -x$. What are the coordinates of the reflected vertices?
5) A point $Q(5,7)$ is reflected over the horizontal line $y = 2$. What are the coordinates of the reflected point?

Dilations

Do the following problems.

1) A point $B(6,9)$ is dilated by a scale factor of $\frac{1}{3}$ with the center of dilation at the origin. What are the coordinates of the dilated point?
2) A triangle has vertices at $C(1,2)$, $D(3,4)$, and $E(5,1)$. Dilate the triangle by a scale factor of 3 with the center of dilation at the origin. What are the coordinates of the dilated vertices?
3) A square has vertices at $M(2,2)$, $N(2,4)$, $O(4,4)$, and $P(4,2)$. Dilate the square by a scale factor of 2 with the center of dilation at $(1,1)$. What are the coordinates of the dilated vertices?
4) The original triangle has the following vertices: $A(-2,0), B(-2,-2), C(-4,-2)$. After dilation with a scale factor, the new vertices are: $A'(-1,0), B'(-1,-1), C'(-2,-1)$, find the scale factor.
5) A rectangle has vertices at $Q(-4,2)$, $R(0,2), S(0,4)$ and $T(-4,4)$. Dilate the triangle by a scale factor of $\frac{1}{4}$ with the center of dilation at $(-1,1)$. What are the coordinates of the dilated vertices?

Scale Factor

Find the answers.

1) A model car is 8 cm long. The actual car is 4 meters long. What is the scale factor of the model car?
2) A rectangle has a length of 5 cm and a width of 3 cm. It is dilated by a scale factor of 4. What is the area of the dilated rectangle?
3) A map has a scale factor of $1{:}\,100{,}000$. If two cities are 5 cm apart on the map, how far apart are they in real life?
4) A city planner is creating a map of a park. The real park is 500 meters long and 300 meters wide. On the map, the park is represented with dimensions $10\,cm \times 6\,cm$. What scale factor was used to create the map?
5) A cube has side lengths of 2 cm. It is dilated by a scale factor of 3. What is the volume of the dilated cube?

Sequences of Congruence Transformations

Do following problems about sequences of congruence transformations.

1) A point $A(2,3)$ is first translated by $(x+3, y-2)$ and then reflected over the x-axis. What are the coordinates of the final point?
2) A point $C(1,2)$ is first rotated 90° counterclockwise about the origin and then translated by the vector $\vec{v} = (-2,-1)$. What are the coordinates of the final point?
3) A point $F(6,8)$ is first rotated 270° counterclockwise about the origin and then reflected over the y-axis. What are the coordinates of the final point?
4) A triangle has vertices at $I(1,1)$, $J(3,4)$, and $K(5,2)$. It is first translated by $\vec{a} = (2,-3)$, then reflected over the line $y = x$, and finally rotated 180° clockwise about the origin. What are the coordinates of the final vertices?
5) A trapezoid $ABCD$ is rotated 90 counterclockwise about the origin, translated 2 units right, and finally reflected across the y-axis. The original vertices are: $A(1,1), B(4,1), C(3,4), D(1,4)$. Determine the coordinates of $A'B'C'D'$ after all transformations.

List all the transformations needed to convert Shape A to Shape C and for questions 9 and 10, Shape A to Shape D in order:

6)

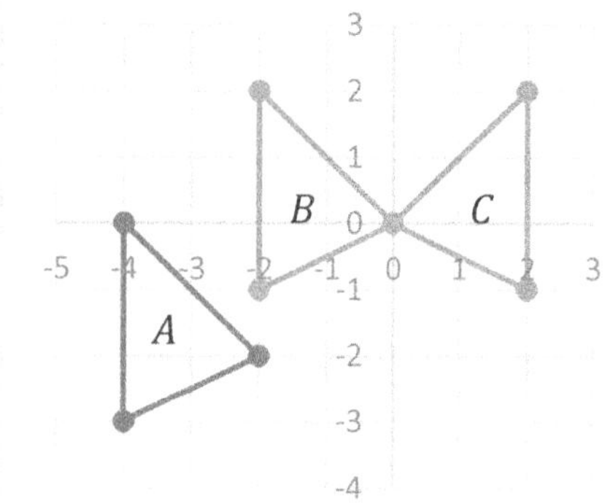

7)

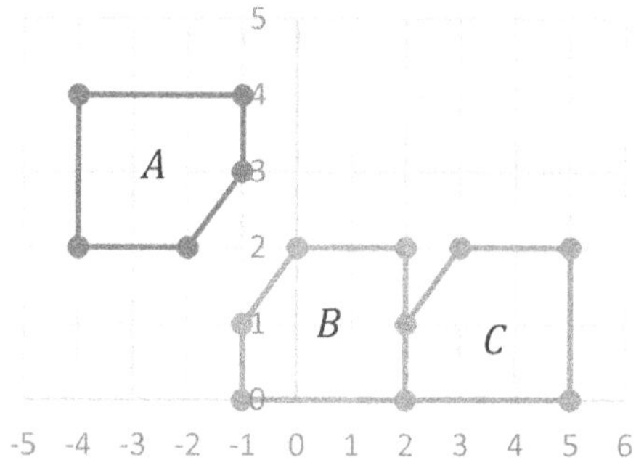

8)

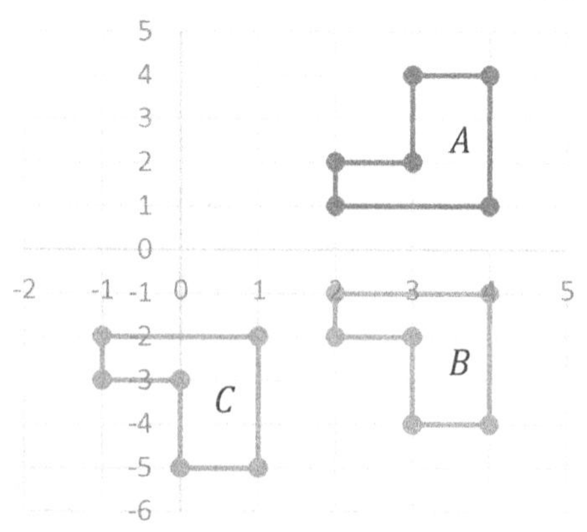

9)

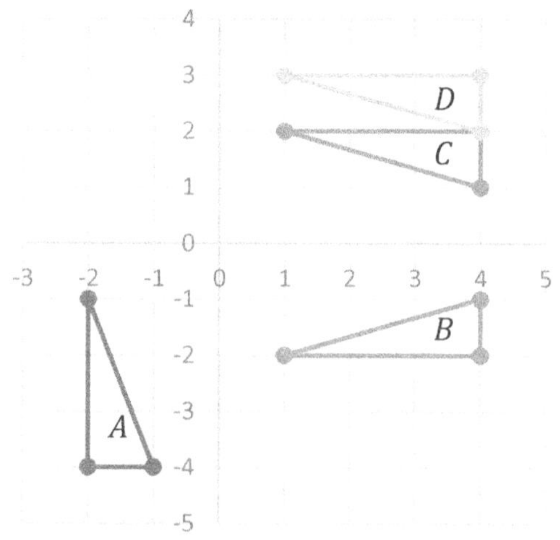

10)

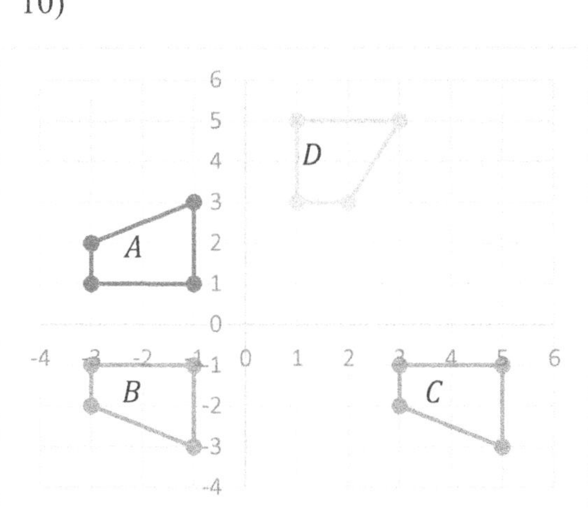

Real World Applications

Do following word problems about transformation:

6) The minute hand of a clock is at the 12 position. It rotates 90° clockwise. What position does it point to now?
7) A map has a scale factor of 1: 20,000. If two cities are 5 cm apart on the map, how far apart are they in real life?
8) A logo is designed as a star inside a circle. If the star is rotated by 72° (one-fifth of a full rotation) it looks the same. How many lines of rotational symmetry does the logo have?
9) A solar panel tracker rotates the panel 45° every hour to follow the sun's position. If the panel starts at an angle of 0° at 8 AM, what will be its angle at noon?
10) A campsite is located 20 meters from a river, and across the river, at the exact opposite point, there's a signal tower. A rescue team standing at the campsite sees the reflection of the signal tower in the river. How far from the campsite is the reflection?

Answer of Worksheets

Transformations

1) Congruence transformation
2) None of them
3) Similarity transformation
4) Congruence transformation
5) Similarity transformation

Translations

1) $(3,2), (5,2), (4,5)$
2) $(-5,4), (-5,6), (-3,6), (-3,4)$
3) $A(5,-4), B(3,-2), C(0,-5)$ and $D(2,-7)$
4) $(-4,8)$
5) $y = 2x - 10$

Rotations

1) $(5,4)$
2) $(-1,-1), (-1,-4), (-4,-4), (-4,-1)$
3) $(3,5)$
4) $A(3,2), B(-1,0), C(2,-4)$
5) $(2,4)$

Reflections

1)

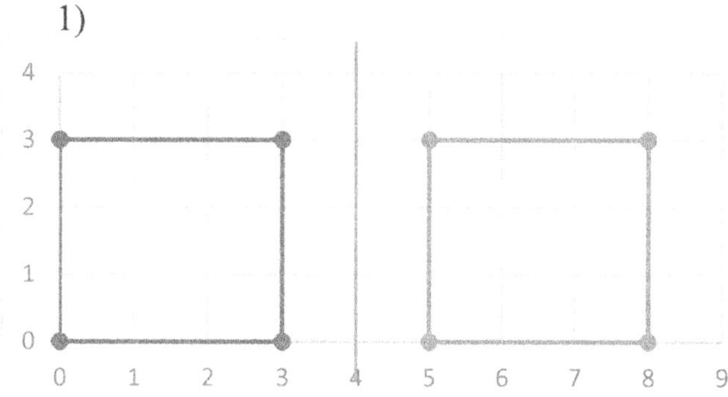

2)

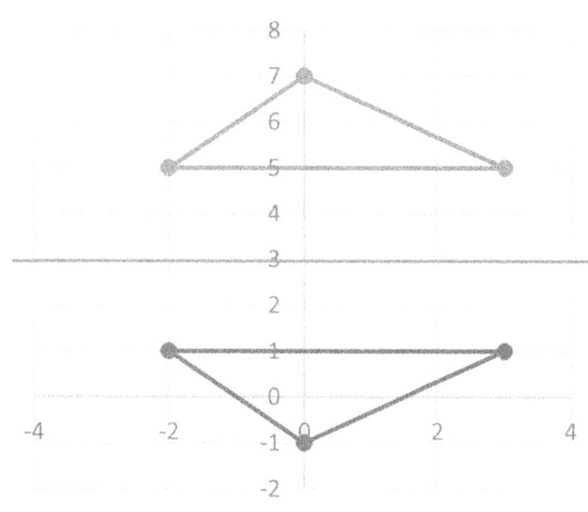

3)

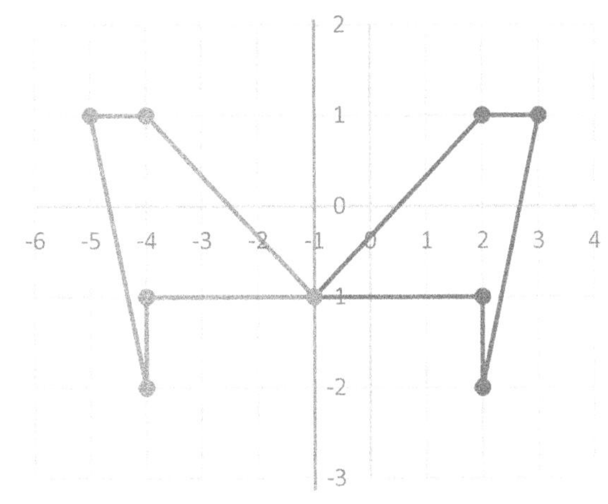

4)

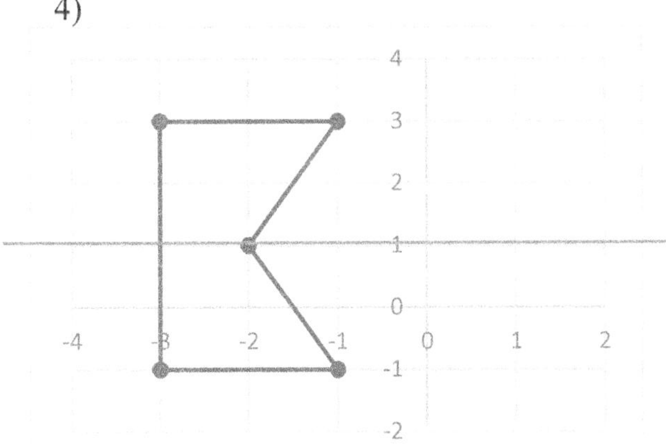

5)

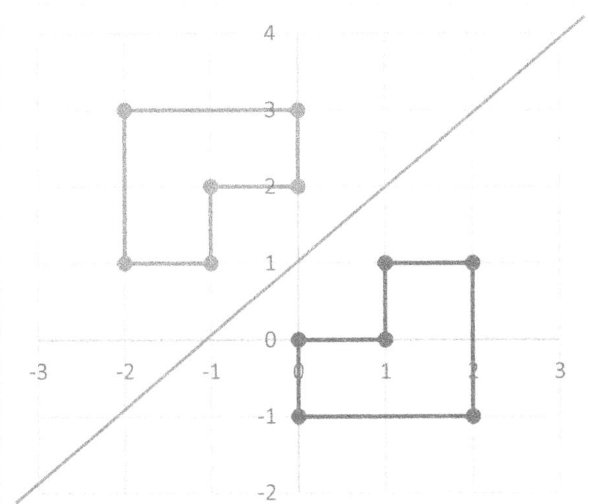

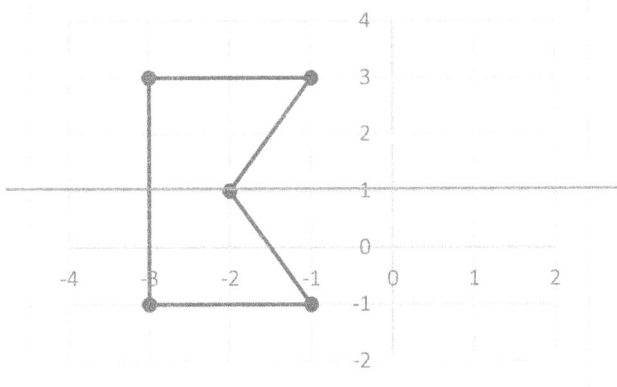

Reflections Over Common Axes

1) $(2,5)$
2) $(6,1)$
3) $D(1,-2)$, $E(3,-5)$, and $F(4,-1)$
4) $K(-3,-2)$, $L(-6,-4)$, $M(-2,-5)$, and $N(-1,-3)$
5) $(5,-3)$

Dilations

1) $(2,3)$
2) $C(3,6)$, $D(9,12)$, and $E(15,3)$.
3) $M(3,3)$, $N(3,7)$, $O(7,7)$, and $P(7,3)$
4) $\frac{1}{2}$
5) $Q(-1.75, 1.25)$, $R(-0.75, 1.25)$, $S(-0.75, 1.75)$ and $T(-1.75, 1.75)$

Scale Factor

1) $\frac{1}{50}$
2) $240\ cm^2$
3) $5\ km$
4) $\frac{1}{5,000}$
5) $216\ cm^3$

Sequences of Congruence Transformations

1) $(5,-1)$
2) $(-4,0)$
3) $(-8,-6)$
4) $I(2,-3)$, $J(-1,-5)$, and $K(1,-7)$
5) $A'(-1,1), B'(-1,4), C'(2,3), D'(2,1)$
6) $A \to B$: translated by $\vec{v} = (2,2)$ and $B \to C$: reflected over y-axis
7) $A \to B$: rotated 180° around $(-1,2)$ and $B \to C$: translated by $\vec{v} = (3,0)$
8) $A \to B$: reflected over x-axis and $B \to C$: translated by $\vec{v} = (-3,-1)$
9) $A \to B$: rotated 270° clockwise and $B \to C$: reflected over x-axis and $C \to D$: translated by $\vec{v} = (0,1)$
10) $A \to B$: reflected over x-axis and $B \to C$: translated by $\vec{v} = (6,0)$ and $C \to D$: rotated 90° counterclockwise

Real World Applications

1) The 3 position
2) $1\ km$
3) 5
4) 180°
5) 40 meters

Chapter 12: Geometry

Topics that you'll learn in this chapter:

- ✓ Pythagorean Theorem
- ✓ Converse of the Pythagorean Theorem
- ✓ Complementary and Supplementary Angles
- ✓ Vertical, Adjacent and Congruent Angles
- ✓ Solving Equations Using Angle Relationships
- ✓ Transversals of Parallel Lines
- ✓ Transversals of Triangles
- ✓ Bisected Line Segments and Angles
- ✓ Quadrilaterals
- ✓ Exterior Angle Theorem
- ✓ Interior and Exterior Angle of Polygons
- ✓ Parts of a Circle
- ✓ Area and Circumference of Circles
- ✓ Area and Perimeter of Semicircles and Quarter Circles
- ✓ Area Between Two Shapes
- ✓ Perimeter and Area: Changes in Scale
- ✓ Front, Side and Top View
- ✓ Volume of Cubes, Prisms and Pyramids
- ✓ Surface Area of Cubes, Prisms and Pyramids
- ✓ Volume of Cylinders, Cones and Spheres
- ✓ Surface Area of Cylinders, Cones and Spheres
- ✓ Similar Solids
- ✓ Nets of Three-Dimensional Figures
- ✓ Real World Applications
- ✓ Worksheets
- ✓ Answer of Worksheets

Pythagorean Theorem

As we read in previous years, The **Pythagorean Theorem** is one of the most fundamental principles in geometry. It relates to the sides of a right triangle (a triangle with one 90-degree angle). Here's what it states:

In a right triangle, the square of the length of the hypotenuse (the side opposite the right angle) is equal to the sum of the squares of the lengths of the other two sides (called the legs).

Mathematically, it is written as:

$$a^2 + b^2 = c^2$$

Where:

- a and b are the lengths of two legs,
- c is the length of the hypotenuse.

Example: Find the value of x in the triangle below.

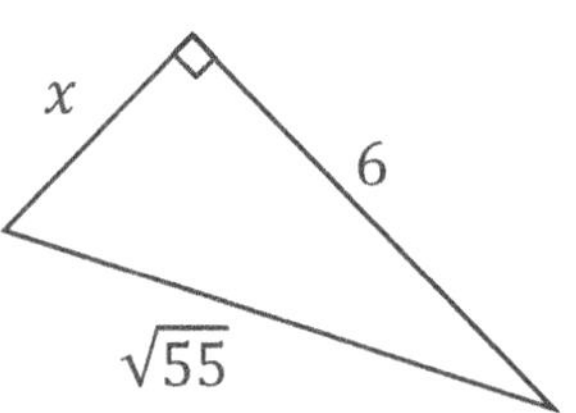

Solution: Using Pythagorean Theorem we have:

$$(\sqrt{55})^2 = 6^2 + x^2 \rightarrow 55 = 36 + x^2 \rightarrow x^2 = 55 - 36 \rightarrow x^2 = 19 \rightarrow x = \sqrt{19}$$

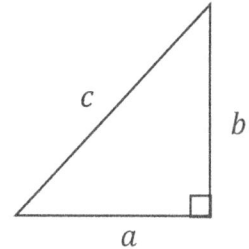

Converse of the Pythagorean Theorem

The Converse of the Pythagorean Theorem is like working the theorem in reverse. It states:

If the square of the length of the longest side of a triangle is equal to the sum of the squares of the other two sides, then the triangle is a right triangle.

This is a very useful tool for checking if a triangle is the right triangle when you're only given the lengths of its sides.

Example: Is the triangle in front a right triangle?

Solution:

1. Identify the sides of triangle: Label the sides as a, b and c, with c being the longest side (potential hypotenuse): $a = 6, b = 9$ and $c = 10$
2. Apply the Pythagorean Theorem:
   $a^2 + b^2 = c^2 \rightarrow 6^2 + 9^2 = 36 + 81 = 117$ but $10^2 = 100 \rightarrow 117 \neq 100$
   So, the triangle is not the right triangle.

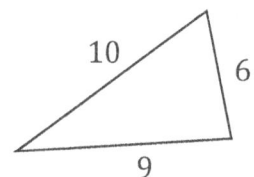

Complementary, Supplementary Angles

Complementary and supplementary angles are types of angle relationships in geometry. Here's what they mean:

- **Complementary Angles:** Two angles are complementary if the sum of their measures is 90 degrees. For example, if one angle is $40°$, the other angle must be $50°$ to make them complementary (since $40° + 50° = 90°$).

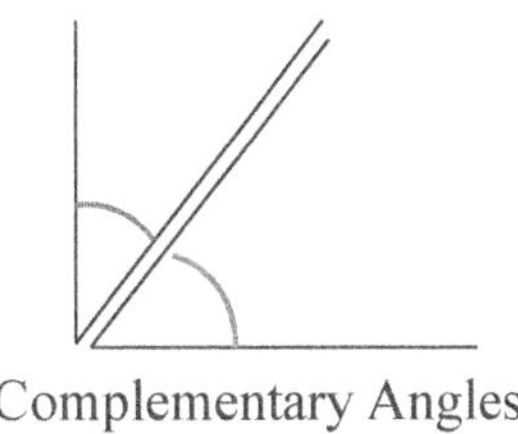

Complementary Angles

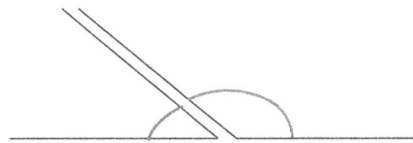

- **Supplementary Angles:** Two angles are supplementary if the sum of their measures is 180 degrees. For example, if one angle is $110°$, the other angle must be $70°$ to make them supplementary (since $110° + 70° = 180°$).

Supplementary Angles

Key Points:

- It does not matter whether complementary or supplementary angles share a vertex or a side.
- The difference between the supplement and complement of an angle is always 90 degrees:

$$Supplement - Complement = 90°$$

Examples:

1) Find the complement and supplement of a $80°$ angle.

   Solution:

   - Complement of an $80°$ angle$= 90° - 80° = 10°$
   - Supplement of an $80°$ angle$= 180° - 80° = 100°$

2) If the complement of an angle is $15°$, what is the supplement of this angle?

   Solution:

   The difference between the supplement and complement of an angle is $90°$, so, we have:

   $\text{Supplement}-\text{Complement}= 90° \rightarrow \text{Supplement}-15° = 90° \rightarrow \text{Suppiement}= 90° + 15° = 105°$

Vertical, Adjacent and Congruent Angles

Vertical Angles:

- Vertical angles are the angles that are opposite each other when two lines intersect.
- They are always equal (or congruent) in measure.

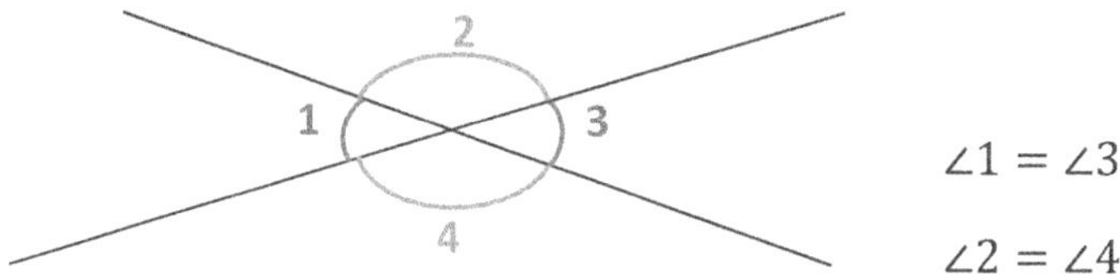

Adjacent Angles:

- Adjacent angles share a **common side** and a **common vertex** (corner point), but they do not overlap.
- They sit next to each other, like neighbors.

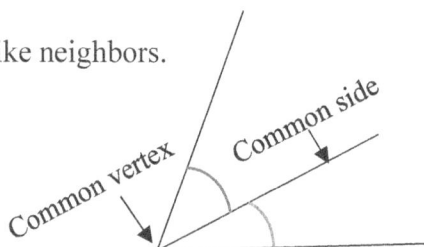

Common side

Common vertex

Congruent Angles:

- Congruent angles are angles that have the **same measure**, regardless of their orientation or position.

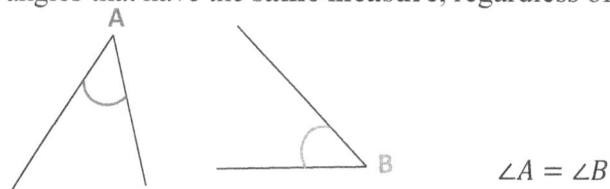

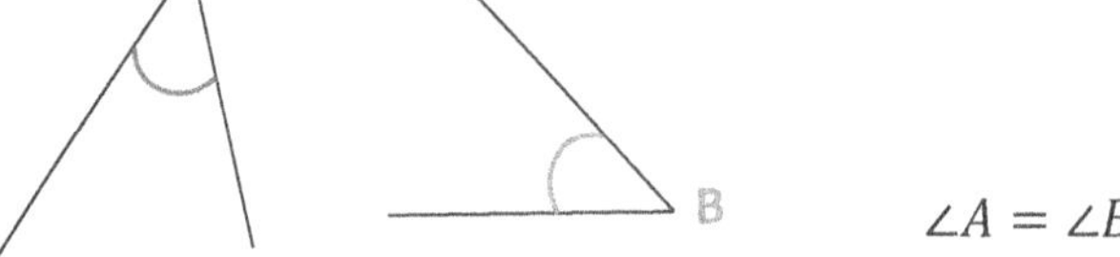

Example:

If $\angle DOC = 35°$ Find the value of angles $\angle AOE$ and $\angle AOB$.

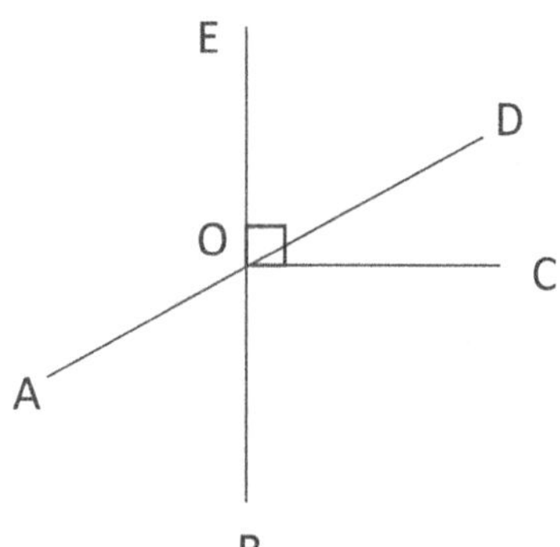

Solution:

- The two angles $\angle DOC$ and $\angle DOE$ are adjacent and complementary because their sum is 90° degrees so we have: $\angle DOC + \angle DOE = 90° \rightarrow 35° + \angle DOE = 90° \rightarrow \angle DOE = 90° - 35° = 55°$
- The two angles $\angle DOE$ and $\angle AOB$ are vertical so, they are equal: $\angle AOB = 55°$
- The two angles $\angle AOB$ and $\angle AOE$ are supplementary because they are adjacent angles on a straight line, so we have: $\angle AOB + \angle AOE = 180° \rightarrow 55° + \angle AOE \rightarrow \angle AOE = 180° - 55° = 125°$

Solving Equations Using Angle Relationships

Solving equations using angle relationships involves identifying the type of relationship between angles (such as complementary, supplementary, vertical, or adjacent angles) and applying the relevant geometric principles. Here's how you can do it step by step:

1. **Understand the Angle Relationship:** Determine if the angles are complementary angles, supplementary angles, vertical angles, adjacent on a straight line or angles in a triangle.
2. **Set Up the Equation:**
   - Write an equation based on the given relationship and substitute the known values.
   - If an angle is represented by a variable (e.g., x), include it in your equation.
3. **Solve for the Variable:**
   - Simplify the equation by combining like terms.
   - Use algebraic techniques (like addition, subtraction, multiplication, or division) to isolate the variable.
4. **Find the Required Angles:**
   - Substitute the value of the variable back into the expressions for the angles, if needed.
   - Confirm that the solution satisfies the angle relationship.

Examples:

1) Two angles are supplementary, and one angle is x, while the other is $2x - 30$. Find the angles.

   Solution: The angles are supplementary: $x + (2x - 30) = 180$

   Simplify the equation: $3x - 30 = 180$; Solve for x: $3x = 210 \rightarrow x = 70$

   Find the angles:

   First angle: $x = 70°$; Second angle: $2x - 30 = 2(70) - 30 = 140 - 30 = 110°$

2) If the interior angles of a triangle are $2x$, $2x - 25$, and $x + 10$ in order, find the value of all three angles.

   Solution:

   1. Set up the equation: Sum of angles $= 2x + (2x - 25) + (x + 10) = 180$
   2. Simplify the equation and solve for x: Combine like terms:

      $2x + (2x - 25) + (x + 10) = 180 \rightarrow 5x - 25 + 10 = 180 \rightarrow 5x - 15 = 180$

      $5x = 195 \rightarrow x = \frac{195}{5} = 39$
   3. Verify the solution: Substitute x back into the expressions for the angles:
      - First angle: $2x = 2 \times 39 = 78°$
      - Second angle: $2x - 25 = 2 \times 39 - 25 = 53°$
      - Third angle: $x + 10 = 39 + 10 = 49°$

Transversals of Parallel Lines

A transversal is a line that crosses or intersects two or more lines at distinct points. When the lines being crossed are parallel, the transversal creates special angle relationships that are very useful in geometry.

1. **Angles Formed:** When a transversal intersects two parallel lines, it creates 8 angles. These angles are grouped into specific pairs with unique relationships:

   ➢ **Corresponding Angles:**

   - Corresponding angles are found on the same side of the transversal, with one angle above a parallel line and the other below the other parallel line.
   - When the lines being crossed are parallel, corresponding angles are always equal in measure.
   - For example, if the transversal cuts the parallel lines and creates angles labeled as $1, 2, 3,$ and 4 on one side, and $5, 6, 7,$ and 8 on the other side, angles 1 and 5, as well as angles 2 and 6 would be corresponding angles. Similarly, angles 3 and 7, along with angles 4 and 8 form other pairs of corresponding angles.

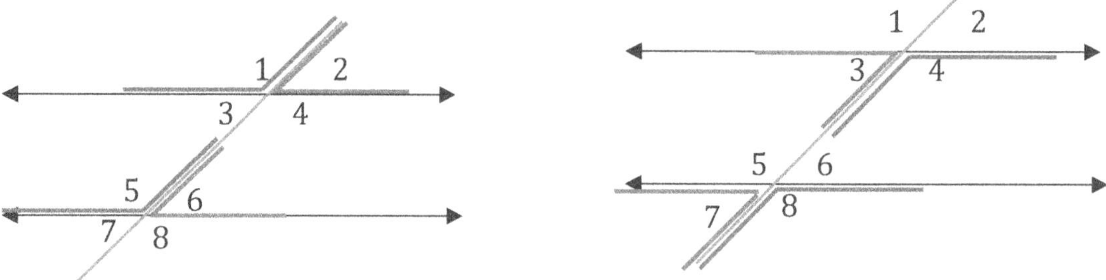

   ➢ **Consecutive Interior Angles (Same-Side Interior Angles):**

   - Consecutive Interior Angles, also known as Same-Side Interior Angles, are pairs of angles that are located on the same side of the transversal and inside the two parallel lines.
   - They lie between the two parallel lines, on the same side of the transversal.
   - Using the same labeling example, angles 3 and 5 form one pair of consecutive interior angles, and angles 4 and 6 form another pair.
   - consecutive interior angles are supplementary. This means their measures add up to $180°$.

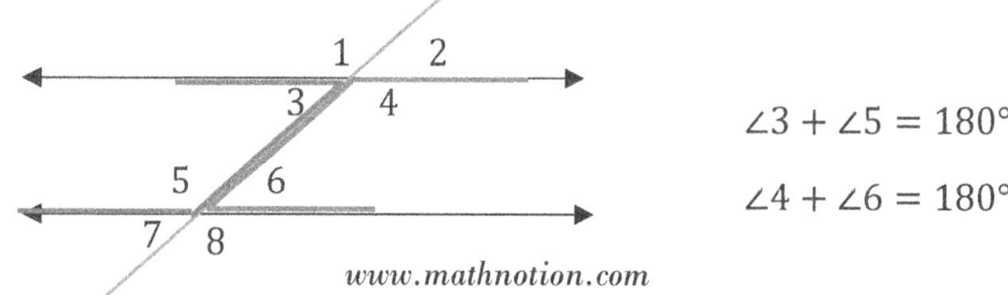

$\angle 3 + \angle 5 = 180°$

$\angle 4 + \angle 6 = 180°$

- **Alternate Interior Angles:**
  - These angles are found inside the two parallel lines.
  - They are on opposite sides of the transversal.
  - Using the same labeling example, angles 3 and 6 would be alternate interior angles. Similarly, angles 4 and 5 would form another pair.
  - Alternate interior angles are always equal when the lines are parallel.
- **Alternate Exterior Angles:**
  - These angles are found outside the two parallel lines.
  - They are also on opposite sides of the transversal.
  - Using the same labeling example, angles 1 and 8 form one pair of alternate exterior angles, and angles 2 and 7 form another pair.
  - Alternate exterior angles are also equal when the lines are parallel.

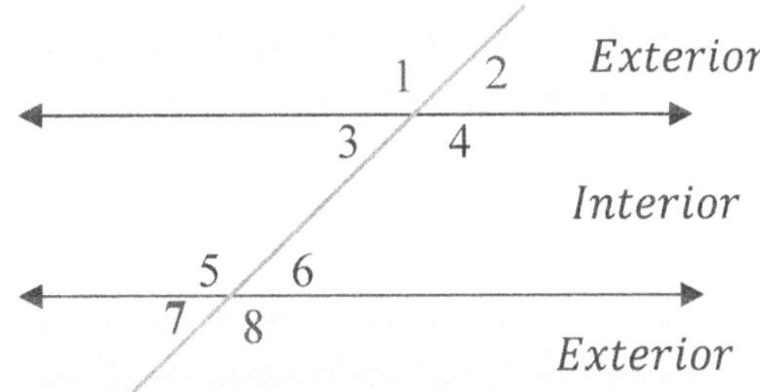

Examples:

1) The two line d_1 and d_2 are parallel, find the value of x.

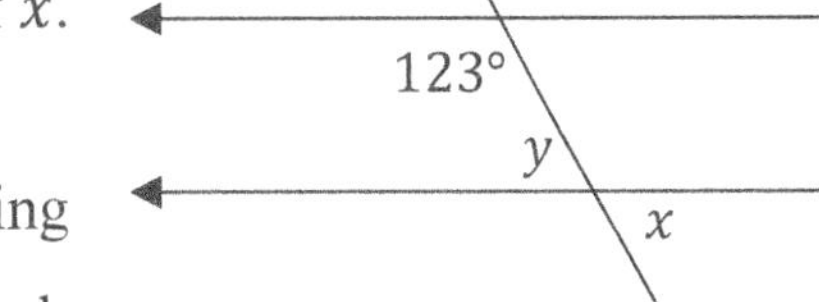

Solution:

- The line t crosses the two parallel lines, creating interior angles on the same side of transversal.
  The angle (123°) and y (we call it y) are consecutive interior angles so their sum is 180°: $y + 123° = 180° \rightarrow \ y = 180° - 123° = 57°$
- The angle y and x are vertical angles so they are equal: $x = y \ \rightarrow x = 57°$

2) The two lines l_1 and $l_2$are parallel, find the value of x.

Solution:

The angles $10x - 6$ and $5x + 24$ are alternate interior angles so they are equal:

$10x - 6 = 5x + 24 \rightarrow 10x - 5x = 24 + 6 \rightarrow$

$5x = 30 \rightarrow x = \frac{30}{5} = 6$

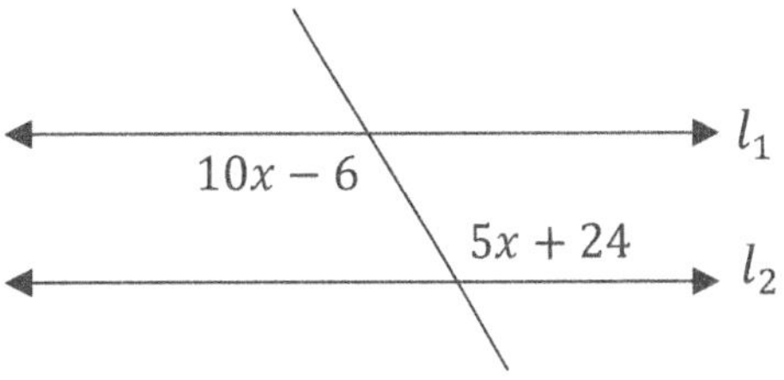

Transversals of Triangles

A transversal of triangles refers to a line that intersects either the sides or the extensions of the sides of a triangle. This concept is often used in geometry to explore relationships between angles, proportionality, and similarity.

Key Points About Transversals in Triangles:

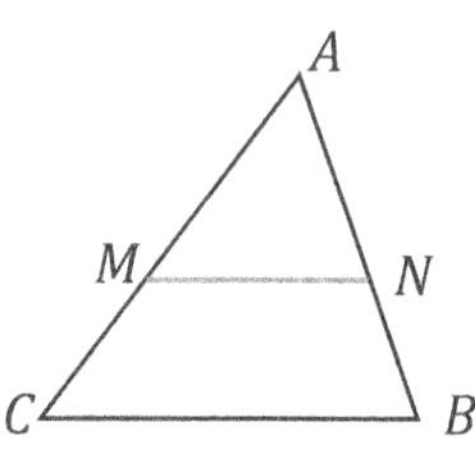

1. **Intersecting Sides:** A transversal can cut across two sides of a triangle, potentially forming smaller segments on those sides. If the transversal is parallel to the third side, it creates similar triangles. ($\Delta ABC \sim \Delta ANM$)
2. **Angle Relationships:** When a transversal intersects two sides of a triangle, it forms angles with those sides. These angles often satisfy certain relationships, such as:
   - Corresponding angles (if the transversal is parallel to the third side).
   - Alternate interior angles (with parallel sides): ($\widehat{M} = \hat{C}$ and $\widehat{N} = \hat{B}$)
3. **Connecting Points Outside the Triangle:** Sometimes, the transversal extends beyond the triangle, forming external angles that relate to the internal geometry of the triangle

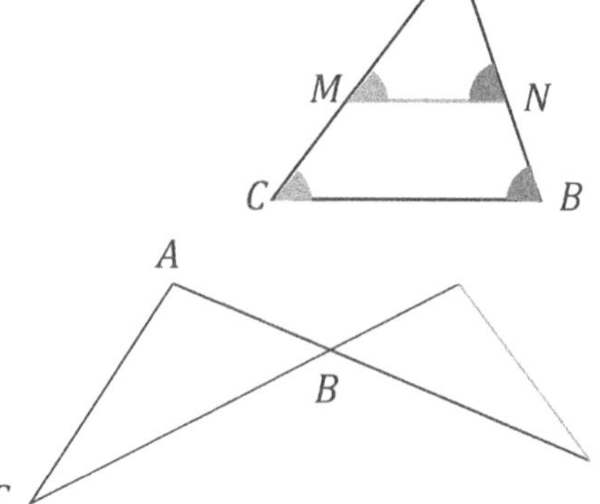

Example:

Find the value of x in following shape. ($d||CB$)

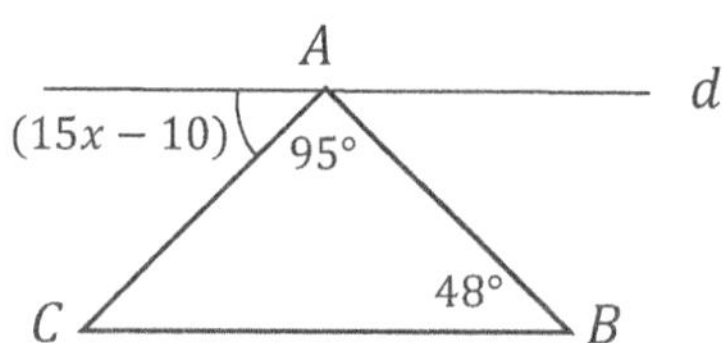

Solution:

- By using the transversal of parallel lines and alternate interior angles we have:

$$\angle C = 15x - 10$$

- Calculating the angle C: The sum of interior angles in ΔABC is 180 so:

  $\angle A + \angle B + \angle C = 180° \rightarrow 95° + 48° + \angle C = 180° \rightarrow \angle C = 180° - 95° - 48° \rightarrow \angle C = 37°$
- Calculating x: $15x - 10 = 37 \rightarrow 15x = 37 + 10 \rightarrow 15x = 47 \rightarrow x = \frac{47}{15}$

Bisected Line Segments and Angles

The word **"bisect"** means **"to cut into two equal parts."** So, when something is **bisected**, it is divided into two equal halves.

Bisecting a Line Segment:

A line segment is a straight line that connects two points. When a line segment is bisected, it is divided into two equal parts at its midpoint.

Example: If you have a line segment AB and point M is the midpoint, then $AM = MB$ because M bisects the segment.

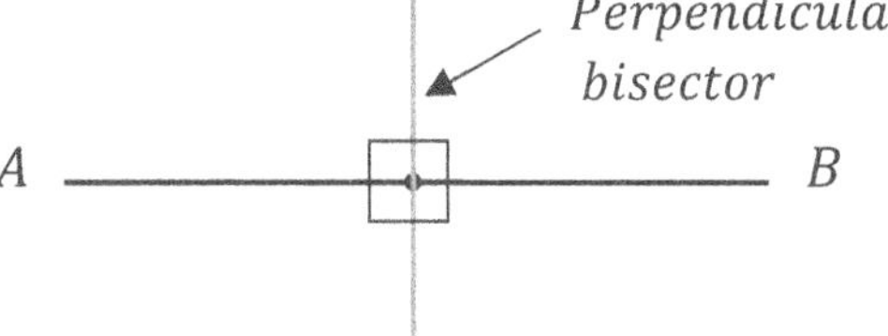

Perpendicular Bisector: A perpendicular bisector is a special kind of bisector that divides a segment into two equal parts at a 90° angle.

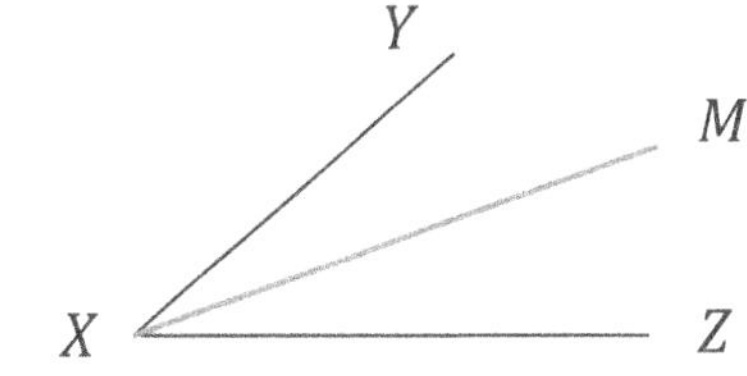

Bisecting an Angle:

When an angle is bisected, it is divided into two smaller angles that are equal.

Example: If you have an angle $\widehat{YXZ}$ and a line XM splits it into two equal parts, then: $\widehat{YXM} = \widehat{MXZ}$ because XM is the angle bisector.

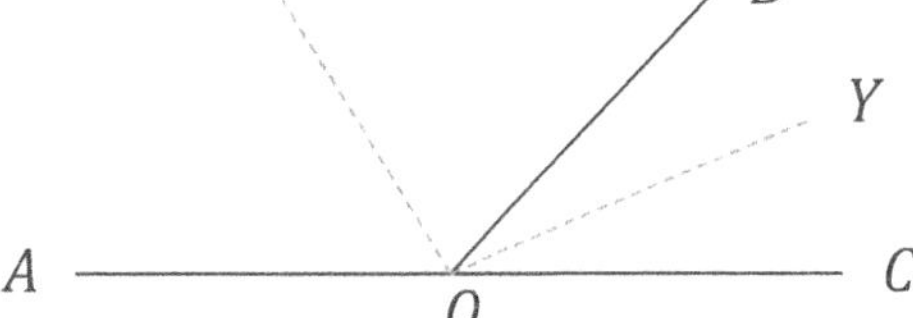

Example:

If OX is the bisector of angle $\widehat{AOB}$ and OY is the bisector of angle $\widehat{BOC}$, find the measure of angle $\widehat{XOY}$:

X B Y A C O

Solution:

The angles $\widehat{AOB}$ and $\widehat{BOC}$ are supplementary because they are adjacent angles on a straight line so: $\widehat{AOB} + \widehat{BOC} = 180°$

The angle $\widehat{XOY}$ is equal to the sum of half of angles $\widehat{AOB}$and $\widehat{BOC}$:

$\widehat{XOY} = \frac{\widehat{AOB}}{2} + \frac{\widehat{BOC}}{2} \rightarrow \widehat{XOY} = \frac{\widehat{AOB}+\widehat{BOC}}{2}$ →From the result of the previous part, we have: $\widehat{XOY} = \frac{180}{2} = 90°$

Quadrilaterals

A **quadrilateral** is a polygon with exactly four sides, four vertices, and four angles. Quadrilaterals are diverse and can take many forms, but they all share some common properties as well as unique characteristics depending on their type.

General Properties of Quadrilaterals:

1. **Angle Sum Property**: The sum of the interior angles of any quadrilateral is always 360°.
2. **4 Sides and 4 Angles**: A quadrilateral always has four sides and four interior angles.
3. **Diagonals**: Quadrilaterals have two diagonals, which are line segments connecting opposite vertices.

The Main Types of Quadrilaterals and Their Properties:

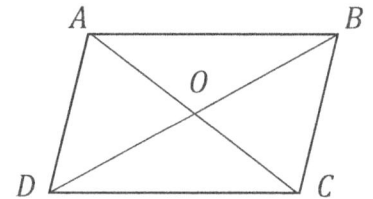

Parallelogram:

- Opposite sides are parallel and equal in length: $AB||DC, AD||BC$ and $AB = DC, AD = BC$
- Opposite angles are equal: $\hat{A} = \hat{C}$ and $\hat{B} = \hat{D}$
- Diagonals bisect each other: $OA = OC$ and $DO = OB$
- Adjacent angles are supplementary (add up to 180°): $\hat{A} + \hat{D} = 180°, \hat{C} + \hat{D} = 180°, \hat{C} + \hat{B} = 180°$ and $\hat{A} + \hat{B} = 180°$

Rectangle:

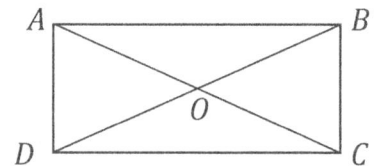

- It is a kind of parallelogram.
- The opposite sides are parallel and equal: $AB||DC, AD||BC$ and $AB = DC, AD = BC$
- All interior angles are 90°.
- Diagonals are equal in length and bisect each other. $AC = BD$ and $OA = OC, DO = OB$

Square:

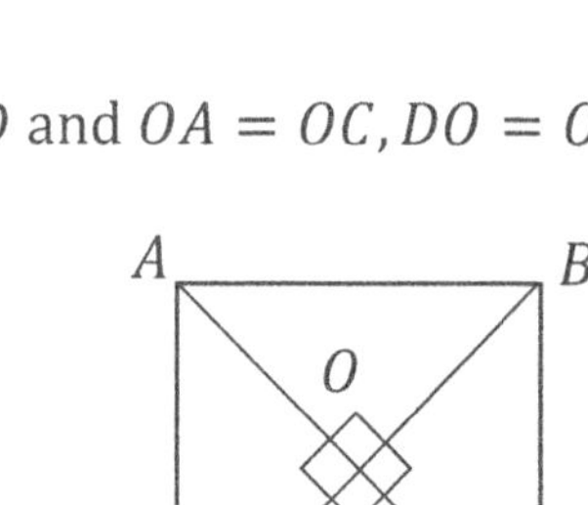

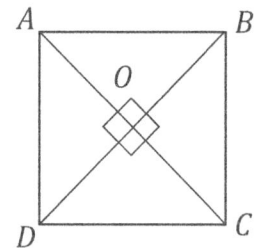

- It is a kind of rectangle.
- All sides are equal: $AB = BC = DC = AD$
- All interior angles are 90°.
- Diagonals are equal, bisect each other at right angles, and divide the square into four right triangles. $AC = BD, OA = OC$ and $DO = OB$

Rhombus:

- It is a kind of parallelogram.
- All four sides are equal in length:
  $AB = BC = DC = AD$
- Opposite angles are equal:
  $\hat{A} = \hat{C}$ *and* $\hat{B} = \hat{D}$
- The diagonals bisect each other at right angles; they are lines of symmetry and also bisect the interior angles:

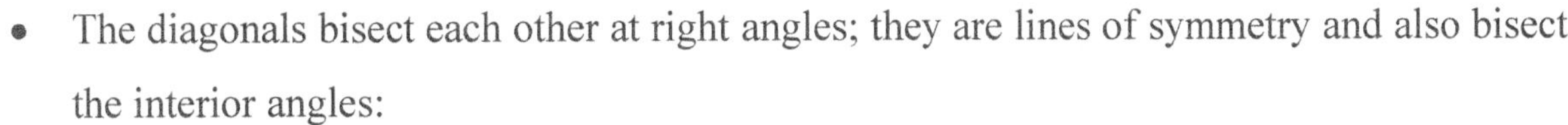

  $AO = OC, DO = OB$ *and* $A_1 = A_2, B_1 = B_2, C_1 = C_2, D_1 = D_2$

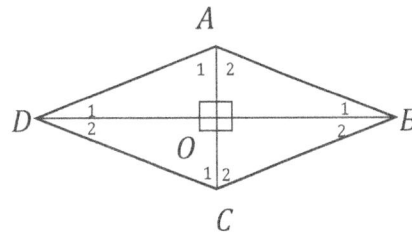

Trapezium (or Trapezoid):

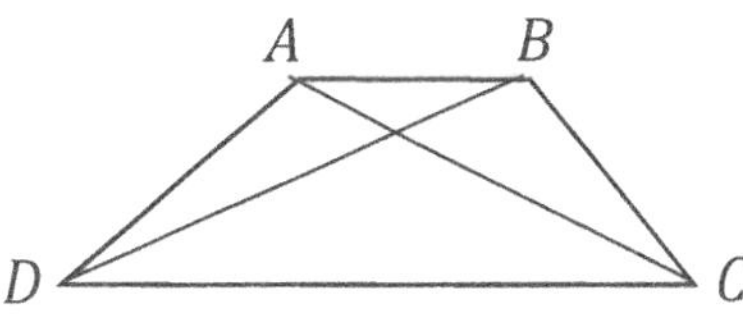

- Only one pair of opposite sides is parallel: $AB||DC$
- AThe angles along the same non-parallel side are supplementary (add up to 180°):
  $\hat{A} + \hat{D} = 180°$*and* $\hat{C} + \hat{B} = 180°$

Kite:

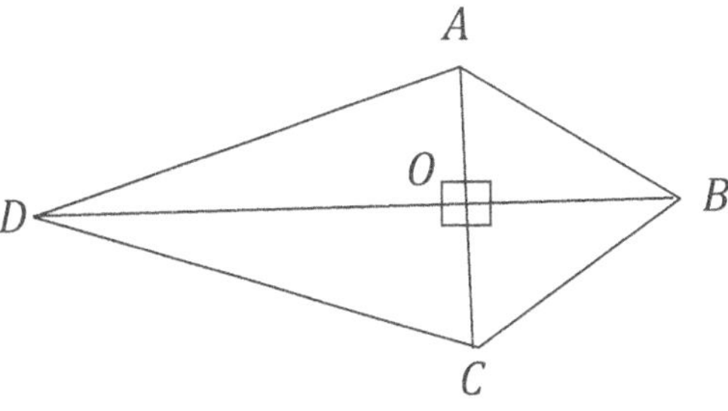

- Two pairs of adjacent sides are equal in length:
  $AB = BC$ *and* $AD = DC$
- One pair of opposite angles is equal: $\hat{A} = \hat{C}$
- The diagonals intersect at right angles, with one diagonal bisecting the other: $AO = OC$

Examples:

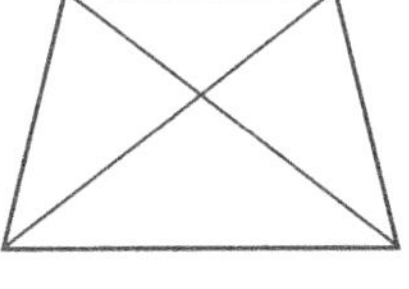

1) Can we say that any quadrilateral whose diagonals are equal is a rectangle?

   Solution:

   No, because an isosceles trapezium also has equal diagonals, but it is not a rectangle.

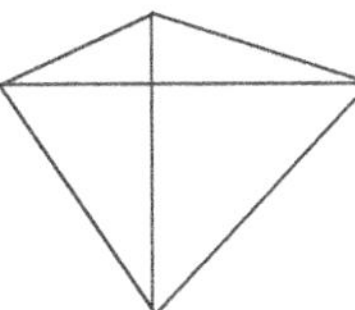

2) The diagonals of a quadrilateral are equal and perpendicular. Can we say that the quadrilateral is a rhombus?

   Solution:

   No, in this quadrilateral the diagonals are equal and perpendicular, but the quadrilateral is not a rhombus.

Exterior Angle Theorem

The **Exterior Angle Theorem** is a key concept in triangle geometry. It states:

"The measure of an exterior angle of a triangle is equal to the sum of the measures of the two non-adjacent interior angles (also called remote interior angles)".

Explanation:

- When you extend one side of a triangle, it creates an exterior angle with the adjacent side.
- The two interior angles that are not directly connected to this exterior angle are called the remote interior angles.

Mathematically:

$$Exterior\ Angle = Interior\ Angle\ 1 + Interior\ Angle\ 2$$

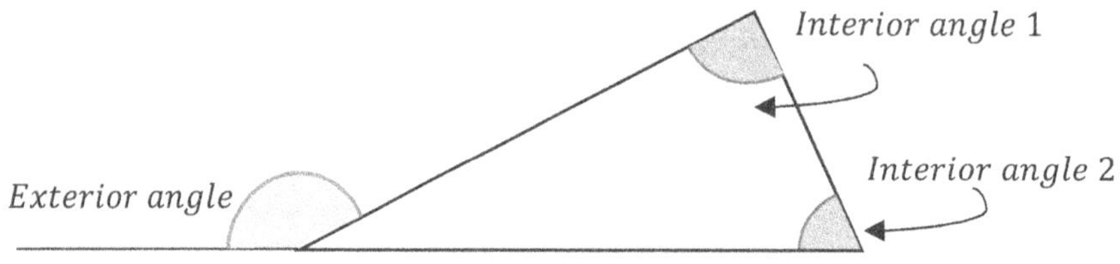

Examples:

3) Find the value of x in following shape:

Solution:

Using the Exterior Angle Theorem, the angle x is equal to the sum of remote interior angles ($37°$ and $90°$):

$x = 90 + 37 = 127°$

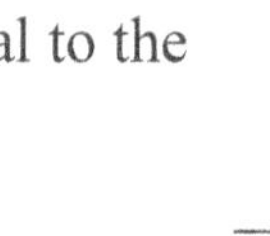

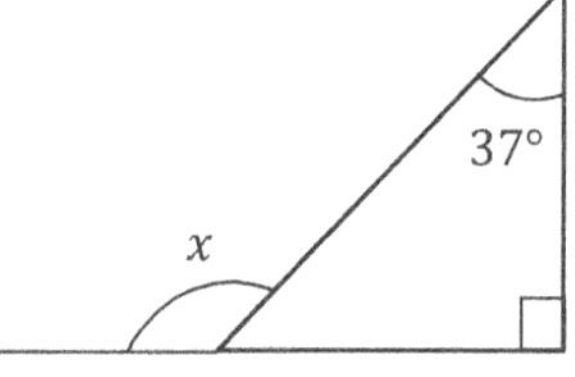

4) Find the value of $x + y - z$.

Solution:

- if we consider x as one of the exterior angles in triangle, we have: $x = \angle 1 + \angle 2$
- The angles $\angle 1$ and z are vertical so they are equal to each other: $\angle 1 = z$
- The angles $\angle 2$ and y are adjacent angles on a straight line: $\angle 2 + y = 180° \rightarrow \angle 2 = 180 - y$
- Using the previous results we have:

  $x = \angle 1 + \angle 2 \rightarrow x = z + 180 - y \rightarrow x + y - z = 180°$

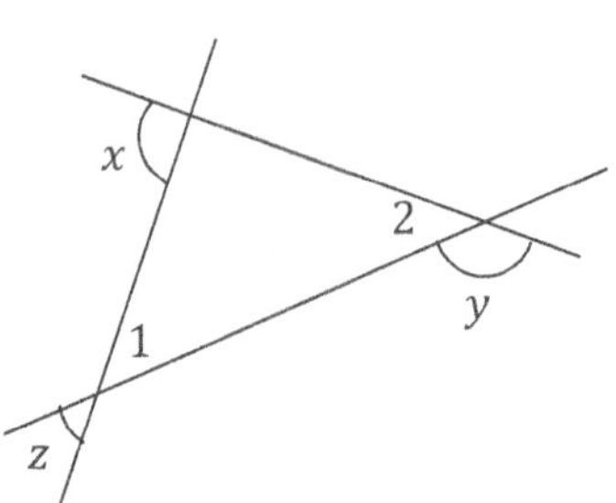

Interior and exterior Angle of Polygons

Interior angles of a polygon are the angles formed inside the polygon at each vertex (corner) where two sides meet. Here are some key points about interior angles:

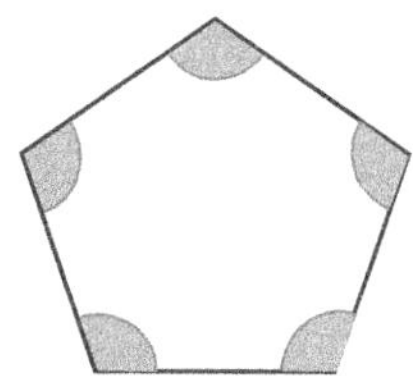

- For any polygon, the sum of its interior angles depends on the number of sides it has.
- The formula to calculate the sum of the interior angles of a polygon is: $(n - 2) \times 180°$, where n is the number of sides of the polygon.

For example:

- A triangle (3 sides) has a sum of interior angles: $(3 - 2) \times 180° = 180°$
- A quadrilateral (4 sides) has a sum of interior angles: $(4 - 2) \times 180° = 360°$
- A pentagon (5 sides) has a sum of interior angles: $(5 - 2) \times 180° = 540°$

Note: If you want to find the measure of each interior angle of a *regular polygon* (where all angles and sides are equal), you divide the sum by the number of sides. The formula is:

Each interior angle = $[(n - 2) \times 180°] / n$.

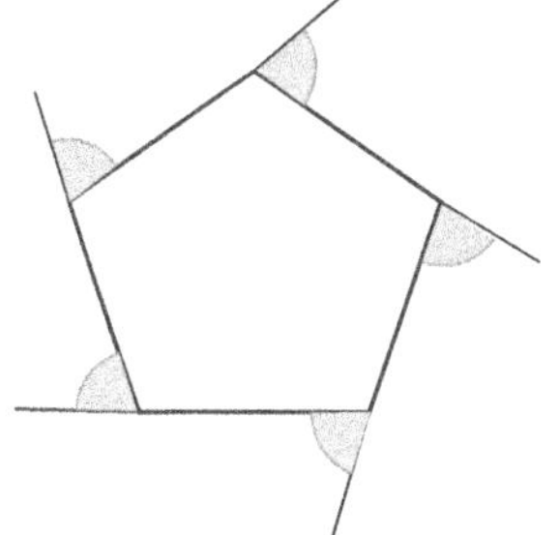

Exterior angles of a polygon are the angles formed when one side of the polygon is extended outward, creating an angle with the adjacent side. These angles are measured outside the polygon. Here are some key points about exterior angles:

- **Sum of exterior angles**: The sum of the exterior angles of any polygon is always $360°$, regardless of the number of sides.
- **Regular polygons**: In a regular polygon (where all sides and angles are equal), each exterior angle can be calculated using the formula: Each exterior angle = $360° / n$, where n is the number of sides.
- **Relationship with interior angles**: Each exterior angle and its corresponding interior angle at a vertex are supplementary, meaning they add up to $180°$.

Example:

Find the measure of each interior angle and each exterior angle of a regular 12-sided polygon.

Solution:

- Each interior angle: $n = 12$ (number of sides): Interior angle $= [(12 - 2) \times 180°] / 12 = (10 \times 180°) / 12 = 1800° / 12 = 150°$
- Each exterior angle: The exterior and interior angles of a vertex are supplementary: Each exterior angle= $180° - 150° = 30°$

So, each interior angle is $150°$ and each exterior angle is $30°$.

Parts of a Circle

A circle is made up of several key parts, each with its own unique role and definition. Here's an overview of the main parts of a circle:

1. **Center**: The fixed point that is equidistant from all points on the circle.
2. **Radius**: The distance from the center to any point on the circle. It's one of the key measures used in calculations.
3. **Diameter**: A line segment that passes through the center of the circle, connecting two points on the boundary. It is equal to twice the radius (Diameter $= 2 \times$ Radius).
4. **Chord**: A line segment connecting any two points on the circle without necessarily passing through the center.
5. **Arc**: A portion of the circle's circumference, like a "curve" between two points on the circle.
6. **Sector**: The area enclosed by two radii and the arc they connect (like a "slice of pizza").
7. **Tangent**: A line that touches the circle at exactly one point and does not cross into its interior.

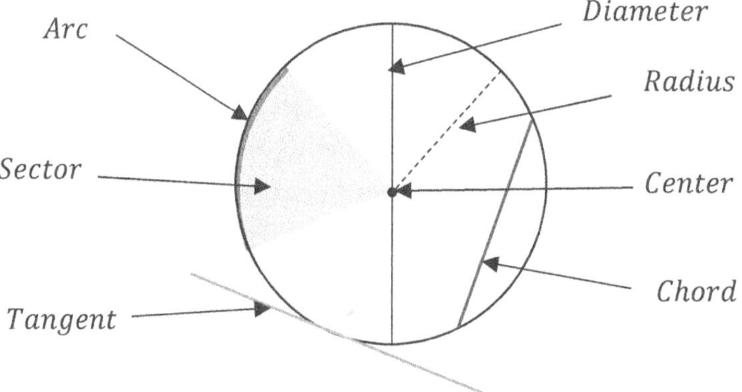

Example:

1) How many possible relationships can there be between a line and a circle?

Solution:

1. The line does not intersect the circle: In this case, the line is completely outside the circle and does not touch it at any point. This is called a non-intersecting line.

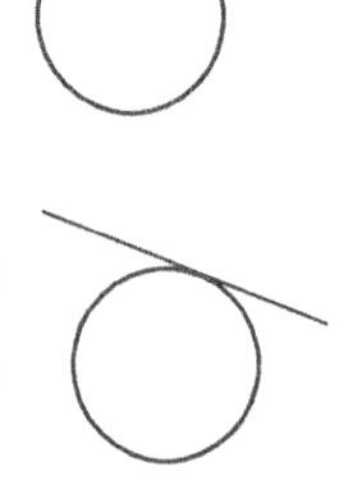

2. The line touches the circle at exactly one point: This is when the line is a tangent to the circle. The tangent line just grazes at the circle at one single point.
3. The line intersects the circle at two points: Here, the line cuts through the circle and crosses it at two distinct points. This is called a secant line.

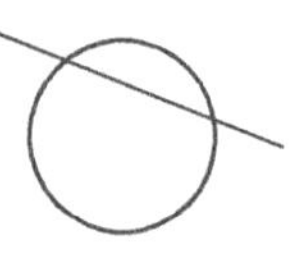

Area and Perimeter of Semicircles and Quarter Circles

Circles:

From past years, we know that the area and circumference of a circle are calculated as follows:

- **Area of a Circle:** Area = $\pi \times r^2$
- **Circumference of a Circle:** Circumference = $2 \times \pi \times r$

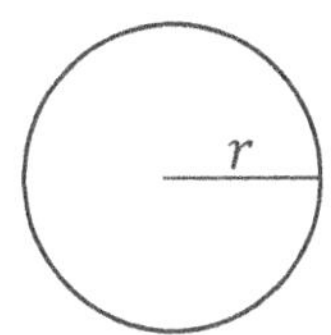

Semicircles:

- **Area of a Semicircle:** Since the semicircle is half of a circle, its area is:

  $Area = (1/2) \times \pi \times r^2 = (\pi \times r^2) / 2$ Where r is the radius.

- **Perimeter of a Semicircle:** The perimeter includes the curved part (half of the circumference of the circle) and the diameter:

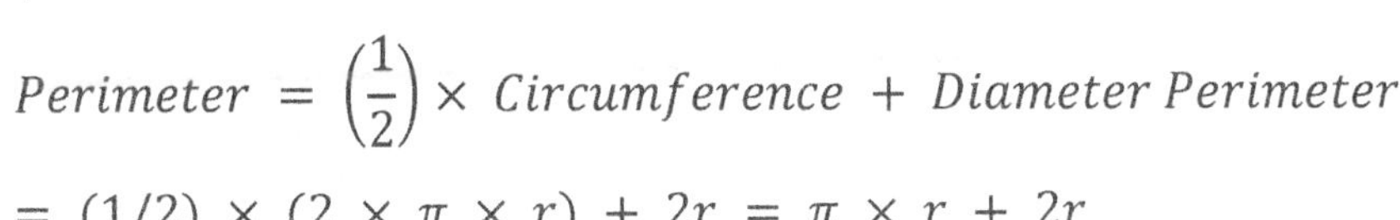

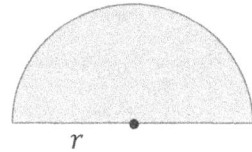

  $Perimeter = \left(\frac{1}{2}\right) \times Circumference + Diameter\ Perimeter$

  $= (1/2) \times (2 \times \pi \times r) + 2r = \pi \times r + 2r$

Quarter Circles:

- **Area of a Quarter Circle:** Since it is one-fourth of a circle, its area is:

  $Area = (1/4) \times \pi \times r^2 = (\pi \times r^2) / 4$

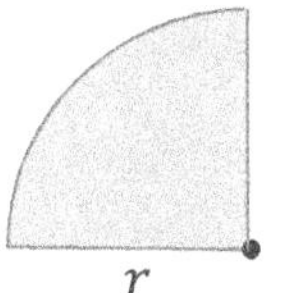

- **Perimeter of a Quarter Circle**: The perimeter includes the curved part (one-fourth of the circumference of the circle) and the two radii:

  $Perimeter = \left(\frac{1}{4}\right) \times Circumference + 2r\ Perimeter$

  $= (1/4) \times (2 \times \pi \times r) + 2r = (\pi \times r) / 2 + 2r$

Examples:

1) If the radius of a semicircle is 4 cm, find its area and perimeter.

   Solution:

   - Area of the semicircle: $\left(\frac{1}{2}\right) \times \pi \times r^2 = \frac{(\pi \times r^2)}{2} = \frac{\pi \times 4^2}{2} = 8\pi$
   - Perimeter of the semicircle: $\pi \times 4 + 2r = 4\pi + 8$

2) If the radius of a quarter circle is 10 cm, find its area and perimeter.

   Solution:

   - Area of the quarter circle: $\frac{(\pi \times r^2)}{4} = \frac{\pi \times 10^2}{4} = \frac{100\pi}{4} = 25\pi$
   - Perimeter of the quarter circle: $\frac{(\pi \times r)}{2} + 2r = \frac{\pi \times 10}{2} + 2 \times 10 = 5\pi + 20$

Area Between Two Shapes

The **area between two shapes** refers to the space that lies inside one shape but outside the other. The calculation depends on the relationship between the two shapes and whether one is fully inside the other or they overlap.

1. **When One Shape is Fully Inside the Other**:

Subtract the area of the smaller shape from the area of the larger shape.

$$Area\ between = Area\ of\ larger\ shape - Area\ of\ smaller\ shape$$

2. **When the Shapes Overlap**:

   If the shapes overlap, the area between them is more complicated to calculate. You typically

   - Determine the areas of the two shapes.
   - Subtract the overlapping area.

Examples:

1) Find the area of grey part.

   Solution:

   Using the formula for calculating the area between two areas:

   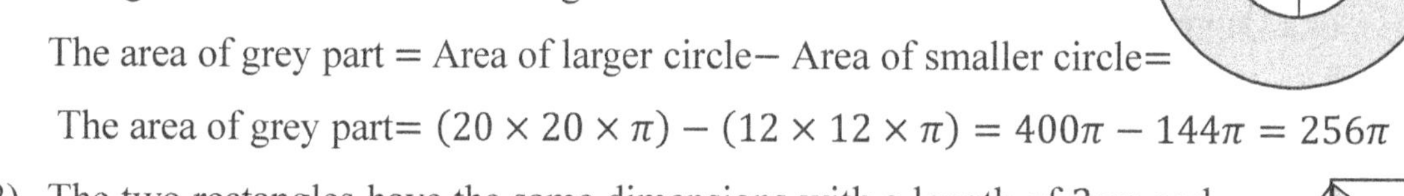

   The area of grey part = Area of larger circle− Area of smaller circle=

   The area of grey part= $(20 \times 20 \times \pi) - (12 \times 12 \times \pi) = 400\pi - 144\pi = 256\pi$

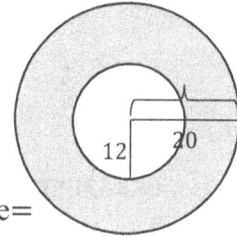

2) The two rectangles have the same dimensions with a length of 2cm and a width of 1cm, find the area of grey part (hint: all four right triangles have the same dimensions):

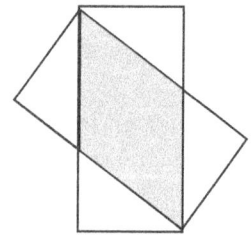

   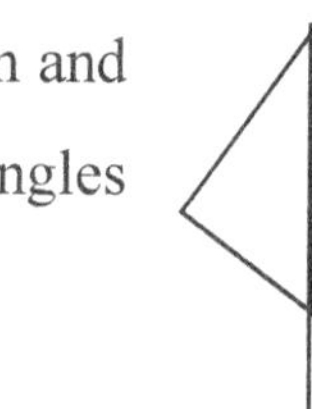

   Solution:

   - If we subtract the area of the two triangles from the area of one of the rectangles, we will get the area of the grey part. We consider one side of the triangle as 1 and the other side as x. Therefore, the top side of the rectangle, which is equal to the hypotenuse of the triangle, is equal to 2 minus x. Using the Pythagorean theorem, we will have:

     $(2-x)^2 = x^2 + 1^2 \rightarrow 4 - 4x + x^2 = x^2 + 1^2$

     $\rightarrow 4 - 4x = 1 \rightarrow 4x = 3 \rightarrow x = \frac{3}{4}$

   - The area of triangle: $\left(1 \times \frac{3}{4}\right) \div 2 = \frac{3}{8}$
   - The area of grey part:

     The area of rectangle−The area of two triangles= $(2 \times 1) - \frac{3}{8} \times 2 = 2 - \frac{3}{4} = \frac{5}{4}$

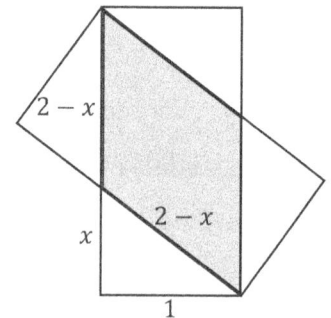

Perimeter and Area: Changes in Scale

When a shape gets scaled up or scaled down (made bigger or smaller while keeping its proportions the same), the perimeter and area change in the following ways:

3. **Scaling the Perimeter**:
    - If you multiply the side lengths of a shape by a certain number, called the "scale factor," the perimeter also gets multiplied by that same number.
    - For example:
        - If a square has a side length of 3 units, its perimeter is $3 \times 4 = 12$ units.
        - If the side length is doubled (scale factor $= 2$), the new side length is 6 units, and the perimeter becomes $6 \times 4 = 24$ units.
2. **Scaling the Area**:
    - The area changes by the square of the scale factor. This means if you double ($\times 2$) the side lengths, the area increases by $2^2 = 4$ times!
    - For example:
        - If a square has a side length of 3 units, the area is $3 \times 3 = 9$ square units.
        - If the side length is doubled, the new side length is 6 units, and the area is $6 \times 6 = 36$ square units (which is 4 times bigger!).

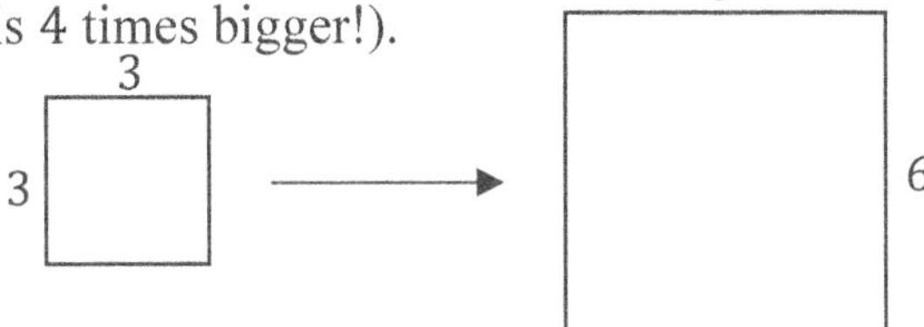

Example:

If we triple the radius of a circle, by how many times will its circumference and area increase?

Solution:

- Circumference: If you triple the radius ($r \to 3r$), the new circumference will be $C = 2\pi(3r) = 3(2\pi r)$. So, the circumference becomes 3 times larger.
- Area: If you triple the radius ($r \to 3r$), the new area will be $A = \pi(3r)^2 = \pi(9r^2) = 9(\pi r^2)$. So, the area becomes 9 times larger.

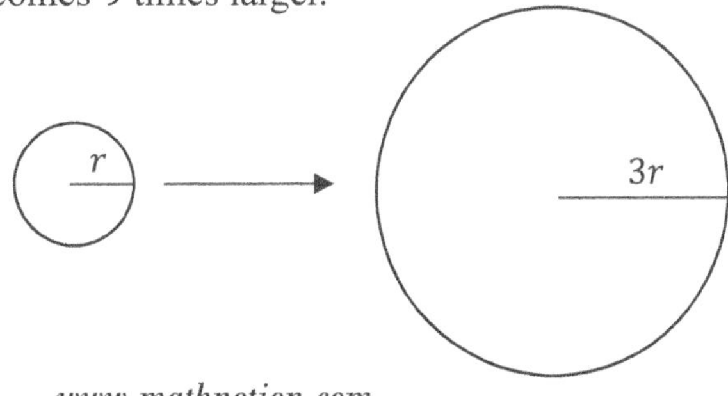

Front, Side and Top View

The front, side, and top views of 3D figures are 2D representations of what a 3D object looks like from different perspectives. These views are used in fields like design, engineering, and architecture to better understand the shape and structure of an object.

1. **Front View**
   - This shows the object as seen from the front.
   - It highlights the height and width of the figure.
   - Think of standing directly in front of an object and capturing what you see.

Example: Looking at a house from the front will show you the door, windows, and the roofline (but you won't see the sides or back).

2. **Side View**
   - This shows the object as seen from either the left or right side.
   - It highlights the depth and height of the figure.
   - Imagine standing to the side of an object and observing what's visible from there.

Example: From the side view of a car, you would see the doors, wheels, and windows on one side (but not the front or rear).

3. **Top View**
   - This shows the object as seen from directly above.
   - It highlights the width and depth of the figure.
   - Imagine you're looking down at the object from above.

Example: The top view of a table would show its tabletop shape and size, but not the legs or underside.

Example:

Identify how the three-dimensional shape is seen from the front, side, and top views.

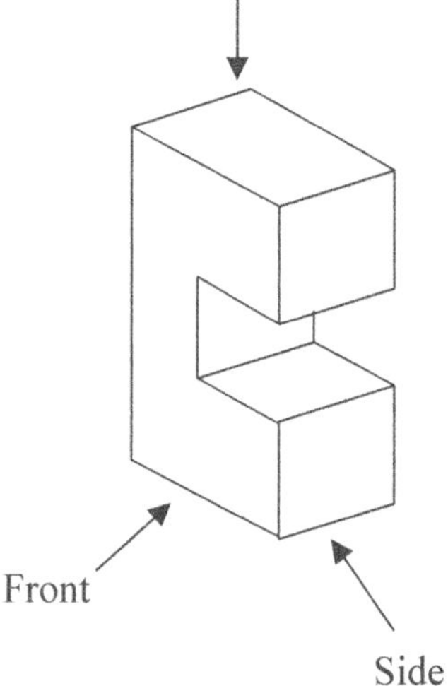

Solution:

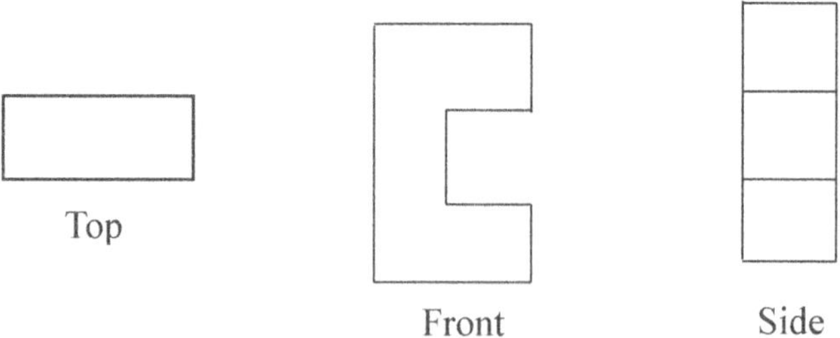

Volume of Cubes, Prisms and Pyramids

1. **Volume of a Cube:** A cube is a three-dimensional shape with all sides of equal length.

$$Volume = a^3$$

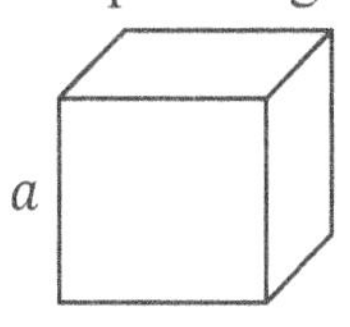

Where a is the length of a side.

2. **Volume of a Prism:** A prism has two identical bases connected by vertical faces. Common examples are rectangular and triangular prisms.

Base area

Height

$$Volume = Base\ Area \times Height$$

where the "$Base\ Area$" depends on the shape of the base.

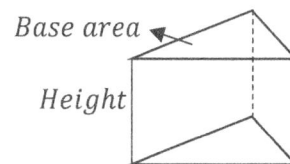

3. **Volume of a Pyramid:** A pyramid has a polygon as its base and triangular faces that meet at a single point (the apex).

$$Volume = \frac{1}{3} \times Base\ Area \times Height$$

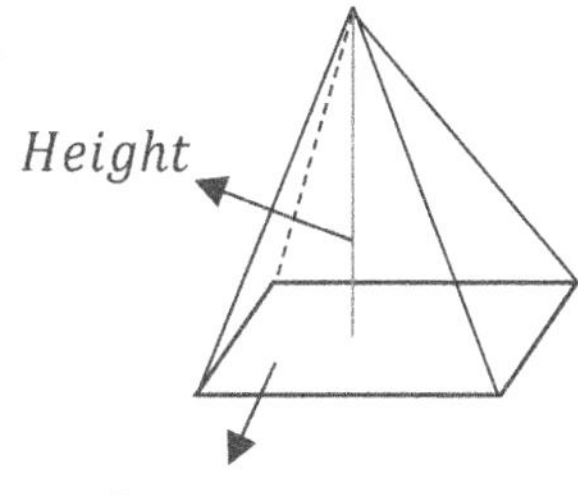

where the "$Base\ Area$" depends on the shape of the base, and the "$Height$" is the perpendicular distance from the base to the apex.

Examples:

1) If each side of a cube is 4 meters, find its volume.

Solution:

Substituting this value into the cube volume formula: $Volume = 4^3 = 4 \times 4 \times 4 = 64\ m^3$

2) If the base of a prism is a rectangle with length 5 cm and width 3 cm, and the height is 10 cm, find its volume:

Solution:

Using the prism volume formula: $Volume = (5 \times 3) \times 10 = 150\ cm^3$

3) If the base of a pyramid is a square with a side length of 6 inches, and the height is 9 inches, find its volume.

Solution:

Using the pyramid volume formula:

$Volume = \frac{1}{3} \times (6 \times 6) \times 9 = 108$ cubic inch

Surface Area of Cubes, Prisms and Pyramids

1. **Surface Area of a Cube:** A cube has 6 equal square faces.

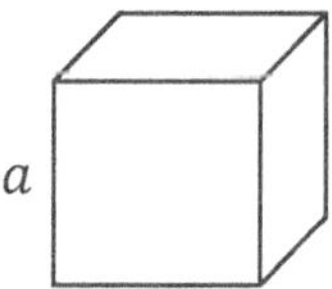

$$Surface\ Area = 6a^2$$

Where a is the length of a side.

2. **Surface Area of a Prism:** The total surface area is the sum of the areas of the two bases and the lateral (side) faces.

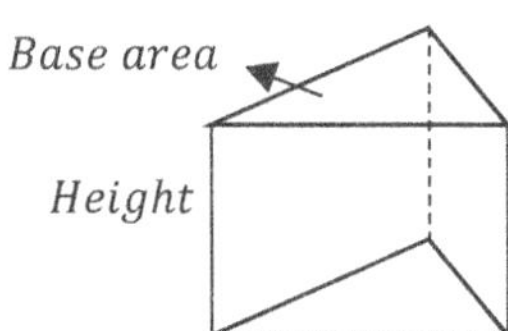

$$Surface\ Area = 2 \times Base\ Area + Lateral\ Area$$

The lateral area depends on the perimeter of the base and the height of the prism.

$$Lateral\ Area = Perimeter\ of\ Base \times Height$$

3. **Surface Area of a Pyramid:** The surface area includes the base area plus the area of the triangular faces.

$$Surface\ Area = Base\ Area + Lateral\ Area$$

The lateral area is the sum of the areas of all the triangular faces. For a square pyramid, this can be calculated using the slant height (l).

$$Lateral\ Area = \frac{1}{2} \times Perimeter\ of\ Base \times Slant\ Height$$

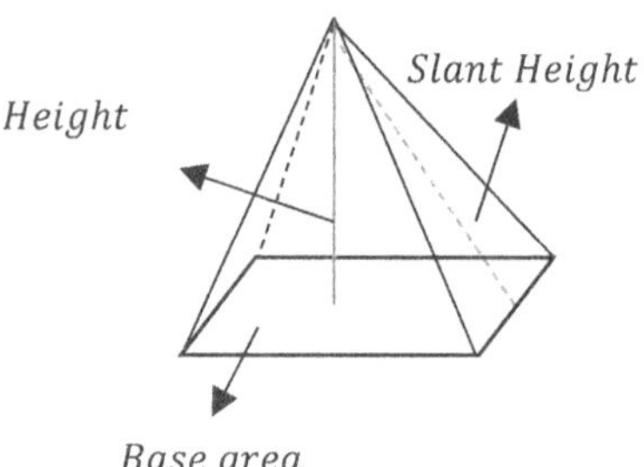

Examples:

1) Find the surface area of a prism whose bases are rhombuses with diagonals of 6 ft and 8 ft, and side length of 5 ft. Additionally, the height of the prism is 7 ft.

Solution:

- The area of base: $\frac{1}{2} \times d_1 \times d_2$ where d_1 and d_2 are the diagonals of rhombus : $\frac{1}{2} \times 8 \times 6 = 24\ ft^2$
- The lateral area: $Perimeter\ of\ rhombus \times height = (4 \times 5) \times 7 = 140ft^2$
- Total surface area: $Lateral\ area + Area\ of\ two\ bases = 140 + 2 \times 24 = 188ft^2$

2) In a pyramid if the base is an equilateral triangle with side length of $\sqrt{3}$ cm and the height and slant height are both $\frac{3}{2}$ cm, find its surface area.

Solution:

- Area of the base: $\frac{1}{2} \times side \times height = \frac{1}{2} \times \sqrt{3} \times \frac{3}{2} = \frac{3\sqrt{3}}{4}\ cm^2$
- Area of one face: $\frac{1}{2} \times side\ of\ triangle \times slant\ height = \frac{1}{2} \times \sqrt{3} \times \frac{3}{2} = \frac{3\sqrt{3}}{4} cm^2$
- Lateral area: Since there are 3 faces, the total lateral area is: $3 \times \frac{3\sqrt{3}}{4} = \frac{9\sqrt{3}}{4}\ cm^2$
- Total surface area: $Base\ Area + Lateral\ Area = \frac{3\sqrt{3}}{4} + \frac{9\sqrt{3}}{4} = 3\sqrt{3}cm^2$

Volume of Cylinders, Cones and Spheres

1. **Volume of a Cylinder**: A cylinder has two circular bases and a height. The formula for its volume is:

$$Volume = \pi r^2 h$$

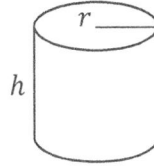

Where r is the radius of the base and h is the height of cylinder.

2. **Volume of a Cone:** A cone has a circular base and a pointed top. The formula for its volume is:

$$Volume = \frac{1}{3}\pi r^2 h$$

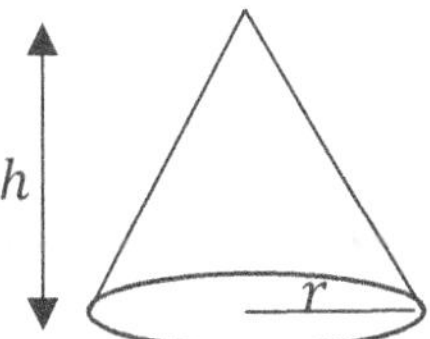

This is one-third of the volume of a cylinder with the same base and height.

3. **Volume of a Sphere:** A sphere is a perfectly round 3D object. The formula for its volume is:

$$Volume = \frac{4}{3}\pi r^3$$

Where r is the radius of the sphere.

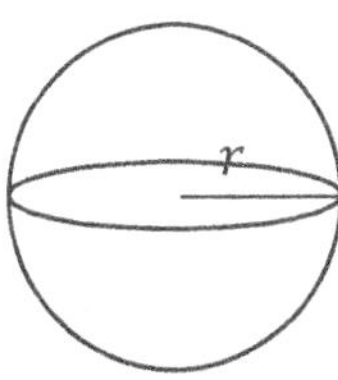

Examples:

1) Find the volume of a cylinder with the radius of 3 cm and a height of 10 cm.

Solution:

Using the formula mentioned above: $Volume = \pi \times 3^2 \times 10 = 90\pi\ cm^3$

2) If a cone has a radius of 4 inches and a height of 9 inches find the volume of this cone.

Solution:

For this cone, radius is 4 inches and the height is 9 inches. Substituting the values into the formula: $Volume = \frac{1}{3} \times \pi \times (4)^2 \times 9 = \frac{1}{3} \times \pi \times 16 \times 9 = \frac{1}{3} \times \pi(144) = 48\pi$ cubic inches

3) If a spere has a radius of 0.5 meters, what is its volume?

Solution:

In this case, the radius $r = 0.5$ meters. Substituting into the formula:

$Volume = \frac{4}{3} \times \pi \times 0.5^3 = \frac{4}{3} \times \pi \times 0.125 = 0.1667\pi\ m^3$

Surface Area of Cylinders, Cones and Spheres

1. **Surface Area of Cylinders:** A cylinder has two circular bases and a curved surface. The total surface area is given by:

$$Surface\ Area = 2\pi r^2 + 2\pi rh$$

Where, r is the radius of the base, h is the height, $2\pi r^2$ represents the area of the two circular bases and $2\pi rh$ represents the area of the curved surface.

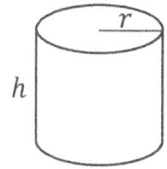

2. **Surface Area of Cones:** A cone has a circular base and a slanted curved surface. The surface area is given by:

$$Surface\ Area = \pi r^2 + \pi rl$$

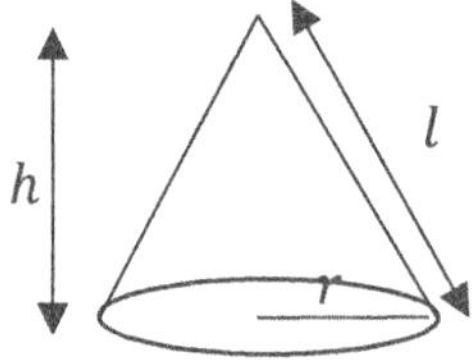

Where, r is the radius of the base, l is the slant height, πr^2 represents the area circular bases and πrl represents the area of the curved surface.

Note: The slant height (l) can be found using the Pythagorean Theorem: $l = \sqrt{r^2 + h^2}$

3. **Surface Area of Sphere:** A sphere has no flat surfaces, and its surface area is given by:

$$Surface\ Area = 4\pi r^2$$

Where, r is the radius.

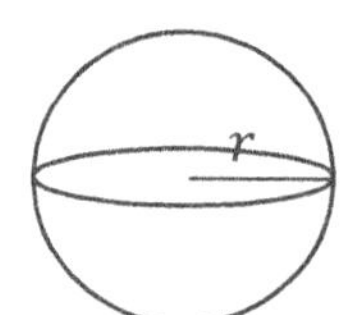

Examples:

1) A cylinder has a radius of 3 cm and a height of 8 cm. What is its total surface area?

Solution:

The radius is 3 cm and the height is 8 cm. substitue the values into the formula:

Substitute the values into the formula:

$Surface\ area = 2\pi(3)^2 + 2\pi(3)(8) = 18\pi + 48\pi = 66\pi$

2) A cone has a base radius of 5 m and a slant height of 10 m. Calculate its total surface area.

Solution:

The radius is 5 m and slant height is 10 m. Substitute the values into the formula:

$Surface\ area = \pi(5)^2 + \pi(5)(10) = 25\pi + 50\pi = 75\pi$

3) The surface area of a basketball is 600 square inches. What is the approximate radius of the basketball? (Use $\pi \approx 3$).

Solution:

To find the radius of the basketball, we use the formula for the surface area of a sphere:
$A = 4\pi r^2$ where $A = 600\ in^2$, Rearranging the formula to solve for r^2:
$r^2 = \frac{A}{4\pi} \rightarrow r^2 = \frac{600}{4\times3} = \frac{600}{12} = 50 \rightarrow r = \sqrt{50} \approx 7.07$ inches

Similar Solids

Similar solids are three-dimensional shapes that have the same shape but are scaled versions of each other. In other words, their corresponding dimensions (such as length, width, height, and radius) are proportional, but the solids themselves can be of different sizes.

Key Characteristics of Similar Solids:

1. **Proportional Corresponding Dimensions:** If two solids are similar, the ratio of their corresponding linear dimensions (such as height, width, and radius) is constant. This ratio is called the scale factor. For example, if the scale factor between two similar cubes is 2, this means each edge of the larger cube is twice as long as the corresponding edge of the smaller cube.
2. **Surface Area Ratio:** The ratio of the surface areas of two similar solids is the square of the scale factor. For example, if the scale factor is k, then:
$$Surface\ Area\ Ratio = k^2$$
3. **Volume Ratio:** The ratio of the volumes of two similar solids is the cube of the scale factor. For example, if the scale factor is k, then:
$$Volume\ Ratio = k^3$$
4. **Same Shape, Different Sizes:** The solids have the same overall geometric shape, meaning their corresponding angles are the same, and their corresponding faces are proportional.

Examples:

1) Two cubes are similar, and the side length of the smaller cube is 3 cm, while the side length of the larger cube is 6 cm. If the surface area of the smaller cube is 54 cm², what is the surface area of the larger cube?

Solution:

- The scale factor: The scale factor between the two similar cubes is determined by the ratio of their corresponding side lengths: Scale factor= $\frac{side\ length\ of\ larger\ cube}{side\ length\ of\ smaller\ cube} = \frac{6}{3} = 2$
- Surface area ratio: Surface area ratio= $(scale\ factor)^2 = 2^2 = 4$
- Surface area of larger cube= $Surface\ Area\ of\ Smaller\ Cube \times Surface\ Area\ Ratio$
Surface Area of Larger Cube= $54 \times 4 = 216\ cm^2$

2) Two cones are similar, and the height of the smaller cone is 5 inches, while the height of the larger cone is 15 inches. If the volume of the smaller cone is 100 in³, what is the volume of the larger cone?

Solution:

- The scale factor: Scale factor= $\frac{height\ of\ larger\ cone}{height\ of\ smaller\ cone} = \frac{15}{5} = 3$
- The volume ratio: Volume ratio= $(scale\ factor)^3 = 3^3 = 27$
- The Volume of the larger cone: $= Volume\ of\ Smaller\ Cone \times Volume\ Ratio$
Volume of Larger Cone= $100 \times 27 = 2{,}700\ in^3$

Nets of Three-Dimensional Figures

A **net** of a three-dimensional figure is a two-dimensional pattern that, when folded along its edges, forms the $3D$ shape. It essentially "unwraps" the $3D$ figure into its flat components, allowing you to see all the individual faces and how they connect.

How Nets Work:

Think of a net as "cutting open" and laying flat on the outer surfaces of a $3D$ object:

- Each face of the $3D$ figure appears as a flat shape (e.g., squares, rectangles, triangles, circles, etc.).
- The arrangement of these flat shapes in the net shows how they fit together in the $3D$ form.

Nets of Common $3D$ Figures:

Cube	Rectangular Prism	Cylinder	Cone	Pyramid
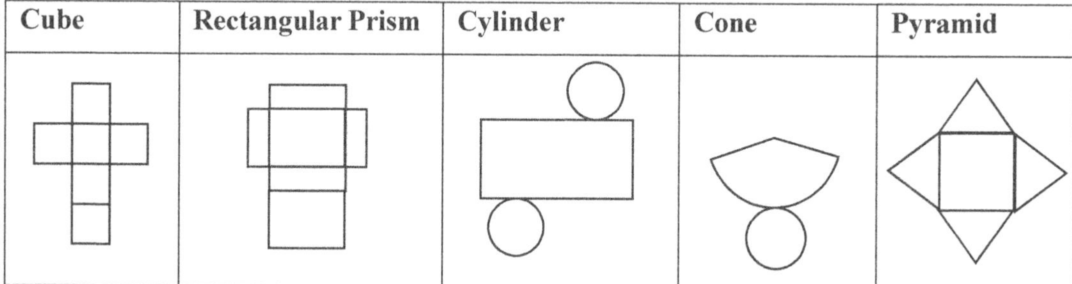				

Examples:

1) A cylinder's net includes two circles and a rectangle. If the radius of the base is 4 cm and the height is 10 cm, what are the dimensions of the rectangle in the net?

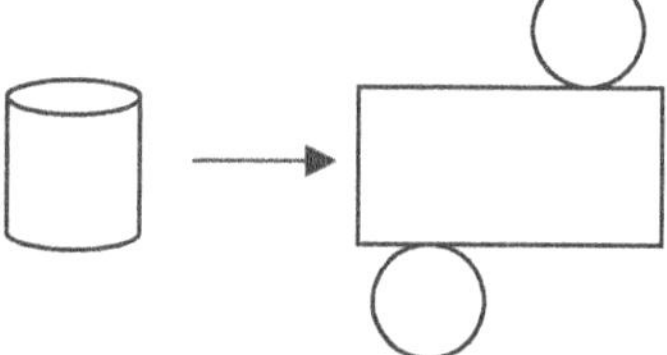

Solution:

- Width of the rectangle: The width of the rectangle corresponds to the height of the cylinder, which is already given as: Height$= 10\ cm$
- Length of the rectangle: The length of the rectangle is equal to the circumference of the circular base. The formula for the circumference of a circle is:

  Circumference$= 2\pi r = 2\pi(4) = 8\pi \approx 8 \times 3.14 = 25.12\ cm$

2) Design the net of a rectangular prism with a length of 8 cm, width of 5 cm, and height of 3 cm. Label all parts of the net

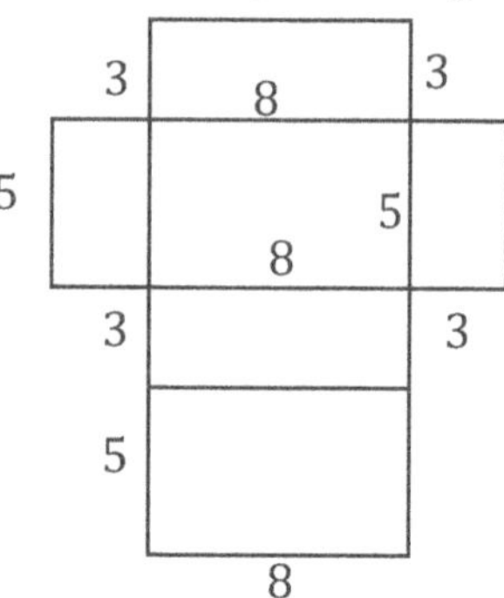

Solution:

- Top and bottom faces: Both are $length \times width$: 8×5
- Front and back faces: Both are $length \times height$: 8×3
- Left and right faces: Both are $width \times height$: 5×3

Real World Applications

Geometry has a vast range of general applications that influence our daily lives, technology, and industries. Here's an overview of its broader uses:

1. **Architecture and Construction**:
   - Used to design buildings, bridges, and other structures by calculating dimensions, angles, and stability.
   - Helps in creating aesthetically pleasing and functional designs.
2. **Art and Design**:
   - Geometry plays a role in creating patterns, mosaics, sculptures, and modern graphic designs.
   - Artists use symmetry, proportions, and spatial relationships for visual appeal.
3. **Navigation and Mapping**:
   - Essential for GPS systems, map creation, and understanding distances and directions.
   - Used in planning travel routes on land, air, and sea.
4. **Technology and Robotics**:
   - Plays a crucial role in 3D modeling, coding visual interfaces, and designing AI systems.
   - Shapes the design and functionality of robots.
5. **Engineering**:
   - Helps in designing mechanical parts, machines, circuits, and more.
   - Engineers rely on geometry for calculations involving volumes, areas, and precise angles.

Example:

You want to paint the walls of a rectangular room. The room has: Length: 5 meters, width: 4 meters and height: 3 meters.

1. Calculate the total area of the walls (to find out how much paint is needed).
2. Exclude the area of one door (2 meters high and 1 meter wide) and two windows (1.5 meters high and 1 meter wide each).

Solution:

1. Area of two longer walls: $2 \times (5 \times 3) = 30\ m^2$
2. Area of two shorter walls: $2 \times (4 \times 3) = 24\ m^2$
3. Total wall area: $30 + 24 = 54\ m^2$
4. Subtract the area of the door and windows:
   - Area of door: $2 \times 1 = 2\ m^2$
   - The area of two windows: $2 \times (1.5 \times 1) = 3\ m^2$
   - Total area to subtract: $2 + 3 = 5\ m^2$
5. Calculate the net area to paint: Total wall area− area to subtract= $54 - 5 = 49\ m^2$

Worksheets

Pythagorean Theorem

Find the unknown value.

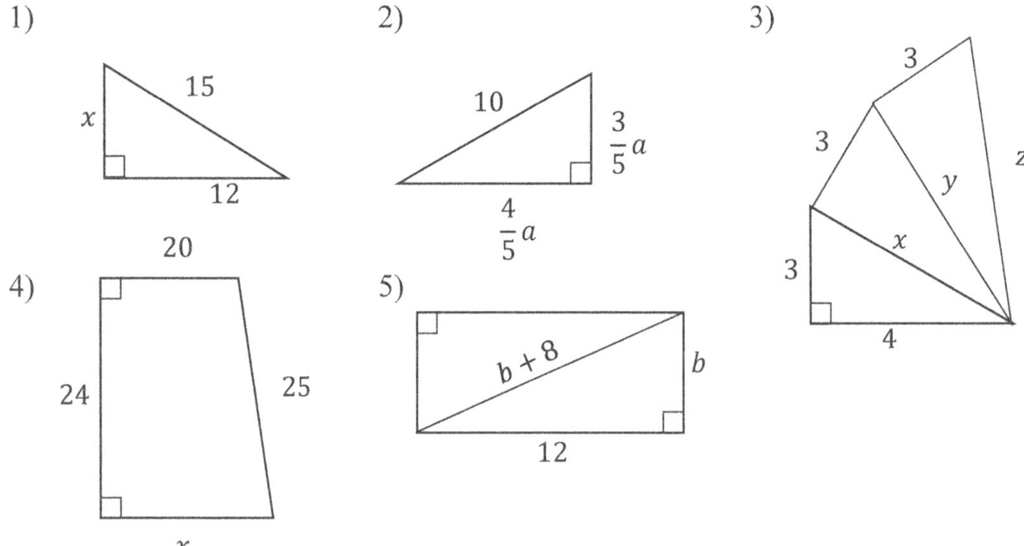

Do the following problems.

6) An ant starts moving from point A. First, it moves 8 meters to the east, then 5 meters to the north, and finally 4 meters to the east until it reaches point B. What is the distance between point A and point B?

7) The side length of an equilateral triangle is 10 centimeters. Find the height of the triangle.

8) Alice has a ladder that is 7.5 meters long. She leans the ladder against a wall. The distance from the bottom of the ladder to the wall is 2.1 meters. How high up the wall can she climb?

9) Calculate the area of following triangle.

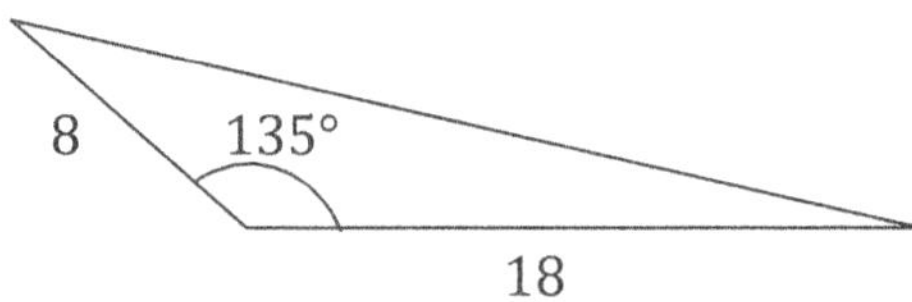

10) In the figure below, how much larger is the area of the big square compared to the small square?

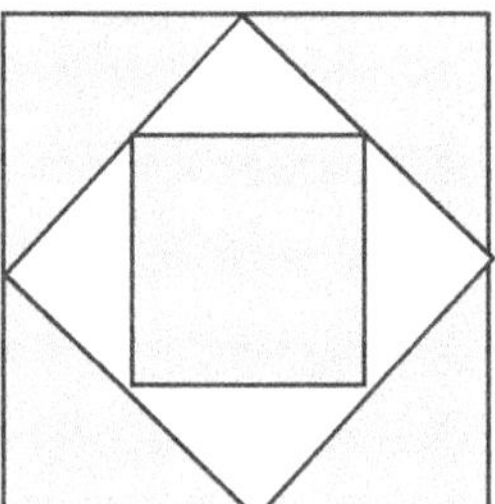

Converse of the Pythagorean Theorem

Determine whether the triangles below are right-angled or not.

1) 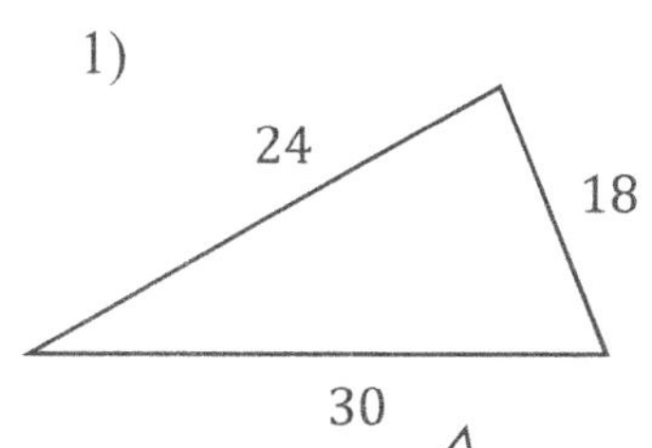

2)  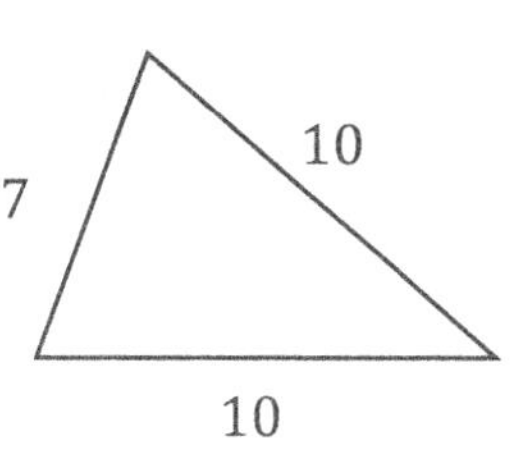

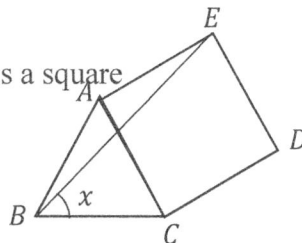

3)  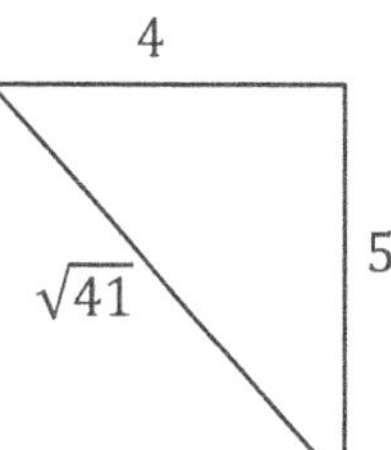

4) 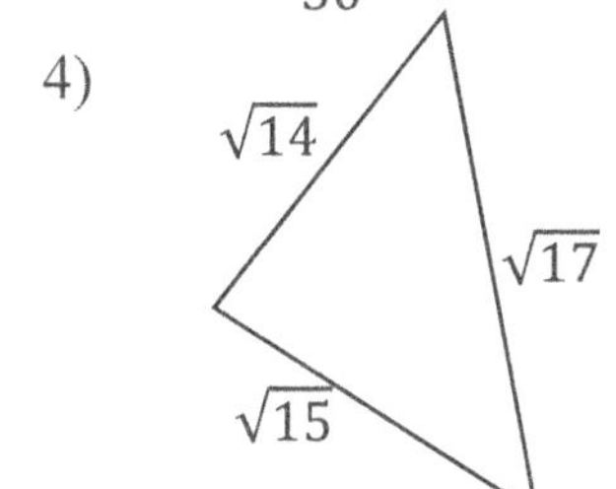

5) 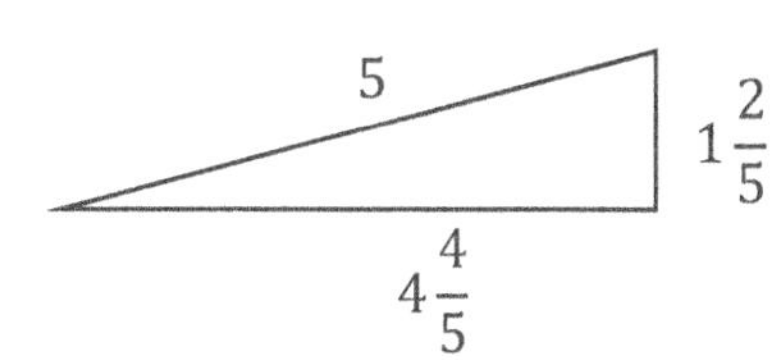

Complementary and Supplementary Angles

Find the answers.

1) Two angles are complementary. One angle is $30°$. What is the measure of the other angle?
2) Two angles are supplementary. One angle is $110°$. What is the measure of the other angle?
3) If two complementary angles are in the ratio $2:3$, find the measures of both angles.
4) An angle is $40°$ more than its complement. What are the measures of both angles?
5) The ratio of the supplement to the complement of an angle is 5 to 3. Find the angle.

Vertical, Adjacent and Congruent Angles

Find the measure of the requested angle.

1) 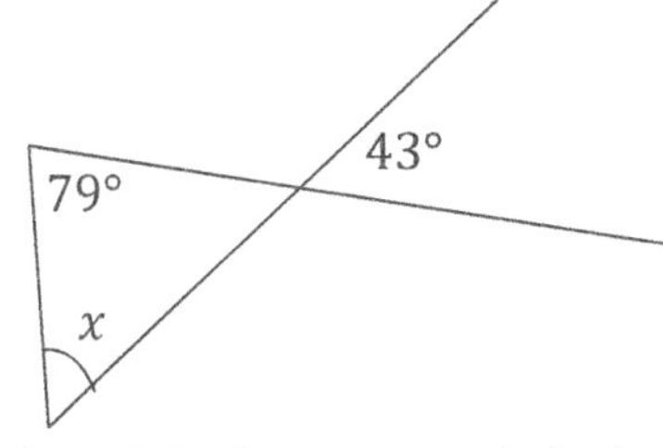

2) The dashed line is the bisector of $\widehat{AOB}$

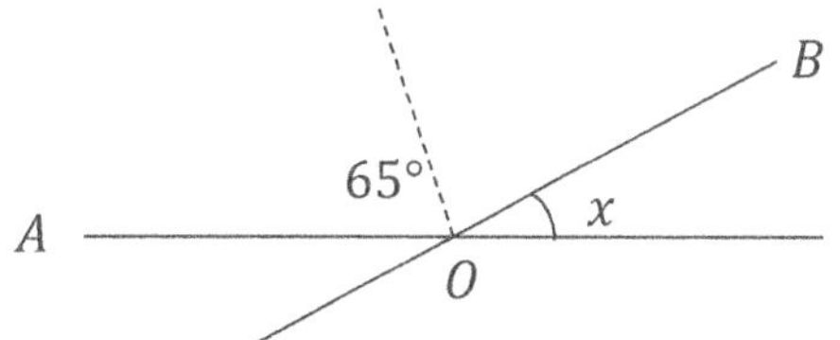

3) The points O is the center of circle.

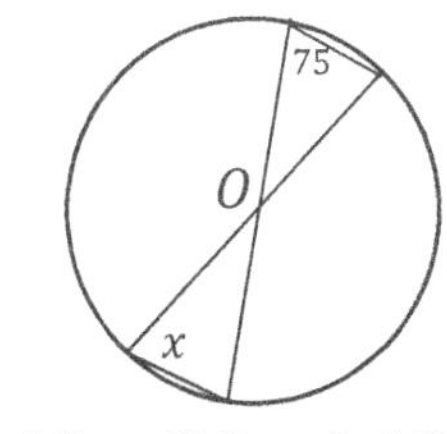

4) $\widehat{BOD} = 70°$, $\widehat{COA} = 105°$ and $\widehat{AOD} = 165°$

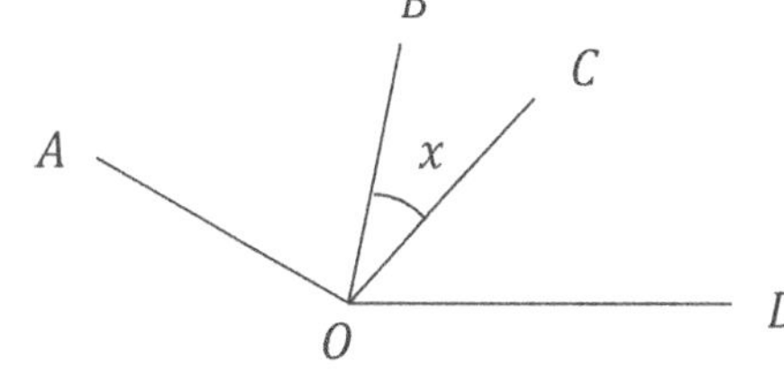

5) $AB = AC = BC$ and $ACDE$ is a square

Solving Equations Using Angle Relationships

Do the following problems.

1) Two supplementary angles are $14x + 42$ and $10x - 6$, find the measure of two angles.
2) Two vertical angles are $10a - 5$ and $8a + 2$, find the measure of two angles.
3) The difference between two complementary angles is 36 degrees. Find the supplement of the smaller angle.
4) determine the value of x in given shape.

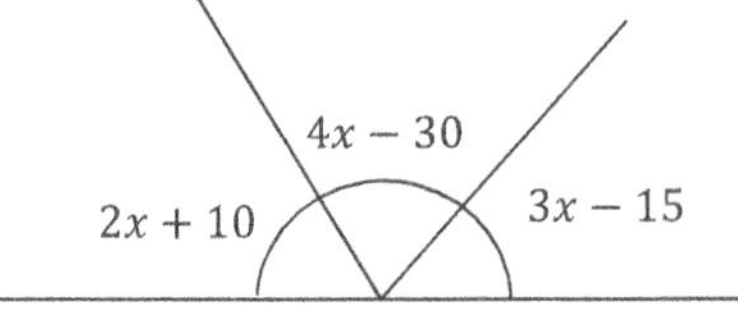

5) Find the value of a in following triangle. $(AC = AB)$

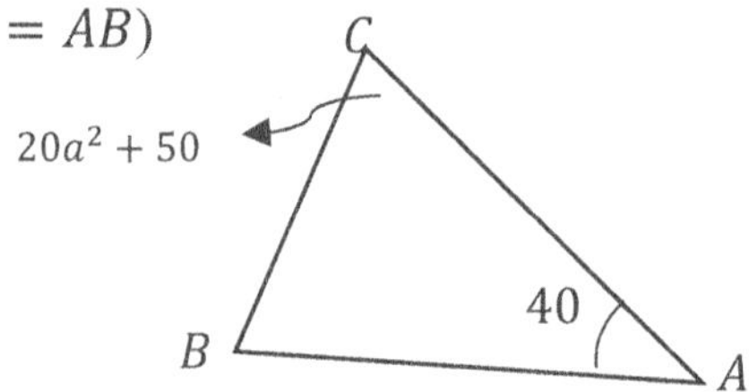

Transversals of Parallel Lines

Find the missing values (In all shapes, the two directed lines are parallel) .

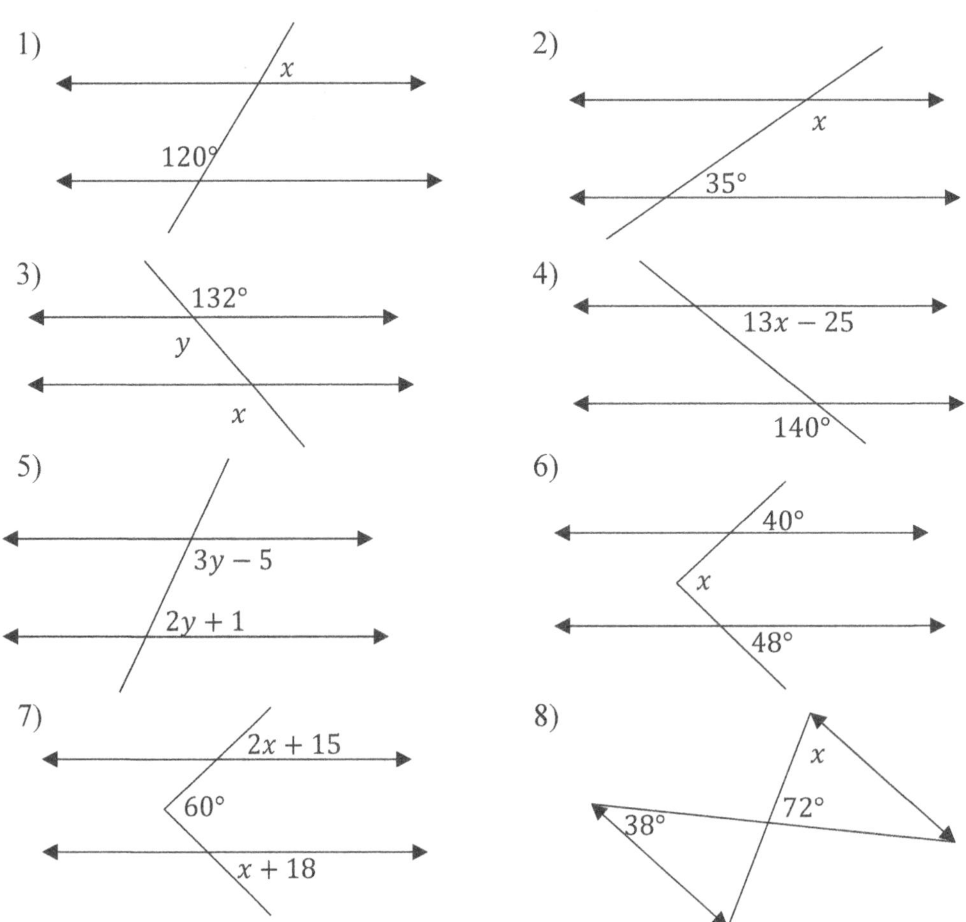

9) AO is the bisector of $\widehat{BAm}$ and BO is the bisector of $\widehat{ABn}$, find angle x.

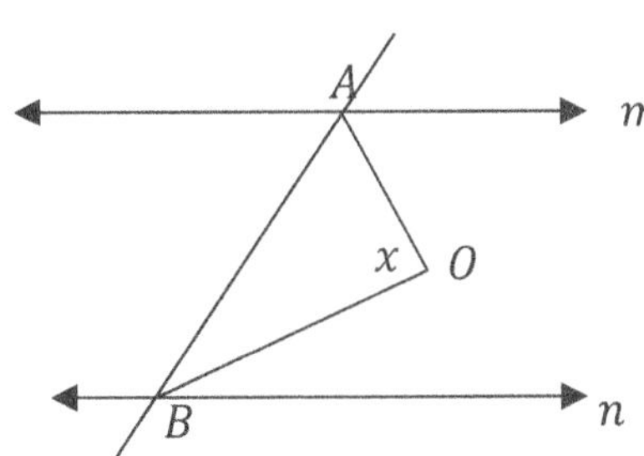

10)

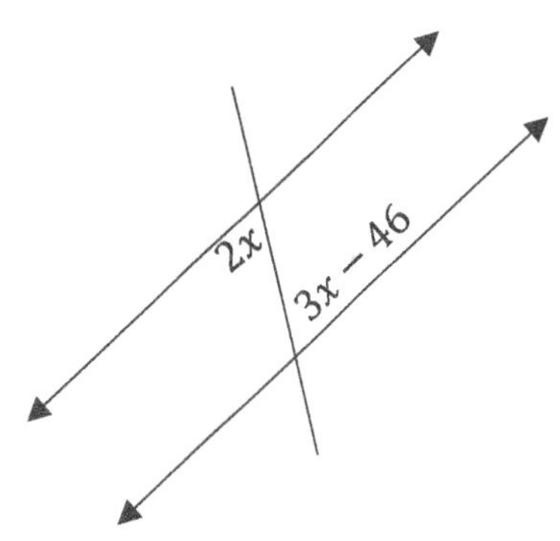

Transversals of Triangles

Do the following problems.

1) In following triangle, the line segment mn is parallel to BC, find the angle x.

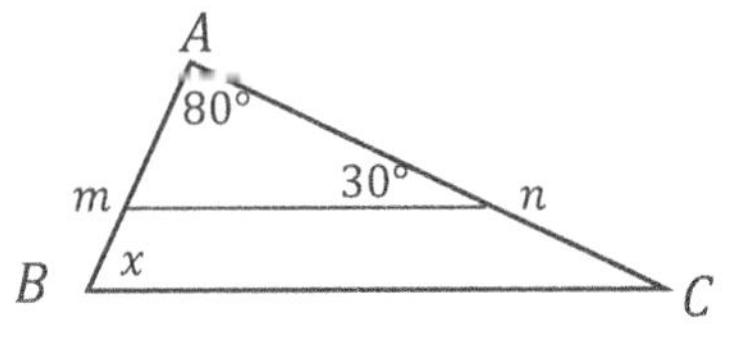

2) Find the angle y.

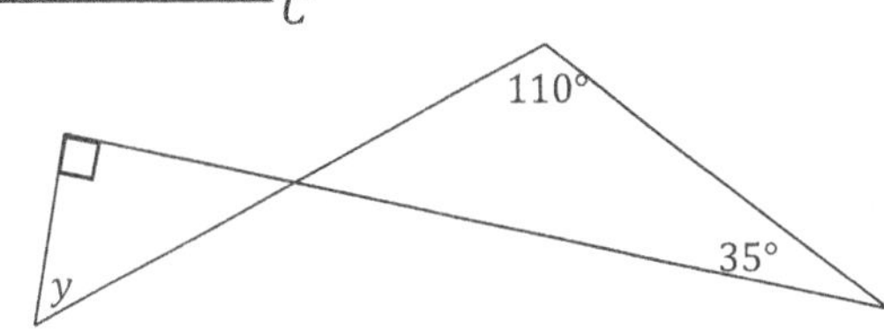

3) Find the values x and y.

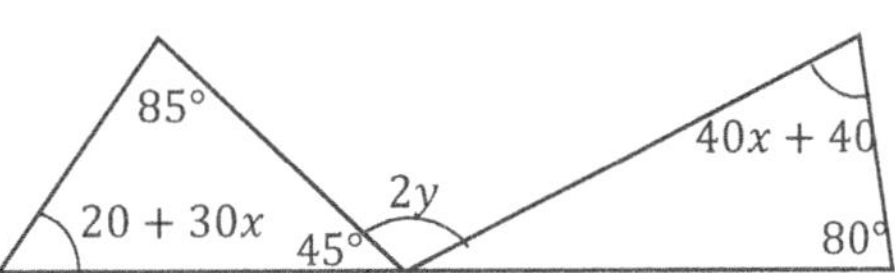

4) If triangle ABC is isosceles and the line d is parallel to AB, find the missing angles.

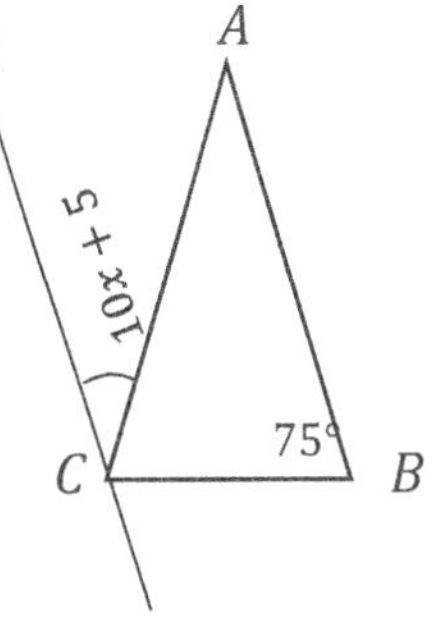

5) In following shape, $d||s$ and $t||p$, find missing values.

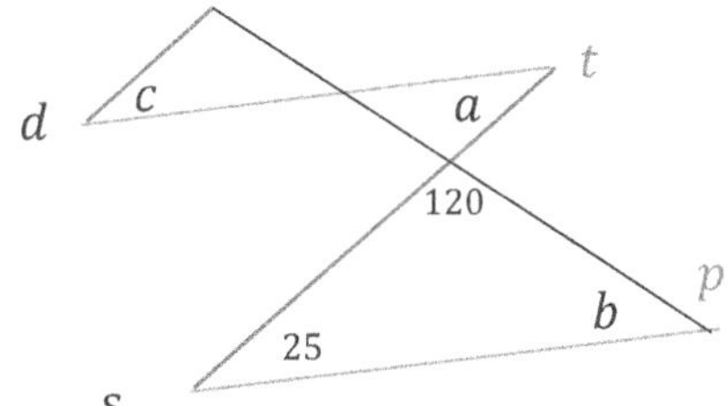

Bisected Line Segments and Angles

Find the answers.

1) A line segment PQ is bisected by point M. If $PM = 3x + 2$ and $MQ = 5x - 4$, find the value of x and the total length of PQ.
2) Let M be the midpoint of the line segment joining $A(3,7)$ and $B(x,11)$. If M has coordinates $(5,9)$, find the value of x.
3) The line segment AB is 7.5 cm long. How many points exist on the perpendicular bisector of this line segment where the distance of these points from the endpoints of the segment is equal to 5?
4) In the figure below, OC is perpendicular to OA, and OD is the bisector of angle $\widehat{AOB}$ and OE is the bisector of angle $\widehat{COB}$. Find the angle $\widehat{DOE}$.

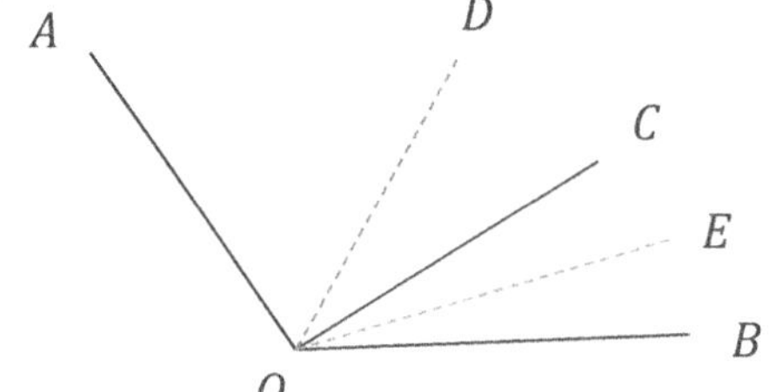

5) In the figure below, BD is the bisector of angle $\widehat{ABC}$ and CD is the bisector of angle $\widehat{ACE}$, find the angle $\widehat{BDC}$.

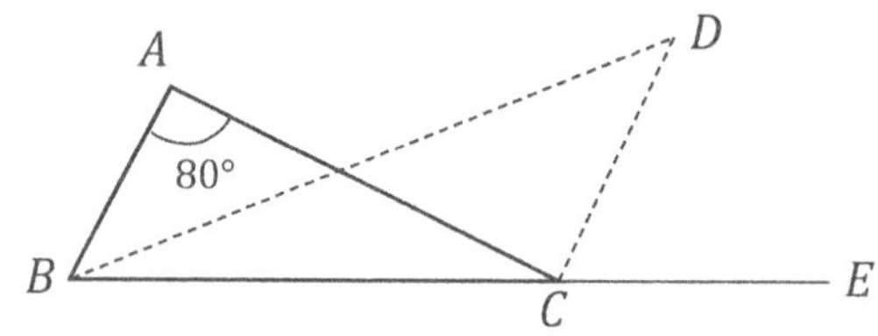

Quadrilaterals

Do the following problems.

1) If the midpoints of the sides of a rhombus are connected, what type of quadrilateral is formed?
2) If the diagonals of a parallelogram are equal, can it be concluded that the parallelogram is a rectangle?
3) If the length and the width of a rectangle are 8 and 6 cm, find the length of its diagonal.
4) In the parallelogram below, $\hat{A} = 2\hat{B} + 30$. Calculate the measures of angles $\hat{A}$ and $\hat{D}$.

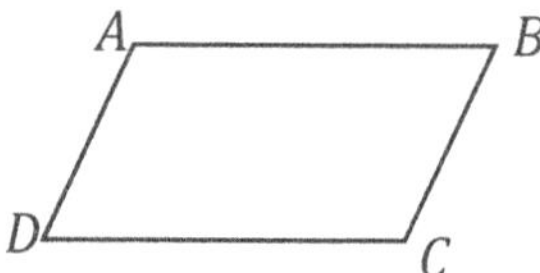

5) In the trapezoid below, AE and DE are bisectors. What is the measure of angle E?

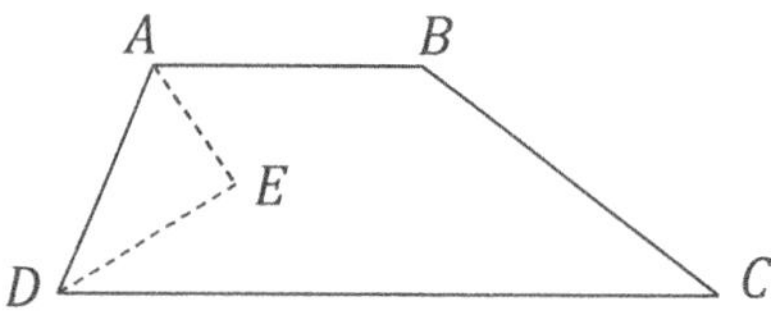

6) In following rhombus, find the value of x and y.

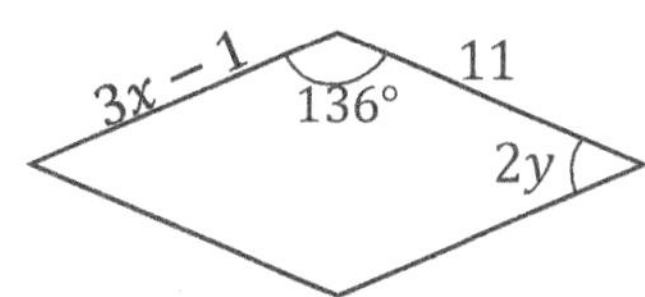

7) The perimeter of the rectangle below is 46 cm, find the lengths of AB and BC.

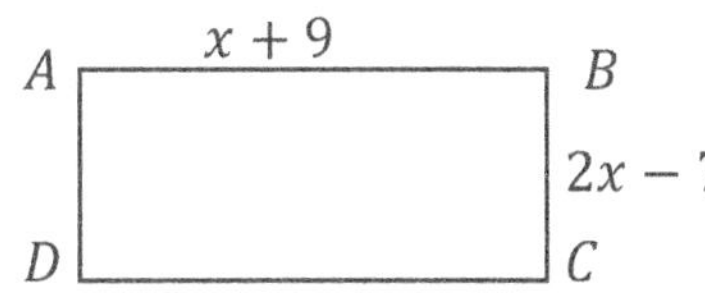

8) In following kite, the diagonal AC is 8 cm, and BD is 12 cm. Calculate the area of kite.

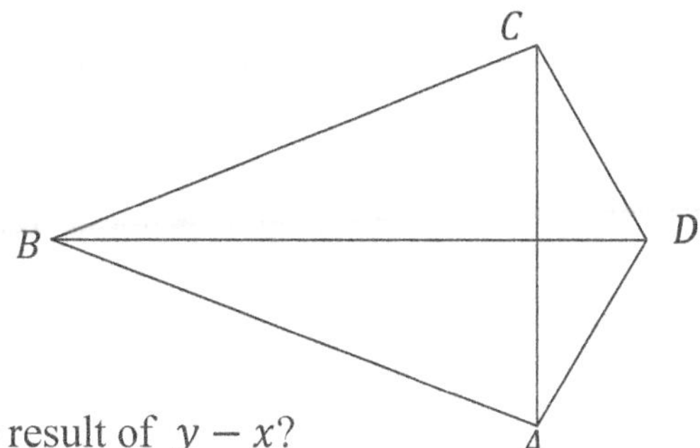

9) The trapezoid shown is isosceles. What is the result of $y - x$?

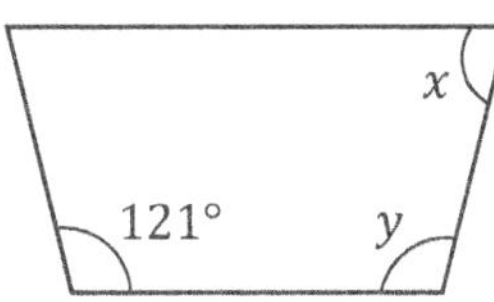

10) In the figure below, all polygons are regular, and the total perimeter of the shape is 357. Find the value of x.

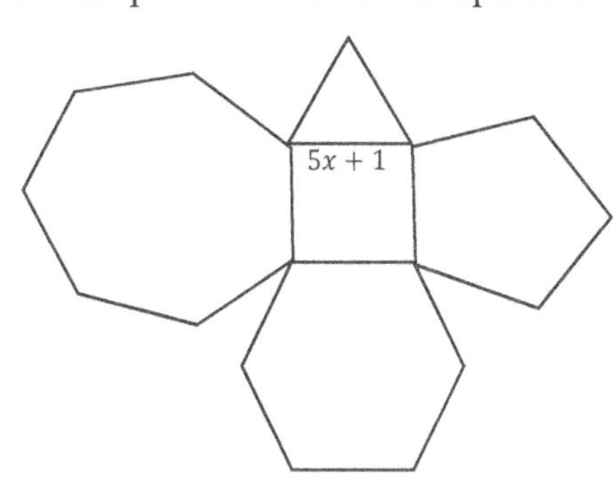

Exterior Angle Theorem

Fill in the blanks according to the figure.

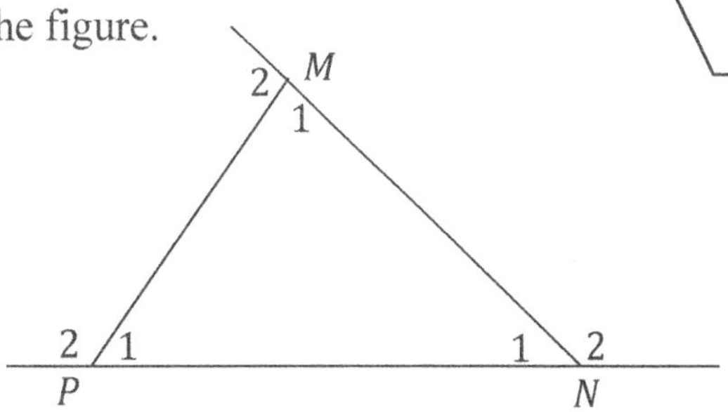

1) $\widehat{M}_1 = 180° - (\ldots + \cdots)$
2) $\widehat{N}_1 = \cdots - \widehat{N}_2$
3) $\widehat{N}_2 = \cdots + \cdots$
4) $\widehat{M}_2 = \widehat{N}_1 + \cdots$
5) $\widehat{P}_2 = \cdots + \cdots$
6) $\widehat{P}_2 + \widehat{M}_2 + \widehat{N}_2 = \cdots$

Find the missing values in following shapes.

7)

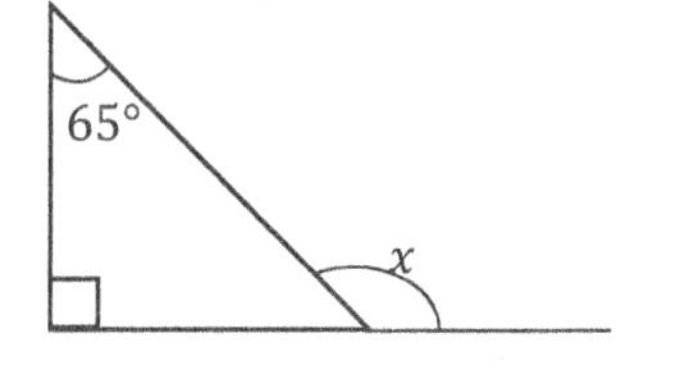

9)

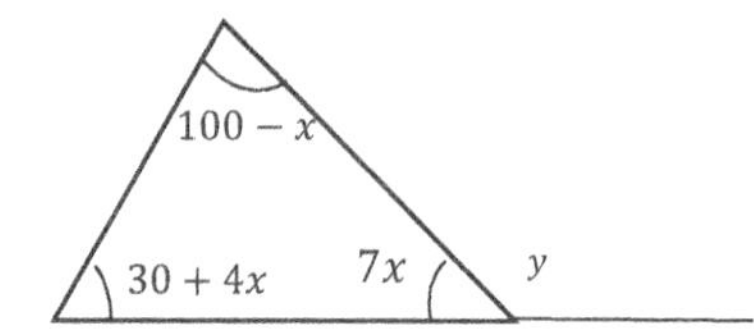

8)

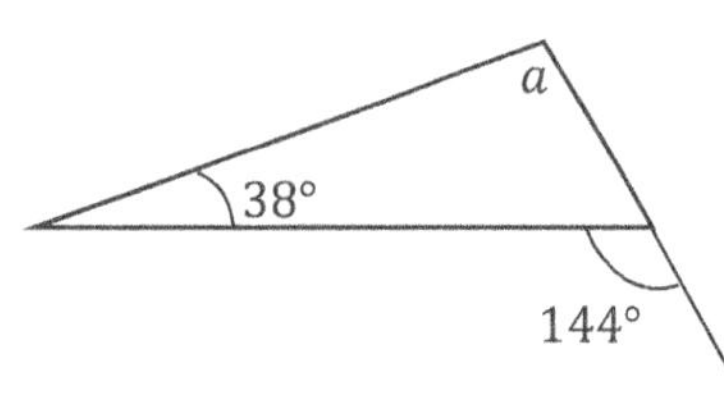

10) $\propto + \beta = 30°$

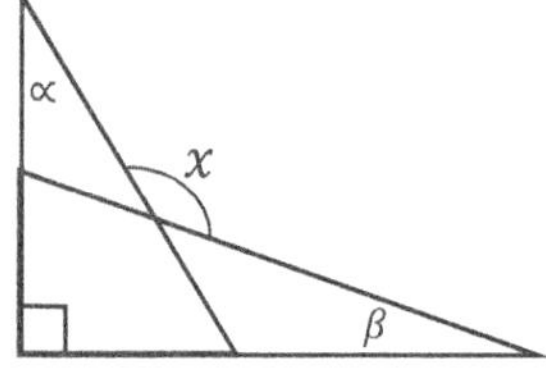

Interior and Exterior Angle of Polygons

Calculate.

1) A regular octagon has 8 sides. Find the measure of one exterior angle of the octagon.
2) The measure of an interior angle of a regular polygon is $162°$. Determine the number of sides of the polygon.
3) A regular polygon has an exterior angle of $30°$. Find the number of sides in the polygon.
4) How many times is the sum of the interior angles of an 18-sided polygon greater than the sum of the interior angles of a 10-sided polygon?
5) We reduced the number of sides of a polygon by two, and the sum of the interior angles was halved. Find the number of sides of this polygon.

Parts of a Circle

Find the answers.

1) A circle has a radius of 10 cm. A central angle of $60°$ subtends an arc. Find:
   a) The length of the arc.
   b) The area of the sector.
2) The chord of a circle is 16 cm long and is 6 cm away from the center of the circle. Find the radius of the circle.
3) A chord of a circle is 8 cm long and the radius of the circle is 5 cm. Find the perpendicular distance from the center of the circle to the chord.
4) In a circle with center O, a radius OB is perpendicular to the tangent AB at the point B. How much is the angle formed between the radius and tangent at the point of contact?
5) In the figure below, two chords AB and CD are equal. Are two arcs $\overset{\frown}{AB}$ and $\overset{\frown}{CD}$ also equal? Why?

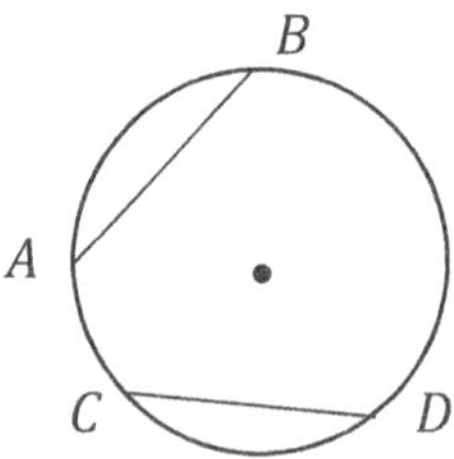

Area and Perimeter of Semicircles and Quarter Circles

Calculate. ($\pi \approx 3$)

1) A semicircle has a diameter of 14 cm. Calculate its area and perimeter.
2) A quarter circle has a radius of 10 cm. Find its area and the length of its curved edge.
3) A quarter circle is cut out from a square with side length 16 cm. Find the remaining area of the square.
4) A semicircle and a quarter circle have equal areas. If the radius of the semicircle is 12 cm, find the radius of the quarter circle.
5) A rectangle has a length of 10 cm and a width of 6 cm. A semicircle is attached to one of the shorter sides. Find the total area of the shape.

Area Between Two Shapes

Find the area of grey part in following shapes. ($\pi \approx 3$)

1)

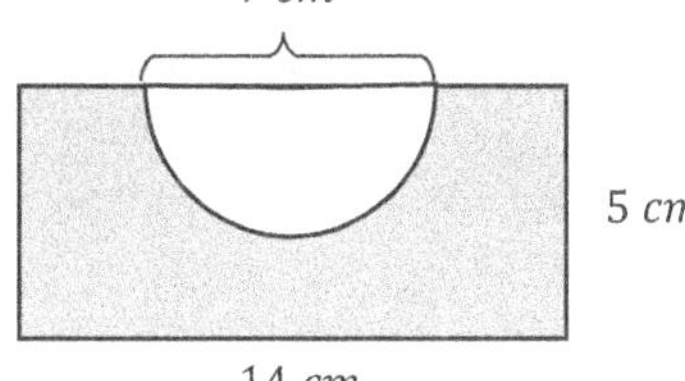

2)

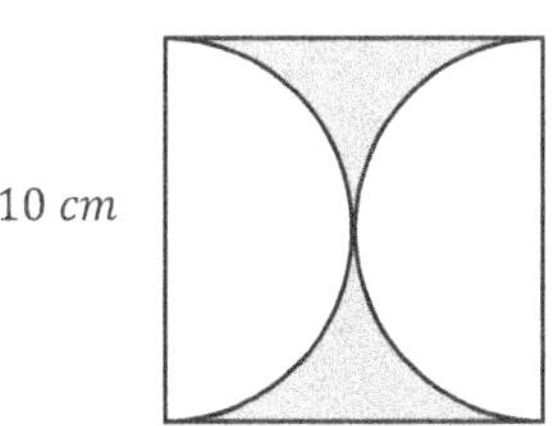

3) The triangle below is a right-angled isosceles triangle.

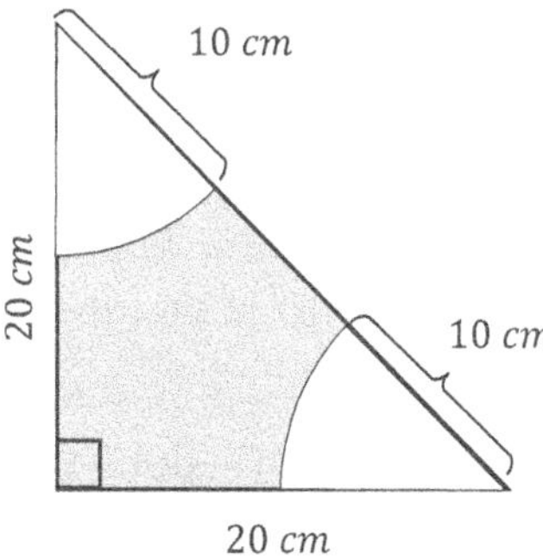

4) The quadrilateral is square.

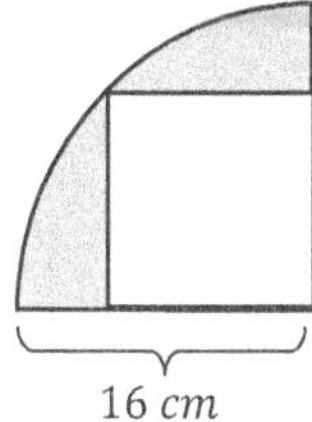

5) The larger circle's diameter is 24 cm.

Perimeter and Area: Changes in Scale

Do the following problems:

1) A square has a side length of 5 cm. If the side length is doubled, how do the perimeter and area change? Calculate the new values.
2) A rectangle has dimensions 6 cm by 4 cm. If both the length and width are scaled up by a factor of 3, what will be the new perimeter and area of the rectangle?

3) A triangle has a base of 10 cm and a height of 8 cm. If the base and height are both increased by 50%, calculate how the area changes and find the new area.
4) The radius of a circle is reduced by half. By what factor does the perimeter (circumference) change? By what factor does the area change?
5) A rectangular garden has a length of 20 m and a width of 12 m. The garden is to be enlarged such that its new area is 1.5 times the original area. If the shape remains a rectangle and the enlargement uses the same scale factor for both the length and width, find the scale factor and the new dimensions of the garden.

Front, Side and Top View

Determine how each shape is seen from the requested angle of view.

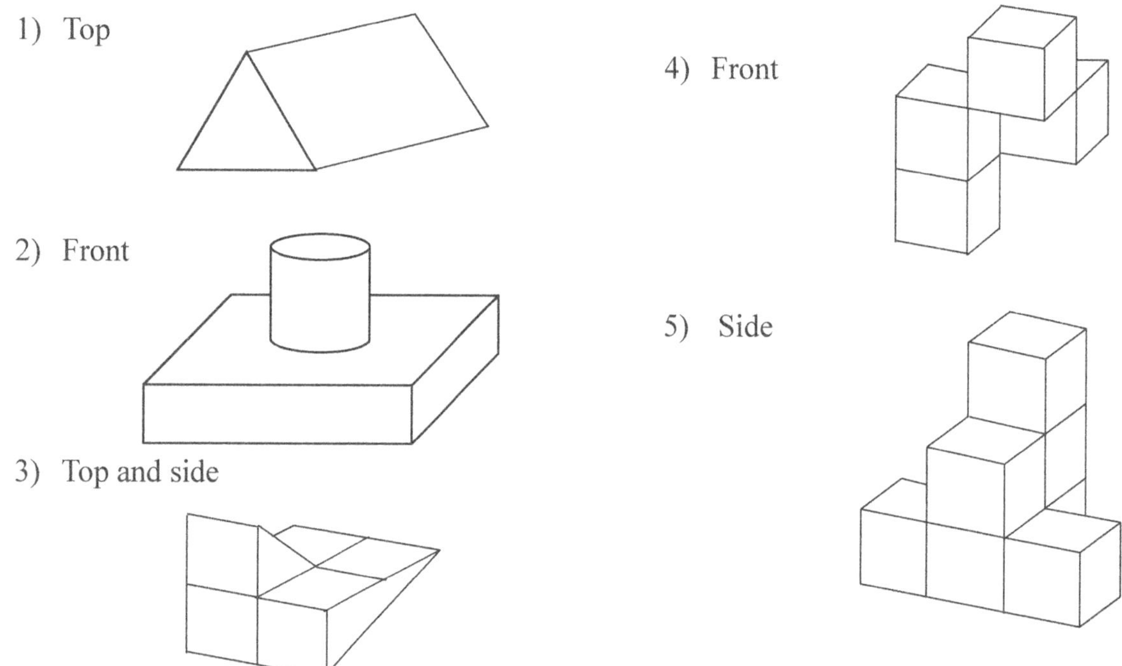

Volume of Cubes, Prisms and Pyramids

Calculate.

1) A square pyramid has a base side length of 6 cm and a height of 9 cm. Find its volume.
2) A rectangular prism has a volume of 240 cm^3. If its length is 10 cm and its width is 8 cm, find its height.
3) A composite shape is made by placing a cube with a side length of 4 cm on top of a square pyramid with the same base dimensions. If the pyramid's height is 6 cm, calculate the total volume of the shape.
4) A pyramid has a rectangular base of dimensions 10 cm by 8 cm and a height of 12 cm. A cube with a side length of 4 cm is carved out from the pyramid. Calculate the remaining volume.
5) A cube and a rectangular prism have the same volume. If the cube's side length is 6 cm, and the prism's base is 9 cm by 4 cm, find the height of the prism.

Surface Area of Cubes, Prisms and Pyramids

Solve.

1) A rectangular prism has dimensions of 8 cm, 4 cm, and 3 cm. Calculate its total surface area.
2) A triangular pyramid has a base triangle with sides 5 cm each and a slant height of 7 cm. Find its total surface area.
3) A square pyramid has a base side length of 6 cm and a slant height of 8 cm. Calculate its total surface area.
4) A composite solid is formed by attaching the right triangular prism to one face of a cube. The cube has a side length of 10 cm, and the triangular prism has:
    - A triangular base with base= $4cm$ and height= $3cm$
    - A length (depth) of 8 cm

    The triangular face of the prism is attached to one face of the cube, find the total surface area of the composite solid.

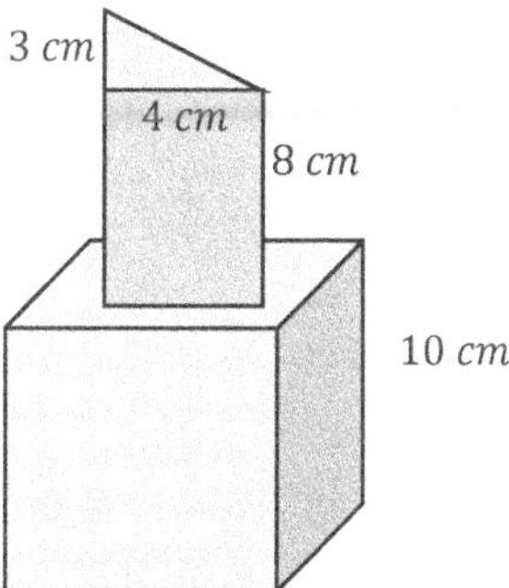

5) A cube and a rectangular prism have equal surface areas. If the cube's side length is 4 cm, and the prism's dimensions are 5 cm by 3 cm by h, find the height h of the prism.

Volume of Cylinders, Cones and Spheres

Find the answers. ($\pi \approx 3$)

1) A cylindrical water tank has a radius of 4 m and a height of 8 m. How many cubic meters of water can it hold?
2) A cone has a volume of 150 cm^3 and a base radius of 5 cm. Find its height.
3) The radius of a sphere is doubled. By what factor does its volume change?
4) A composite solid is made up of a cone and a cylinder. The cylinder has a radius of 6 cm and a height of 12 cm, while the cone has the same radius but a height of 8 cm. Find the total volume of the solid.
5) A cone is inscribed in a cylinder with the same radius and height. How many times is the volume of the cylinder greater than the volume of this cone?

Surface Area of Cylinders, Cones and Spheres

Calculate. ($\pi \approx 3$)

1) A sphere has a radius of 7 cm. Find its surface area.
2) A cylinder has a radius of 4 cm and a height of 8 cm. Find the lateral surface area and the total surface area.
3) A sphere and a cylinder have equal surface areas. If the sphere's radius is 3 cm and the cylinder's height equal its diameter, find the cylinder's radius.

4) A cone is placed on top of a cylinder. Both have the same base radius of 5 cm, and the heights of the cone and cylinder are 8 cm and 12 cm, respectively. Find the total surface area of the combined shape
5) A metal sphere with a radius of 10 cm is melted and recast into a cone with a base radius of 5 cm. Find the slant height of the cone if the curved surface area of the cone equals the sphere's surface area.

Similar Solids

Do the following problems.

1) Two cylinders are similar, with heights in the ratio $2:1$. If the radius of the larger cylinder is 6 cm, what is the radius of the smaller cylinder?
2) wo cones are similar, and their heights are in the ratio of $3:2$. If the volume of the smaller cone is 40 cm^3, find the volume of the larger cone.
3) A solid rectangular prism has dimensions scaled up by a factor of 4. How do its volume and surface area change?
4) Two similar pyramids have base areas of 36 cm^2 and 81 cm^2. If the height of the smaller pyramid is 5 cm, find the height of the larger pyramid.
5) Two cubes are similar, and their surface areas are in the ratio $16:25$. Find the ratio of their volumes.

Nets of Three-Dimensional Figures

Do the following problems.

1) Identify the net of a cube from the following shapes.

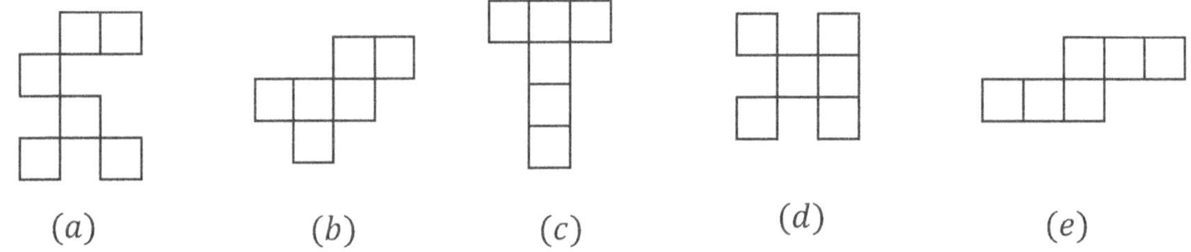

2) Draw the net of a rectangular prism with dimensions $4\ in$ by $2\ in$ by $1\ in$, and then calculate the perimeter of the net.
3) Given the net of a pyramid, calculate its volume and surface area.

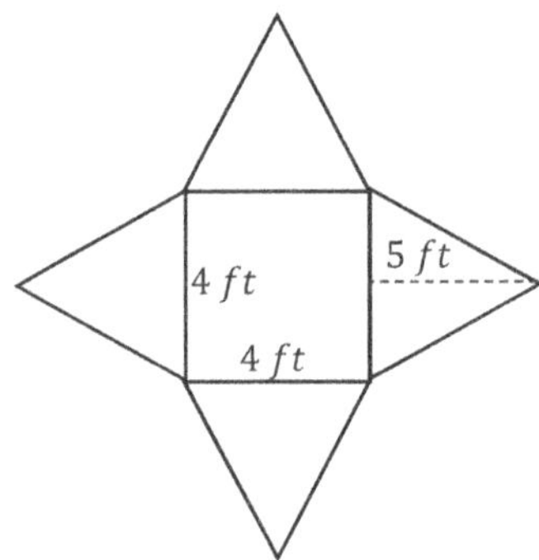

4) If the height of one face of a tetrahedron is 5 cm, what is the height of its corresponding net?
5) We have unfolded a cube with a volume of 27 cubic centimeters into a net. What is the perimeter of the cube's net in centimeters?

Real World Applications

Find the answers to each question.

1) A straight ladder leans against a wall. If the ladder forms a 60° angle with the ground, what is the complementary angle between the ladder and the wall?
2) If the basketball with a radius of 12 cm is completely inflated, what is the volume of air it can hold?
3) On a map, a road crosses two parallel streets, forming angles of 75° and 105° with one street. What are the corresponding angles formed with the other street?
4) A rectangular prism-shaped tank with a length of 3 m, a width of 2 m, and a height of 1.5 m needs to be painted. What is the surface area that needs to be painted?
5) You are designing a custom gift box for a company using a net of a triangular prism. The base of the triangular prism has sides of 6 cm, 8 cm, and 10 cm. The height of the prism is 15 cm. Determine how much material is needed to make it.
6) A circular fountain has a radius of 7 m. Find its area and the length of the fence needed to enclose it (use $\pi \approx 3.14$).
7) We use a roller, as shown below, to paint the wall. Calculate the area of the wall that is painted with one complete roll of the cylinder.

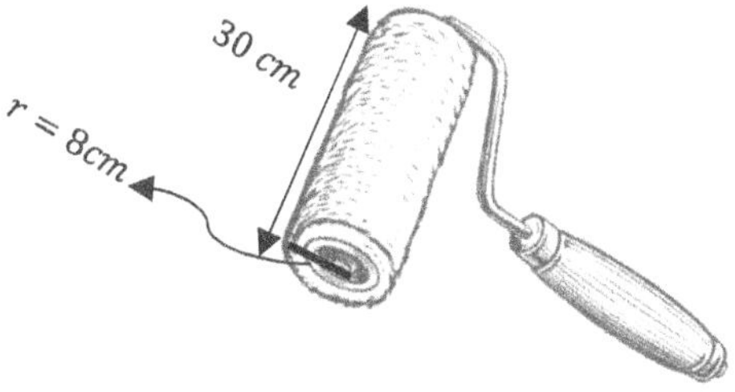

8) David used 36 cubes to build a wall around a square-shaped area, ensuring that each cube was connected to the others. How many more of these cubes are needed to cover the surface inside this area?
9) We have an aquarium with dimensions 17.5 cm by 44 cm by 25 cm, filled with water up to a height of 12.5 cm. A heavy statue weighing 7.7 kilograms and with a volume of 3850 cubic centimeters is placed inside the aquarium. How does the water level in the aquarium change?

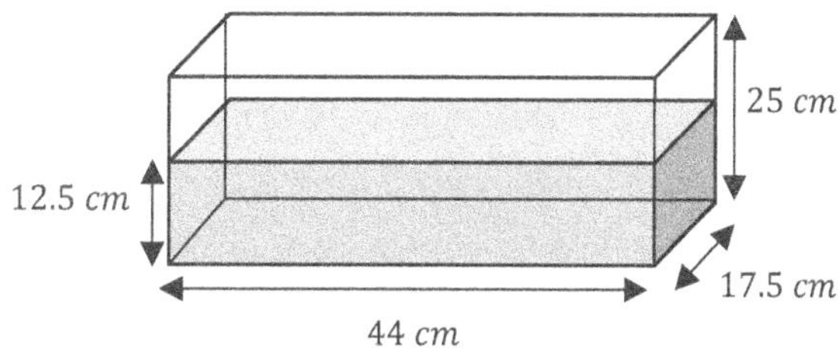

Answer of Worksheets

Pythagorean Theorem

1) 9
2) 10
3) $x = 5, y = \sqrt{34}$ and $z = \sqrt{43}$
4) 27
5) 5
6) 13 meters
7) $5\sqrt{3}$ cm
8) ≈ 7.2 meters
9) $36\sqrt{2}$ square units
10) 4 times

Converse of the Pythagorean Theorem

1) Right-angled
2) Not right-angled
3) Right-angled
4) Not right-angled
5) Right-angled

Complementary and Supplementary Angles

1) $60°$
2) $70°$
3) $36°$ and $54°$
4) $25°$ and $65°$
5) $45°$

Vertical, Adjacent and Congruent Angles

1) $58°$
2) $50°$
3) $75°$
4) $10°$
5) $45°$

Solving Equations Using Angle Relationships

1) $126°$ and $54°$
2) Both angles are $30°$
3) $153°$
4) ≈ 23.8
5) $a = \pm 1$

Transversals of Parallel Lines

1) $60°$
2) $145°$
3) $x = y = 132°$
4) 5
5) $36.8°$
6) $84°$
7) $9°$
8) $70°$
9) $90°$
10) $46°$

Transversals of Triangles

1) $70°$
2) $55°$
3) $x = 1$ and $y = \frac{115}{2}$
4) $10x + 5 = A = 30°$ and $x = 2.5$
5) $a = 25, b = 35$ and $c = 25$

Bisected Line Segments and Angles

1) $x = 3$ and $PQ = 22$
2) $x = 7$
3) Two points
4) $45°$
5) $40°$

Quadrilaterals

1) Rectangle
2) Yes
3) 10 cm
4) $\hat{A} = 130°$ and $\widehat{D} = 50°$
5) $E = 90°$
6) $x = 4$ and $y = 22°$
7) $AB = 16$ and $BC = 7$
8) $48\ cm^2$
9) 62
10) $x = 4$

Exterior Angle Theorem

1) $\widehat{M}_1 = 180° - (\widehat{N}_1 + \widehat{P}_1)$
2) $\widehat{N}_1 = 180° - \widehat{N}_2$
3) $\widehat{N}_2 = \widehat{M}_1 + \widehat{P}_1$
4) $\widehat{M}_2 = \widehat{N}_1 + \widehat{P}_1$
5) $\widehat{P}_2 = \widehat{M}_1 + \widehat{N}_1$
6) $\widehat{P}_2 + \widehat{M}_2 + \widehat{N}_2 = 360°$
7) $155°$
8) $106°$
9) $y = 145°$ and $x = 5°$
10) $x = 120°$

Interior and Exterior Angle of Polygons

1) 45°
2) 20 sides
3) 12 sides
4) 2 times
5) 6 sides

Parts of a Circle

1) The arc length is $\frac{10\pi}{3}$ and the sector area is $\frac{50\pi}{3}$
2) 10 cm
3) 3 cm
4) 90°
5) Yes, by connecting points A and B to the center of the circle, two congruent triangles are formed. Since the angles opposite the arcs are equal, the arcs will also be equal.

Area and Perimeter of Semicircles and Quarter Circles

1) Area: 73.5 cm^2 and perimeter:35 cm
2) Area: 75 cm^2 and length of the curved edge: 15 cm
3) Remaining area: 64 cm^2
4) $\approx 17\ cm$
5) $114\ cm$

Area Between Two Shapes

1) $51.625\ cm^2$
2) $25\ cm^2$
3) $125\ cm^2$
4) $64\ cm^2$
5) $72\ cm^2$

Perimeter and Area: Changes in Scale

1) The perimeter doubles (perimeter: 40 cm) and the area increases by a factor of 4 (area: 100 cm^2)
2) Perimeter: 60 cm and area: 216 cm^2
3) The area increases by a factor of 2.25 (area: 90 cm^2)
4) The circumference is reduced by a factor of 2 and the area is reduced by a factor of 4.
5) Scale factor is $\sqrt{1.5}$, length: $20\sqrt{1.5}$ and width: $12\sqrt{1.5}$

Front, Side and Top View

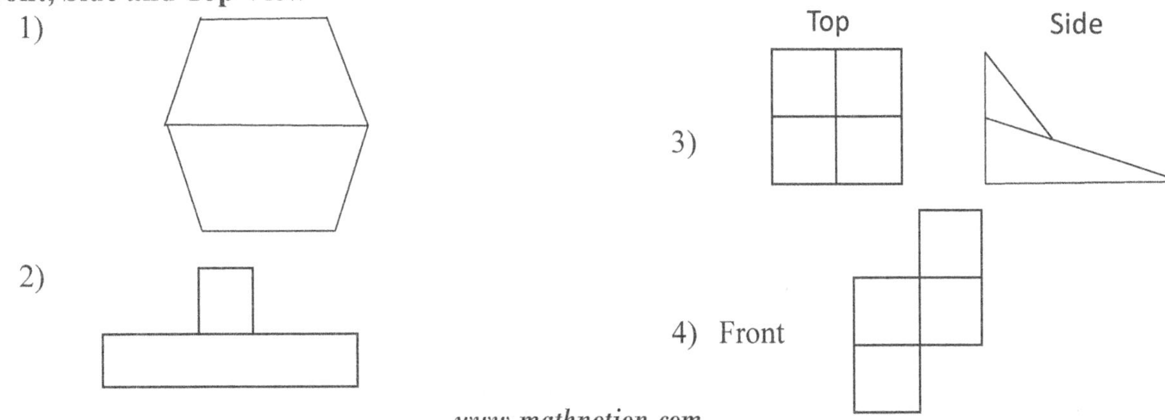

5)  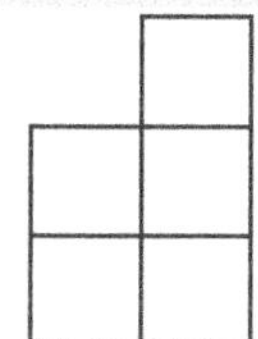

Volume of Cubes, Prisms and Pyramids

1) $108\ cm^3$
2) $3\ cm$
3) $96\ cm^3$
4) $256\ cm^3$
5) $6\ cm$

Surface Area of Cubes, Prisms and Pyramids

1) $136\ cm^2$
2) $\approx 63.33\ cm^2$
3) $132\ cm^2$
4) $312\ cm^2$
5) $\approx 4.125\ cm$

Volume of Cylinders, Cones and Spheres

1) $384\ m^3$
2) $6\ cm$
3) 8
4) $1584\ cm^3$
5) 3 times

Surface Area of Cylinders, Cones and Spheres

1) $588\ cm^2$
2) Lateral surface area: $192\ cm^2$, and the total surface area: $288\ cm^2$
3) $\sqrt{6}\ cm$
4) $\approx 576.45\ cm^2$
5) $80\ cm$

Similar Solids

1) $3\ cm$
2) $135\ cm^3$
3) Volume:64 times and Surface area:16 *times*
4) $7.5\ cm$
5) $64:125$

Nets of Three-Dimensional Figures

1) b, c and e
2) The perimeter of the net: $28\ in$

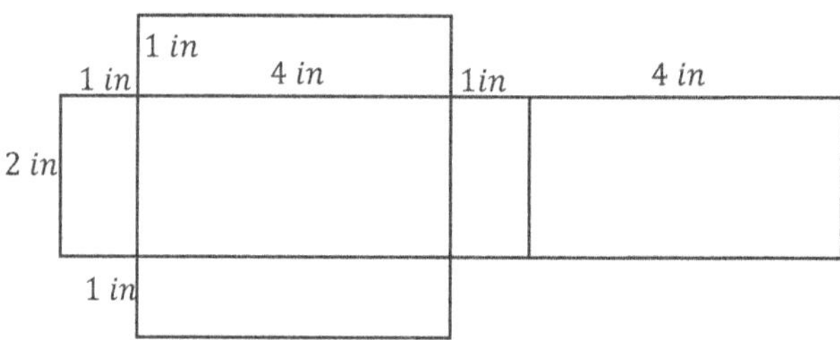

3) The volume: $\frac{16\sqrt{21}}{3}\ ft^3$, and the surface area: $56\ ft^2$
4) $10\ cm$
5) $42\ cm$

Real World Applications

1) $30°$
2) $\approx 7238.44\ cm^3$ of air
3) The angles formed with the other street are $75°$ and $105°$
4) $27\ m^2$
5) $408\ cm^2$ of material
6) $43.96\ m$
7) $1507.2\ cm^2$
8) 64 more cubes
9) $5\ cm$

Chapter 13: Data and Graph

Topics that you'll learn in this chapter:

- ✓ Mean, Median, Mode and Range
- ✓ Mean Absolute Deviation
- ✓ Quartiles and Outlier
- ✓ Box and Whisker Plots
- ✓ Scatter Plots
- ✓ Correlation and Causation
- ✓ Correlation Coefficients
- ✓ Line of Best Fit
- ✓ Equation of a Line of Best Fit
- ✓ Two-Way Frequency Tables
- ✓ Two-Way Relative Frequency Tables
- ✓ Real World Applications
- ✓ Worksheets
- ✓ Answer of Worksheets

Mean, Median, Mode and Range

1. **Mean (Average):** The sum of all values divided by the total number of values.
   - **Effect of Changes**:
     - **Adding/Removing a number**: The mean changes unless the number added/removed is equal to the mean.
     - **Changing one value**: If a number increases or decreases, the mean will shift in the same direction.
     - **Outliers (Extreme values)**: The mean is highly affected by outliers.
2. **Median:** The middle number when data is arranged in order.
   - **Effect of Changes**:
     - **Adding/Removing numbers**: If the middle changes, the median changes.
     - **Shifting one value**: If it's not the middle value, the median remains unchanged.
     - **Outliers**: The median is not affected by extreme values.
3. **Mode:** The number that appears most often.
   - **Effect of Changes**:
     - **Adding/Removing values**: If the most frequent value is removed or another value appears more frequently, the mode changes.
     - **Ties**: If multiple numbers appear with the same highest frequency, there can be more than one mode.
4. **Range:** The difference between the maximum and minimum values.
   - **Effect of Changes**:
     - **Adding/Removing numbers**: The range changes if the smallest or largest number changes.
     - **Outliers**: The range is highly sensitive to extreme values.

Example:

If we add a constant (k) to all numbers, how will the mean, median, mode and range change?

Solution:

Adding a constant k to all numbers, shifts mean, median and mode by k, but the range remains the same.

Mean Absolute Deviation

The **Mean Absolute Deviation (MAD)** is a way to measure how "spread out" or "different" the numbers in a set are from the average (mean). Here's how it works:

1. **Find the Mean**: Add up all the numbers in the data set and divide by how many numbers there are. This gives you the average.
2. **Calculate the Absolute Deviations**: For each number in the set, find the difference between that number and the mean. This difference is called the "deviation." If the deviation is negative, take the absolute value (make it positive), because we're interested in the size of the difference, not the direction.
3. **Find the Average of the Deviations**: Add up all these absolute deviations and divide by how many numbers there are. This final result is the Mean Absolute Deviation (MAD).

Essentially, the MAD tells you, on average, how far each number in the data set is from the mean. It's a way to see how consistent or spread out the numbers are.

Example:

Calculate the mean absolute deviation of this data set: $100, 120, 110, 105, 95, 115, 125$

Solution:

1. Find the mean: Add up all numbers and divide by the 7:
   $(100 + 120 + 110 + 105 + 95 + 115 + 125) \div 7 = 110$
2. Calculate the deviations (differences): For each data point, subtract the mean. Then, take the absolute value of the deviations:
   - $|100 - 110| = 10$
   - $|120 - 110| = 10$
   - $|110 - 110| = 0$
   - $|105 - 110| = 5$
   - $|95 - 110| = 15$
   - $|115 - 110| = 5$
   - $|125 - 110| = 15$
3. Find the mean of the absolute deviation: Add up all the absolute deviations and divide by the 7: $(10 + 10 + 0 + 5 + 15 + 5 + 15) \div 7 \approx 8.57$
   So, the mean absolute deviation is 8.57

Quartiles and Outlier

Quartiles: Quartiles divide a sorted dataset into four equal parts, helping to understand data distribution.

- **Q1 (First Quartile)** → $25th$ percentile (median of the lower half)
- **Q2 (Second Quartile)** → $50th$ percentile (median of the dataset)
- **Q3 (Third Quartile)** → $75th$ percentile (median of the upper half)

Interquartile Range (IQR): The IQR measures the spread of the middle 50% of the data. It's calculated as: $IQR = Q_3 - Q_1$

Outliers: Outliers are data points that are much higher or much lower than most of the data. They don't "fit in" with the rest of the data. Outliers are detected using the IQR rule:

- Lower Boundary: $Q1 - 1.5 \times IQR$
- Upper Boundary: $Q3 + 1.5 \times IQR$

Any value below the Lower Bound or above the Upper Bound is an outlier.

Example:

Consider the following sorted dataset, calculate the quartiles, the interquartile range, and determine if there are any outliers: $2, 5, 7, 10, 12, 15, 18, 22, 25, 30$

Solution:

1. Find quartiles ($Q1, Q2, Q3$):
   - Find Q_2 (Median): The dataset has 10 numbers, so the median is the average of the $5th$ and $6th$ numbers: $Q_2 = \frac{12+15}{2} = 13.5$
   - Find Q_1 (First Quartile - $25th$ Percentile): The lower half: $2, 5, 7, 10, 12$ so, the median of this list is 7
   - Find Q_3 (Third Quartile - $75th$ Percentile): The upper half: $15, 18, 22, 25, 30$, so the median of this list is 22
2. Find Interquartile Range (IQR): $Q_3 - Q_1 = 22 - 7 = 15$
3. Find Outlier Boundaries:
   - Lower Bound: $Q1 - 1.5 \times IQR = 7 - (1.5 \times 15) = 7 - 22.5 = -15.5$. No numbers are less than -15.5, so no upper outliers.
   - Upper Bound: $Q3 + 1.5 \times IQR = 22 + (1.5 \times 15) = 22 + 22.5 = 44.5$. No numbers are greater than 44.5, so no upper outliers.

Box and Whisker Plots

A Box and Whisker Plot is a way to show data visually. It helps us understand how the numbers in a dataset are spread out—where most of the numbers are, if there are any extreme values (outliers), and the range of the data.

Parts of a Box and Whisker Plot:

1. **The Box:**
   - The box shows the middle 50% of the data (between the first quartile Q_1 and the third quartile Q_3).
   - The line inside the box represents the median (the middle number of the data).
2. **The Whiskers:**
   - These are the "lines" extending out from the box.
   - The whiskers show the smallest (minimum) and largest (maximum) values within a certain range of 1.5 times the IQR
3. **Outliers:**
   - If there are data points that are much smaller or larger than the rest of the data, they are called outliers. They are often marked as dots or stars outside the whiskers.

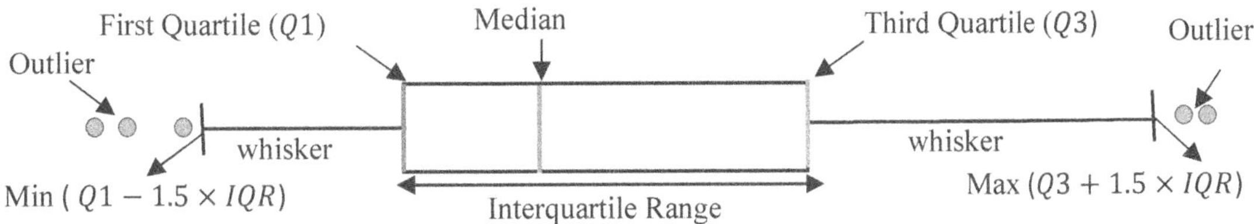

Example: Create a box and whisker plot for following dataset.

$-1, 5, 8, 9, 11, 15, 20, 21, 32, 33, 36, 82$

Solution:

- Find the quartiles: Using the prosses of finding quartiles in previous page we have: $Q_1 = 8.5,\ Q_2 = 17.5$ and $Q_3 = 32.5$
- Calculate the Interquartile Range (IQR): $Q_3 - Q_1 = 32.5 - 8.5 = 24$
- Identify outlier boundaries: Lower boundary: $8.5 - 1.5 \times 24 = -27.5$ and upper boundary: $32.5 + 1.5 \times 24 = 68.5$
- Draw the box and whisker plot:

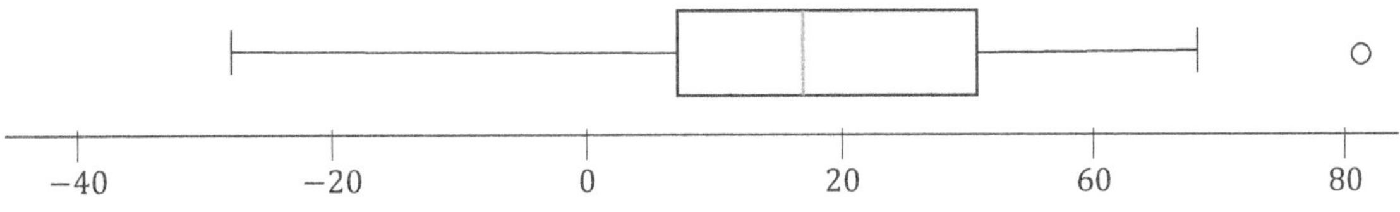

Scatter Plots

A **scatter plot** (also called a scatter diagram or scatter graph) is a type of graph used to display and analyze the relationship between two variables. It's a great tool for spotting trends, patterns, and even correlations in data.

Key Features of a Scatter Plot

1. **Axes**: A scatter plot has two axes:
    - ***X*-axis** (horizontal): Represents one variable.
    - ***Y*-axis** (vertical): Represents the other variable.
2. **Data Points**: Each point on the graph represents a pair of values from the dataset. The position of the point is determined by the values of the two variables.

How to Interpret a Scatter Plot

1. **Positive Correlation:** If the points trend upward from left to right, it means as one variable increases, the other does too. For example, more hours studied → higher exam scores.
2. **Negative Correlation:** If the points trend downward, it means, as one variable increases, the other decreases. For example, more screen time → lower grades.
3. **No Correlation**: If the points are scattered randomly with no clear pattern, there's likely no relationship between the variables.
4. **Clusters and Outliers:**
    - A cluster is when many points are grouped together.
    - An outlier is a point that stands far away from the others, indicating something unusual.

Example:

A teacher is interested in seeing if there's a relationship between the number of hours student's study and their test scores. Draw the scatter plot related to following data:

Hours Studied	1	2	3	4	5	6	7	8	9	10
Test Score	50	55	58	60	61	70	72	73	80	81

Solution:

Once the data is plotted, you should notice the points form a pattern that trends upward from left to right. This suggests a positive correlation—as the number of hours studied increases, test scores tend to increase as well.

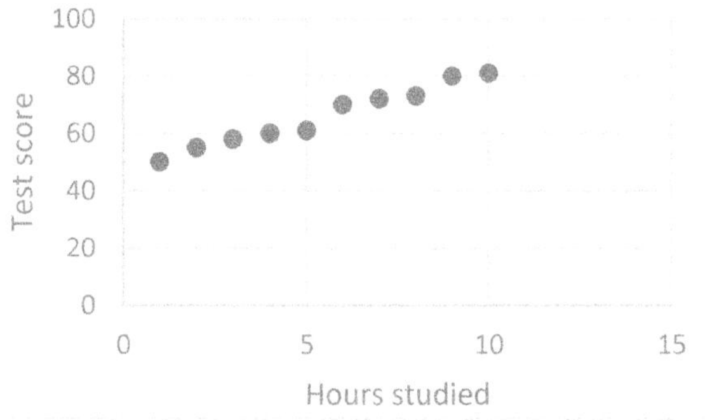

Correlation and Causation

In math and statistics, **correlation** and **causation** are two related but very different ideas when we're looking at relationships between variables.

Correlation means that two variables are related in some way—when one changes, the other tends to change too.

- It could be a **positive correlation** (both go up together); example: The more you exercise, the more calories you burn. Scatter plots for these would show points going upward from left to right.
- Or a **negative correlation** (one goes up, the other goes down); example: The more time you spend on your phone at night, the less sleep you usually get. Scatter plots here would show points going downward from left to right.
- Or **no correlation** (they don't follow any pattern together); example: The color of your shirt and the speed of your internet. Scatter plots for these would look random — the points are just all over.

Causation: Causation means that one variable directly affects or causes a change in another. If A causes $B \rightarrow$ changing A will change B. For example, turning up the heat on a stove causes water to boil faster. That's causation.

Key Difference:

- Correlation = "These things happen together."
- Causation = "This thing makes that happen."

Example:

A school tracks data over a semester and finds a strong correlation between students who bring lunch from home and higher grades. Students who bring lunch from home score an average of 90%, while those who eat cafeteria lunch average 75%. Does this mean that bringing lunch from home causes higher grades? Why or why not?

Solution:

Yes, there is a correlation, students who bring lunch from home tend to have higher grades. That means these two things happen together. However, it is not necessarily a causation and there might be other possible explanations:

- Maybe students who bring lunch from home have parents who are more involved, and that leads to better study habits.
- Maybe students who care more about nutrition also care more about school.
- Maybe students who eat cafeteria food are more rushed or distracted during lunch.

So, while there is a correlation, there is no proof of causation.

Correlation Coefficients

Correlation Coefficients are numbers that measure how strong and in what direction two things are connected.

- The scale goes from -1 to $+1$:
  - $+1$: Strong positive connection (when one goes up, the other also goes up).
  - -1: Strong negative connection (when one goes up, the other goes down).
  - 0: No connection (the two things aren't related).
- Example: If you look at hours spent on video games and grades in school:
  - A negative correlation might show that playing too many video games leads to lower grades.
  - But remember, this doesn't mean games cause bad grades, they're just related.

The following correlation graphs show examples of different range of values for a correlation coefficient:

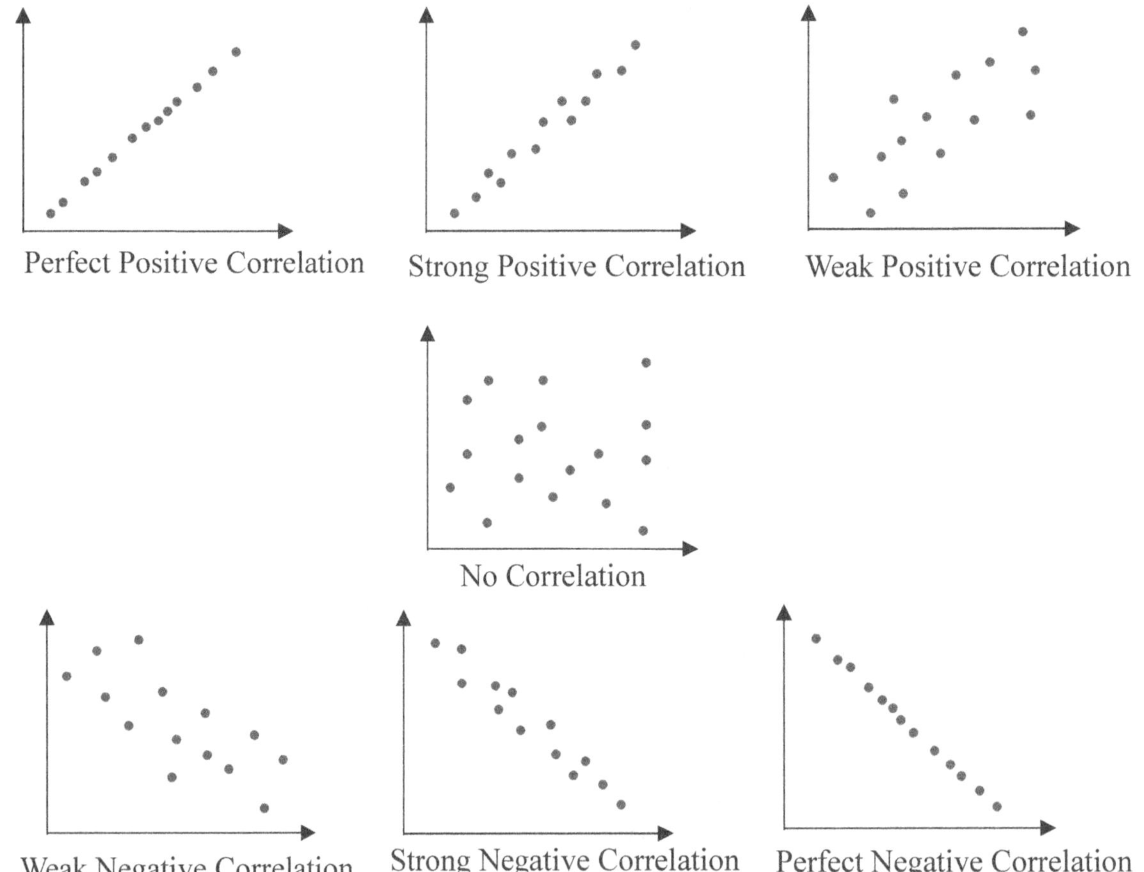

Line of Best Fit

The line of best fit, also called a trend line, is a straight line that best represents the relationship between two variables on a scatter plot. It helps to summarize the data and allows us to make predictions.

How to Place the Line of Best Fit

1. **Draw a Scatter Plot:** Start with your data as a scatter plot (points representing pairs of values).
2. **Look at the Pattern:** Observe the general trend of the points:
   - Are the points going upward? (positive correlation)
   - Are the points going downward? (negative correlation)
   - Are the points scattered randomly? (no correlation)
3. **Place the Line:**
   - The line should follow the trend of the points as closely as possible.
   - Try to have roughly the same number of points above and below the line.
   - For precise placement, use a mathematical formula (least squares regression) to calculate the equation of the line.

How to Interpret the Line of Best Fit

1. **Slope:** The slope tells you how much the dependent variable (y) changes for every one-unit increase in the independent variable (x).
2. **Y-Intercept:** The point where the line crosses the y-axis. It represents the value of y when $x = 0$.
3. **Use for Predictions:** You can use the line's equation ($y = mx + b$) to predict y values for given x values.
4. **Strength of the Fit:**
   - If the points are close to the line, the fit is strong, meaning the data closely follows the trend.
   - If the points are far away from the line, the fit is weaker, meaning there's more variation in the data.

Example:

In the example on the previous page, draw the line of best fit for data.

Solution:

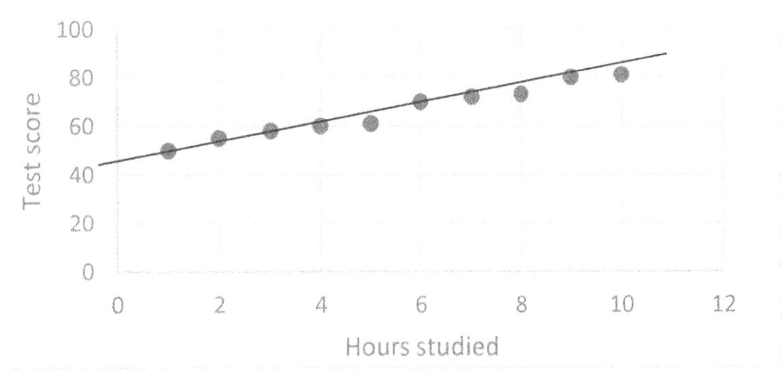

Equation of a Line of Best Fit

Writing the equation of a line of best fit involves finding a straight-line equation that best represents the trend in your data.

Steps to Write the Equation:

To write the equation of the line of best fit (also called the least squares regression line), follow these steps:

1. **Find the Slope (m):** The formula for the slope is: $m = \frac{n\sum xy - \sum x \sum y}{n\sum x^2 - (\sum x)^2}$; Where:

   n= number of data points

   $\sum xy$ = sum of the products of corresponding x and y values

   $\sum x$ = sum of all x values

   $\sum y$ = sum of all y values

   $\sum x^2$ = sum of the squares of x values

2. **Find they-Intercept (b):** The y-intercept is calculated as: $b = \frac{\sum y - m\sum x}{n}$
3. **Write the Equation:** The equation of the line of best fit is written in the form: $y = mx + b$

Example:

A company tracks how much they spend on advertising (x, in \$1,000s) and the corresponding revenue (y, in \$1,000s). The data is shown below. Find the equation of the line of the best fit.

Advertising Budget (\$1,000s)(x)	2	4	6	8	10
Revenue (\$1,000s)(y)	30	45	50	65	80

Solution:

1. Calculate sums needed:
   - $\sum x = 2 + 4 + 6 + 8 + 10 = 30$
   - $\sum y = 30 + 45 + 50 + 65 + 80 = 270$
   - $\sum x^2 = 4 + 16 + 36 + 64 + 100 = 220$
   - $\sum xy = 60 + 180 + 300 + 520 + 800 = 1860$
2. Calculate slope and intercept:

   $m = \frac{(5)(1860)-(30)(270)}{(5)(220)-(30)^2} = \frac{9300-8100}{1100-900} = 6$ and $b = \frac{270-(6\times 30)}{5} = \frac{270-180}{5} = 18$
3. Write the equation: $y = 6x + 18$, it means that for every additional \$1,000 spent on advertising, the company expects an increase of \$6,000 in revenue.

Two-Way Frequency Tables

A **two-way frequency table** is a table that organizes data into two categories, displaying the frequency (count) of occurrences for different combinations of those categories. It helps in analyzing relationships between two categorical variables.

Structure of a Two-Way Frequency Table:

A two-way table typically has:

- **Rows:** Represent one category.
- **Columns:** Represent another category.
- **Cells:** Show the frequency (count) of occurrences.
- **Totals:** Row and column sums (called marginal totals).

Steps to Create a Two-Way Frequency Table:

Example Scenario:

A school surveyed 40 students about their preferred mode of transport (Bus or Car) and grade level (Middle or High School). build the two-way frequency table.

Solution:

Step 1: Identify the Two Categorical Variables: In this case:

1. Mode of Transport → (Bus, Car)
2. Grade Level → (Middle School, High School)

Step 2: Collect Data: Suppose the survey results are:

- Middle school: 15 take the bus, 5 take a car.
- High school: 10 take the bus, 10 take the car.

Step 3: Create the Table Layout: The table should have one category for rows and one for columns, with a "Total" column and row.

	Bus	**Car**	**Total**
Middle School	15	5	20
High School	10	10	20
Total	25	15	40

Step 4: Interpret the Table:

- 25 students take the bus, and 15 take a car.
- 20 Middle school students and 20 high school students were surveyed.

Two-Way Relative Frequency Tables

Two-way relative frequency tables are an extension of two-way frequency tables; except they show proportions or percentages instead of raw counts. These tables help analyze how the frequencies of two categorical variables relate to each other in relative terms. They are especially useful for comparing data when the sample sizes are different or when percentages are easier to interpret than counts.

Key Features:

1. **Relative Frequencies**: Each cell contains a proportion or percentage calculated from the original frequencies.
2. **Types of Relative Frequencies**:
   - **Joint Relative Frequency**: Each cell divided by the grand total (total number of observations). Example: "What percentage of middle school students are those who take the bus?"
   - **Marginal Relative Frequency**: Row or column totals divided by the grand total. Example: "What percentage of all students take the bus?"
   - **Conditional Relative Frequency**: Each cell divided by the row or column total (instead of the grand total). Example: "What percentage of middle school students take the bus?"

Example:

A school surveyed 50 students about their favorite type of movie. The results are:

	Action	Comedy	Total
Male	15	10	25
Female	8	17	25
Total	23	27	50

Convert this into a relative frequency table.

Solution:

Step 1: Joint Relative Frequency (Divide by Grand Total 50): Each value is divided by 50 (grand total):

	Action	Comedy	Total
Male	$15/50 = 0.30 = 30\%$	$10/50 = 0.20 = 20\%$	$25/50 = 0.50 = 50\%$
Female	$8/50 = 0.16 = 16\%$	$17/50 = 0.34 = 34\%$	$25/50 = 0.50 = 50\%$
Total	$23/50 = 0.46 = 46\%$	$27/50 = 0.54 = 54\%$	$1.00 = 100\%$

Interpretation: 30% of all students are males who like action movies.46% of all students prefer action movies overall.

Step 2: Conditional Relative Frequency (Divide by Row Total): To find the percentage within gender groups, divide each value by the row total:

	Action	Comedy	Total
Male	$15/25 = 0.60 = 60\%$	$10/25 = 0.420 = 40\%$	$1.00 = 100\%$
Female	$8/25 = 0.32 = 32\%$	$17/25 = 0.68 = 68\%$	$1.00 = 100\%$

Interpretation: 60% of males prefer action movies. 68% of females prefer comedy movies.

Real World Applications

Data and graphs are used everywhere to help people make informed decisions, spot trends, and communicate complex information visually. Below are some real-world applications in different fields:

1. **Business and Economics:** Businesses use statistics to analyze market trends, consumer behavior, and sales patterns. Tools like scatter plots and regression (line of best fit) help businesses forecast sales based on advertising spend, while measures like interquartile range are used to identify outliers in profit margins.
2. **Healthcare and Medicine:** Doctors and researchers rely on statistics to summarize patient health data, such as average recovery time (mean) and variations (mean absolute deviation). Box-and-whisker plots help identify outliers in clinical trials, and frequency tables aid in understanding correlations, like smoking habits and disease prevalence.
3. **Education:** Schools and policymakers analyze test scores using central tendencies (mean and median) and spread (interquartile range) to assess student performance. Two-way frequency tables might be used to analyze the relationship between study hours and exam grades.
4. **Environmental Science:** Researchers use scatter plots and regression analysis to find relationships between variables like CO2 levels and global temperatures. Quartiles and IQR help categorize data, such as rainfall or pollution levels, while box-and-whisker plots provide a clear visual summary of seasonal variations.
5. **Social Sciences and Surveys:** Surveys benefit from two-way frequency tables to explore relationships between demographic groups and opinions (e.g., age vs. voting preferences). Central tendency and dispersion measures summarize survey data effectively.

Example: A healthcare researcher wants to analyze whether daily exercise (in minutes) affects a person's blood pressure (mmHg). The goal is to determine if more exercise helps lower blood pressure. The researcher collects data from 10 patients as the following table:

Daily Exercise (Minutes) (X)	10	20	30	40	50	60	70	80	90	100
Blood Pressure (mmHg) (Y)	150	145	140	138	135	130	128	125	122	120

Solution

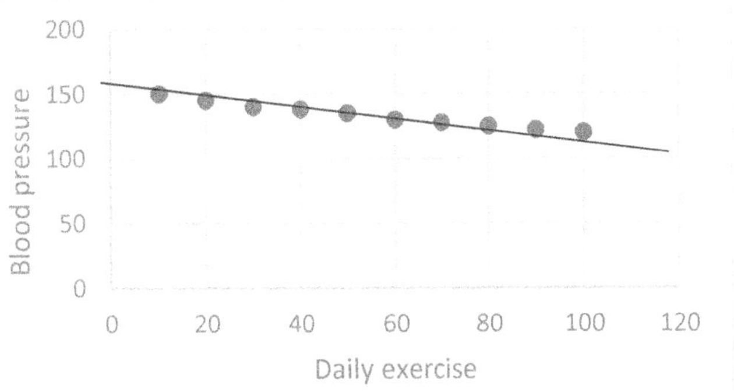

1. Plot the data on a scatter plot: As exercise time increases, blood pressure decreases, this suggests a negative correlation.
2. Find the Line of Best Fit: After calculating the slope and the y-intercept, we get:

   $$y = -0.35x + 153$$

3. Use the Equation to Predict Blood Pressure: If a patient exercise 45 minutes per day, their predicted blood pressure is: $y = -0.35(45) + 153 = 137.25$ so, a person who exercises 45 minutes per day is expected to have a blood pressure of about 137 mmHg.

Worksheets

Change in Mean, Median, Mode and Range

Find the answers.

1) A dataset consists of the following numbers: $[12, 15, 17, 21, 24]$. If three new values—$18, 20$, and 19—are added, what will the new median be?
2) The weights of five dogs in kilograms are: $10, 12, 14, 16$, and 18. If another dog weighing 20 kg is added, how does the mean weight change?
3) If all the numbers in a dataset are multiplied by a number, how will the mean, median mode and range change?
4) A teacher calculates the final grades of students using the following weights:
   - Homework (40%): Scores = $[85, 90, 95]$
   - Quizzes (30%): Scores = $[80, 85]$
   - Final Exam (30%): Score = $[90]$

   What is the overall weighted mean score for the student? If the teacher increases the weight of the Final Exam to 50% while equally reducing the weights of Homework and Quizzes, how does the mean change?
5) If Sara's grades in Geography, Science, Math, and Biology are $76, 98, 92$, and 69, what grade should she get in Chemistry so that the average remains unchanged?

Mean Absolute Deviation

Calculate.

1) Calculate the mean absolute deviation of $15, 21, 11, 7, 19, 23$ and 9.
2) The dataset is: $[10, 10, 10, 10, 10]$. What is the mean absolute deviation?
3) The dataset is: $[-5, -3, 0, 3, 5]$. Calculate the mean absolute deviation.
4) The dataset is: $[100, 150, 240, 280, 310]$. Calculate the MAD. If all values are reduced by 20%, how does the MAD change?
5) A dataset contains the following daily temperatures in degrees Celsius over 7 days: $[15, 18, 20, 22, 25, 30, 35]$. Calculate the MAD. How would the MAD change if the hottest day ($35°C$) were removed from the dataset?

Quartiles and Outlier

Find the answers to the following questions.

1) Find the quartiles (Q_1, Q_2, Q_3) of the following data set: $\{4, 8, 10, 15, 20, 22, 25, 30\}$.
2) Calculate the interquartile range (IQR) of this data set: $\{11, 14, 20, 29, 29, 31, 35, 48\}$.
3) A researcher collects data: $\{4.5, 5.2, 5.8, 6.4, 6.9, 7.3, 7.8, 8.2, 8.9, 9.5\}$. Calculate Q_1, Q_3, IQR, and check if any value exceeds the outlier boundaries.
4) A dataset has $Q_1 = 23, Q_3 = 45$, and $IQR = 22$. An additional value 10 is added. Determine if this new value is an outlier.
5) A teacher records the test scores of 15 students as follows: $42, 55, 60, 61, 65, 70, 72, 75, 78, 80, 85, 88, 90, 92, 95$

   If the teacher gives a 5-point bonus to all students, how will Q_1, Q_3, IQR change?

**Box and Whisker Plots**

Based on the charts below, identify the approximate five-number summary.

1)

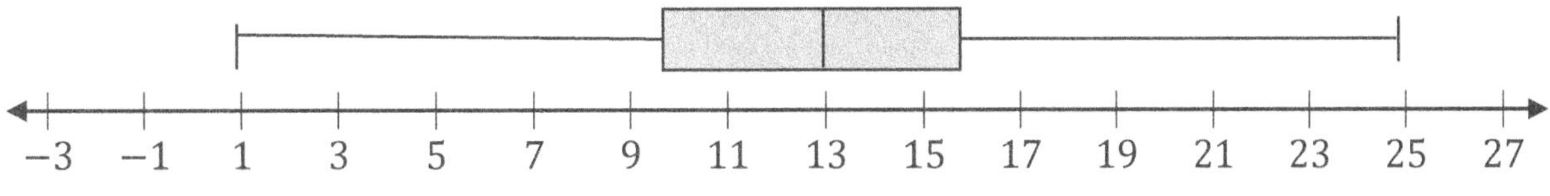

2)

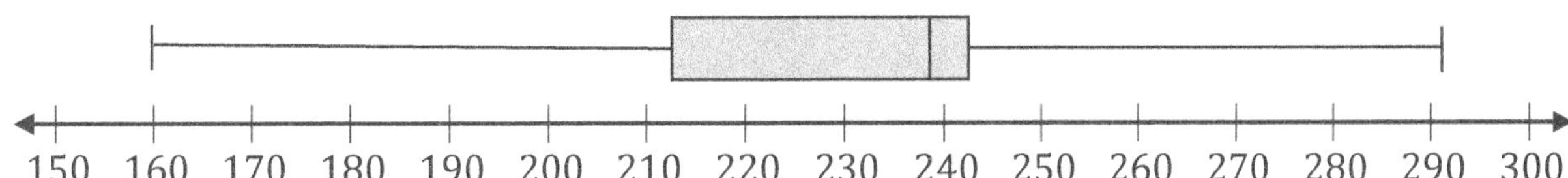

3)

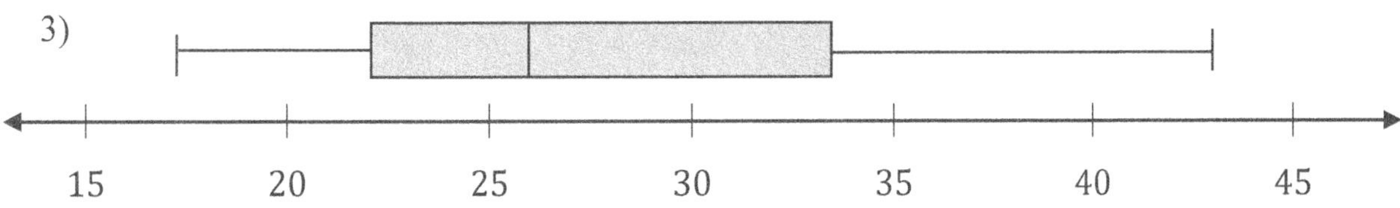

Create the box and whisker plot for following dataset.

4) $1, 15, 23, 35, 40, 46, 128$

5) $120, 139, 642, 792, 1520, 2700$

Scatter Plots

Create the scatter plot to show the following data.

1) The relationship between the amount of water and the times to takes to boil:

Water (cups)	1	2	5	6	8	9	10
Time (min)	2	4	10	12	16	18	20

2) The relationship between the temperature and hot chocolate sales in a café:

Temperature (°C)	−2	0	3	7	12	18	22
Hot chocolate sales (cups)	120	110	85	42	38	25	18

3) The relationship between the height of players and number of baskets in a basketball team:

Height (in)	43	43	48	49	54	54	59	60
Baskets	4	6	5	5	4	6	7	6

Answer the following questions according to the scatter plot.

Maria is learning how to make origami by watching YouTube videos. This scatter plot shows the numbers of videos she watched and how many origamis she finished over the past 6 weeks:

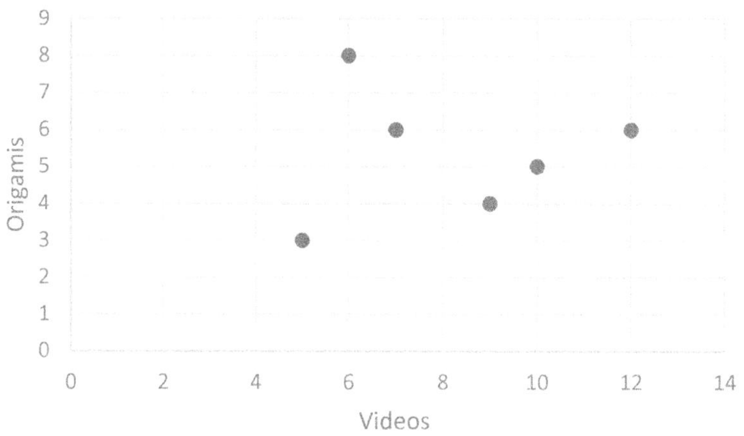

4) During how many weeks did Maria watch more than 8 videos?

Wonderland Amusement Park just opened for the season. This scatter plot shows the number of visitors and the average wait time per ride each day last week.

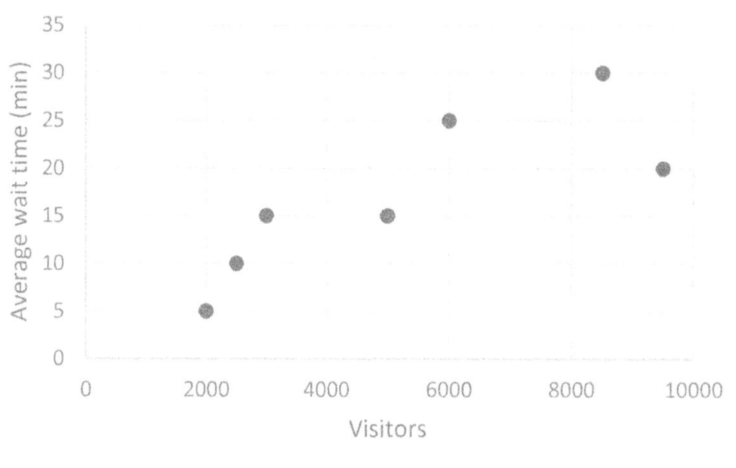

5) How many visitors waited more than 20 minutes for each ride during this week?

Correlation and Causation

Identify the relationship between the variables.

1) A survey found that children who drink more milk tend to grow taller. Does this indicate causation?
2) A graph shows an increase in ice cream sales as outdoor temperatures rise. Is this a correlation or causation?
3) A study finds that people who own cats tend to sleep more hours per night. Suggest a possible third factor that might explain this correlation.
4) Compare the following scenarios: (a) People who eat breakfast regularly perform better in school and (b) People who watch more TV score lower on fitness tests. Are these examples of positive or negative correlation?

5) Two graphs show (a) the increase in global internet users and (b) the increase in global temperatures over the same period. Is there a correlation? Could causation exist?

Correlation Coefficients

1) Which of the following statemets shows a negative correlation?
   a) Studying Hours and test score
   b) Temperature and sales of hot chocolate
   c) Exercise and fitness level
   d) Height and weight
2) A scatter plot shows that as the number of hours watching TV increases, the math test score decreases. What kind of correlation is this?
3) A dataset has a correlation coefficient of -0.87. Which of this best describes the strength and direction of the relationship?
   a) Weak negative
   b) Strong negative
   c) Weak positive
   d) Strong positive
4) A class finds that the correlation between shoe size and reading score is 0.05. What does this tell you about the relationship?
5) If a scatter plot of two variables is shaped like a U-curve (e.g., a parabola), what would the correlation coefficient likely be?
   a) Close to $+1$
   b) Close to -1
   c) Close to 0
   d) Cannot be determined

Line of Best Fit

Do the following problems.

1) In a dataset of shoe sizes and heights, the line of best fit shows a positive slope. What does this suggest about the relationship between shoe size and height?
2) A scatter plot shows the relationship between the number of products sold and advertising budget. The line of best fit has the equation $y = 2x + 5$. If the advertising budget is $\$10$, how many products are expected to be sold?
3) You are given the following dataset for height (cm) vs. weight (kg):

Height (cm)	150	160	170	180	190
Weight (kg)	50	55	65	75	85

   Draw a scatter plot of the points and sketch a reasonable line of best fit.
4) A scatter plot shows years of experience vs. salary with the equation of best fit:
$$y = 5000x + 30000$$
   If someone with 8 years of experience earns $\$60{,}000$, does this fit the trend?
5) A scatter plot is created for the age of a tree and its height. Draw a line of best fit for the following data points: $\{(5,2), (10,4), (15,6), (20,8), (25,10)\}$

Writing the Equation of a Line of Best Fit

Find the equation of the line of best fit for below datasets.

1) $(3,2),(3,3),(4,4),(5,6),(6,5),(7,7),(8,9)$
2) $(1,24),(2,21),(3,21),(4,22),(5,26),(6,17),(7,18),(8,16),(9,19)$

Do the following questions.

3) A dataset has two possible lines of best fit:
   - Model 1: $y = 4x + 15$
   - Model 2: $y = 5x + 10$

   For the actual data point $(6,42)$ which model is more accurate?
4) Write the equation of the line of best fit for following scatter plot.

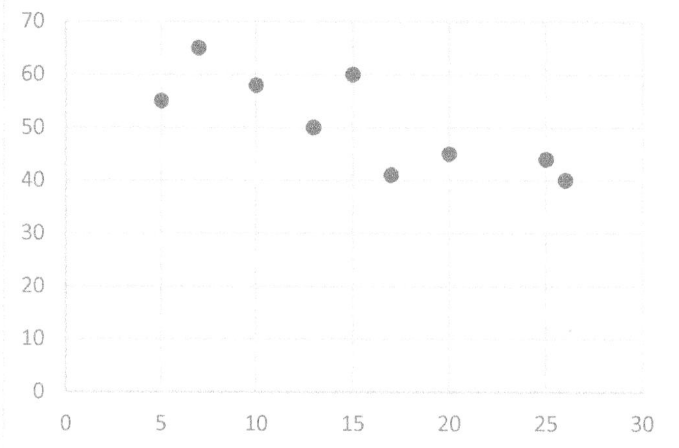

5) What is the equation of the trend line in the scatter plot?

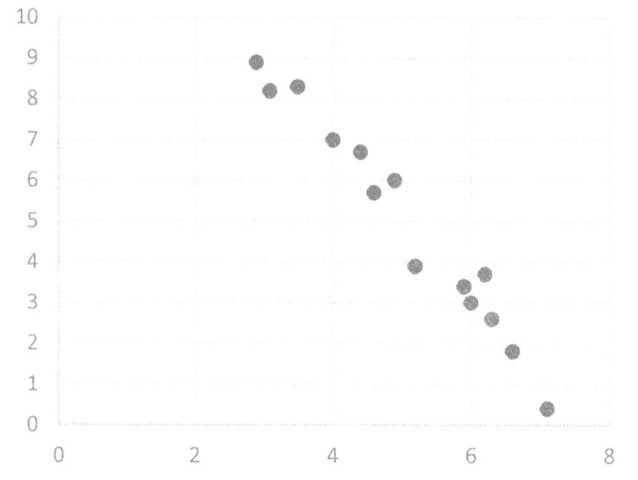

Two-Way Frequency Tables

Do the following problems.

1) A survey asks 30 students about their favorite fruit:
   - 10 likes apples, 5 like bananas and 15 like oranges.
   - 18 are boys, and 12 are girls.

   a) Organize this data into a two-way frequency table.

b) How many girls like bananas

2) A school surveyed 40 students about playing musical instruments:

	Play an Instrument	Does Not Play	Total
Boys	12	8	?
Girls	10	10	?
Total	?	?	?

a) Fill in the missing values.

b) What percentage of students play an instrument?

3) A gym surveyed 250 members to track whether they attend morning or evening workouts based on their age group.

	Morning	Evening	Total
Under 30	?	115	180
30 +	36	?	?
Total	?	?	250

a) Fill in the missing values.

b) If a random member is over 30, what is the probability they prefer evening workouts?

4) A company surveyed 500 employees about their work preference:

	Work from home	Work in office	Total
Managers	80	120	?
Employees	150	150	?
Total	?	?	?

a) Fill in the missing values.

b) If the company hires 200 more employees, how many would you expect to prefer working from home?

5) A university is studying the relationship between students' study habits and their exam performance. They survey 200 students and classify them based on:

1. Study Habits:
   - Consistent Studier (studies regularly)
   - Last-Minute Studier (crams before exams)
2. Exam Performance:
   - Passed (scored 60% or higher)
   - Failed (scored below 60%)

From the survey results:

- 120 students passed the exam.
- 80 students failed the exam.
- 90 students were consistent studiers.
- 50 Last-minute studiers passed the exam.

a) Create a two-way frequency table for this data, filling in the missing values.

b) What percentage of consistent studiers passed the exam?

c) What percentage of students who failed the exam were last-minute studiers?

Two-Way Relative Frequency Tables

Do the following problems.

1) A survey asked students whether they own a pet:

	Has a Pet	No pet	Total
Boys	15	10	25
Girls	20	5	25
Total	35	15	50

   a) Convert this two-way frequency table into a relative frequency table by row.
   b) What percentage of boys have a pet?

2) A survey asks people about coffee or tea preference based on age:

	Coffee	Tea	Total
Under 30	30	20	50
Upper 30	40	10	50
Total	70	30	100

   a) What percentage of people under 30 prefer tea?
   b) What percentage of tea drinkers are 30 or older?

3) A store tracks whether customers pay with cash or card:

	Cash	Card	Total
Weekday	30	70	100
Weekend	50	50	100
Total	80	120	200

   a) Find the row relative frequency for weekday customers.
   b) Find the column relative frequency for card payments.
   c) Are customers more likely to use cash on weekends than on weekdays?

4) A company analyzes customer satisfaction based on response time:

	Satisfied	Not Satisfied	Total
Fast response	60	10	70
Slow response	30	50	80
Total	90	60	150

   Convert this into a relative frequency table by row.

5) A news agency surveys 500 people about their political views:

	Supports Partly A	Support Partly B	Neutral	Total
Urban	120	180	50	350
Rural	80	40	30	150
Total	200	220	80	500

a) Convert this into a relative frequency table.
b) If a random person from an urban area is chosen, what is the probability they support Party B?

Real World Applications

Do the following word problems.

1) A sports team tracks the average number of goals scored per game. If they score 5 extra goals in one game, how does the mean change?
2) A teacher records the math test scores of her class before and after implementing a new study technique.
   Before: $72, 65, 78, 80, 85, 88, 90, 95$
   After: $75, 70, 82, 85, 88, 90, 92, 97$
   - Find the mean of both sets of test scores.
   - How much did the mean change?
3) A group of families in a neighborhood has the following yearly incomes (in thousands of dollars):
   Income: $30, 32, 35, 40, 42, 45, 50, 55, 60, 65$
   a) Find Q_1, Q_2 (median), and Q_3 of the data set.
   b) Compute the interquartile range (IQR).
4) A basketball player records how many hours she practices each week and the number of points she scores in her games.
   Practice Hours: $2, 3, 5, 6, 8, 10$
   Game Points: $10, 12, 20, 22, 28, 30$
   a) Create a scatter plot for this data.
   b) Draw a line of best fit and describe the relationship.
   c) If she practices 7 hours, predict how many points she might score.
5) An online store tracks its ad spending, and the number of purchases made:
   Ad Spend ($\$1000s$):$1, 2, 3, 4, 5, 6$
   Purchases:$10, 18, 25, 35, 40, 50$
   a) Create a scatter plot.
   b) Draw a line of best fit.
   c) Predict how many purchases would occur if the store spent \$7,000 on ads.
6) A survey asks 50 students whether they prefer burgers or pizza and whether they eat fast food weekly or rarely.
   a) How many students prefer burgers?
   b) What percentage of students eat fast food weekly?
   c) What fraction of pizza lovers rarely eats fast food?

	Weekly	Rarely
Burgers	15	10
Pizza	20	5

Answer of Worksheets

Change in Mean, Median, Mode and Range

1) 18.5
2) The mean increases from 14 to 15
3) All of them (mean, median, mode, and range) are multiplied by the same number
4) The mean increases from 87.5 to 88.125
5) 83.75

Mean Absolute Deviation

1) ≈ 5.14
2) 0
3) 3.2
4) MAD reduces from 72.8 to 58.24, It decreases by 20%, just like the data values.
5) Original MAD$\approx$ 5.51, new MAD$\approx$ 4. MAD decreases when the outlier ($35°C$) is removed.

Quartiles and Outlier

1) $Q_1 = 9, Q_2 = 17.5, Q_3 = 23.5$
2) 16
3) $Q_1 = 5.8$, $Q_3 = 8.2$ and $IQR = 2.4$, No outliers
4) 10 is not an outlier
5) Shifting all values adds to Q_1 and Q_3 equally, so IQR stays constant

Box and Whisker Plots

1) $Min \approx 1, Q_1 \approx 9.7, Q_2 \approx 13, Q_3 \approx 15.9$ and $Max \approx 25$
2) $Min \approx 160, Q_1 \approx 213, Q_2 \approx 239, Q_3 \approx 243$ and $Max \approx 291$
3) $Min \approx 18, Q_1 \approx 22, Q_2 \approx 27, Q_3 \approx 34$ and $Max \approx 43$
4) Lower Boundary (Min Boundary)$= -31.5$, $Q_1 = 15, Q_2 = 35, Q_3 = 46$ and Upper Boundary (Max Boundary) $= 92.5$

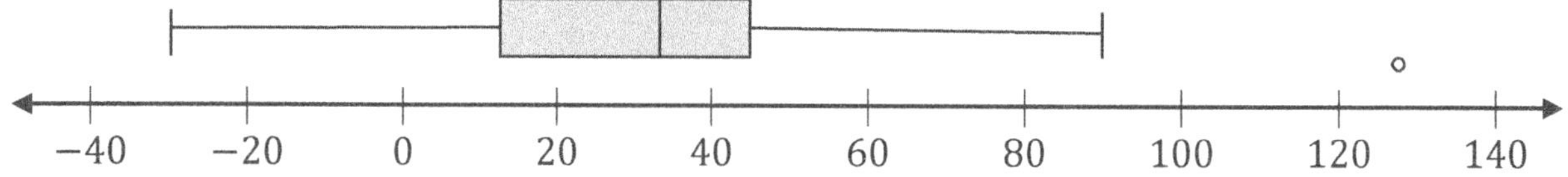

5) Lower Boundary (Min Boundary)$= -1932.5$, $Q_1 = 139, Q_2 = 717, Q_3 = 1520$ and Upper Boundary (Max Boundary) $= 3591.5$

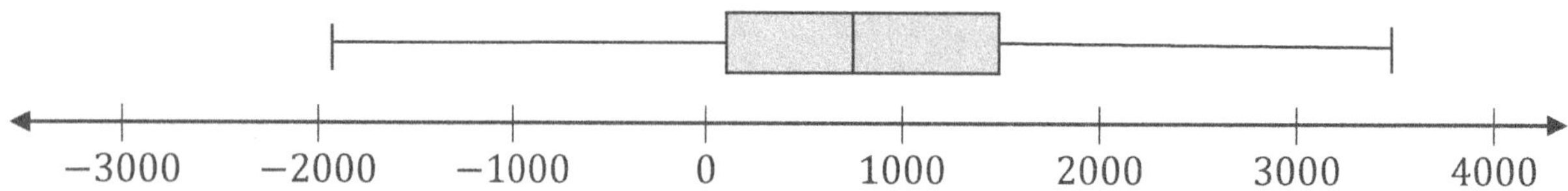

Scatter Plots

1)

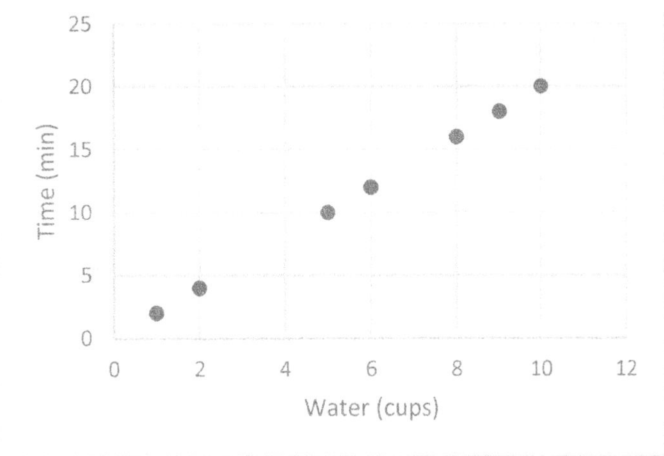

2)

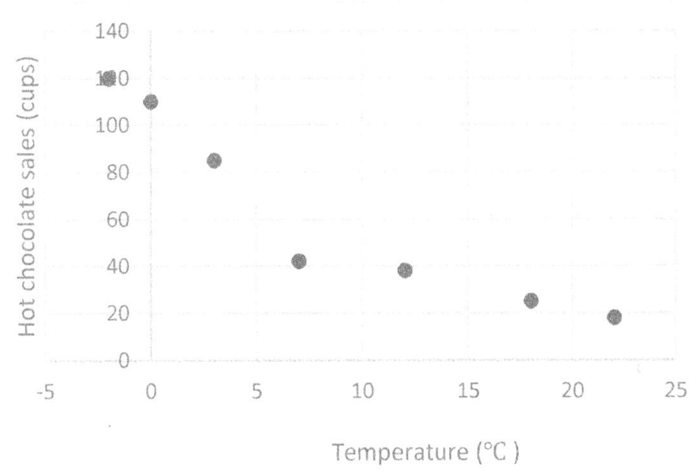

3)

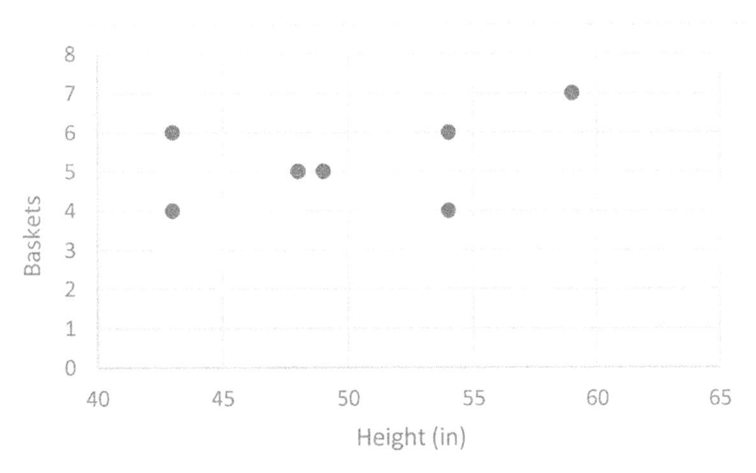

4) 3 weeks
5) 14,500 visitors

Correlation and Causation

1) Drinking more milk and growing taller shows correlation, but not causation. Height is influenced by genetics and overall nutrition, not solely by milk consumption. Milk might be a part of the diet, but it doesn't cause growth.
2) This is a correlation. Warmer temperatures lead to higher ice cream sales, but outdoor temperature doesn't directly cause people to buy ice cream—it's just a preference for cool treats in hot weather.
3) A possible third factor could be lifestyle. Cat owners may prefer a quieter home environment, encouraging better sleep habits.
4) (a) is an example of positive correlation since both variables (eating breakfast and school performance) increase together. (b) is an example of a negative correlation because more TV watching is associated with lower fitness scores.
5) The graphs (internet users vs. global temperatures) show a correlation (both increase over time), but causation doesn't exist. The increase in internet users doesn't affect Earth's climate; instead, both trends happen simultaneously due to independent factors.

Correlation Coefficients

1) Temperature and sales of hot chocolate
2) Negative correlation
3) Strong negative
4) There is almost no relationship
5) Close to 0

Line of Best Fit (Place and Interpreting)

1) There is a positive correlation between shoe size and height. As shoe size increases, height generally increases.
2) 25 products
3)

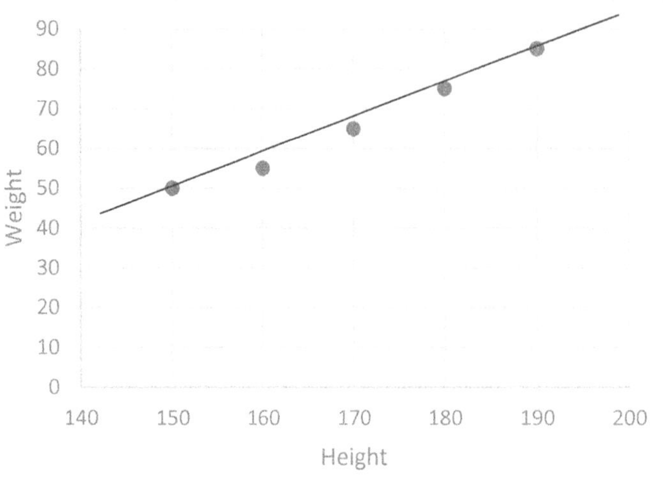

4) No, this salary does not fit the trend.
5) $y = \frac{2}{5}x$

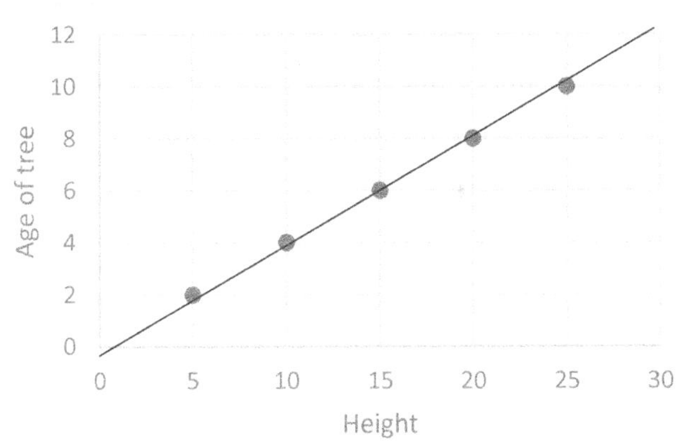

Writing the Equation of a Line of Best Fit

1) $y \approx 1.21x - 1.08$
2) $y \approx -0.85x + 24.69$
3) Model 2
4) We can use two points $(10, 58)$ and $(20, 45)$ for writing the equation: $y = -1.3x + 71$

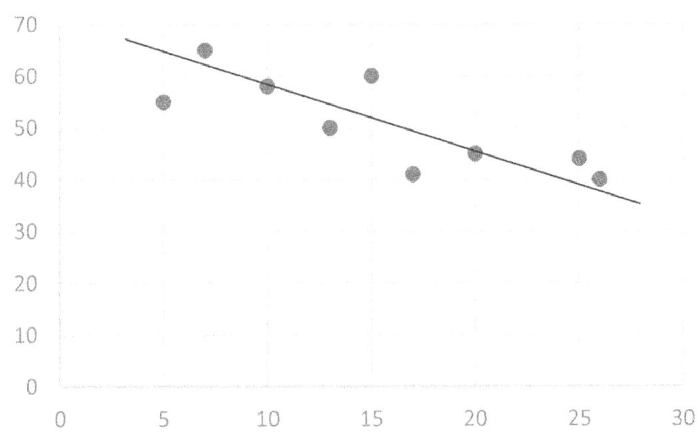

5) We can use two points $(4, 7)$ and $(6, 3)$ for writing the equation: $y = -2x + 15$

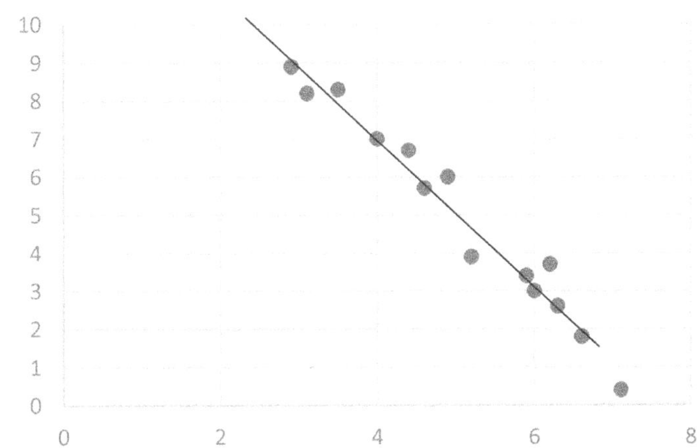

Two-Way Frequency Tables

1) We know the total number of students liking each fruit, but we need to distribute those likes between boys and girls. Let's make reasonable assumptions based on the total number of boys and girls.

a)

	Apples	Banans	Oranges	Total
Boys	6	3	9	18
Girls	4	2	6	12
Total	10	5	15	30

b) 2 girls

2)

a)

	Play an Instrument	Does Not Play	Total
Boys	12	8	20
Girls	10	10	20
Total	22	18	40

b) 55%

3)

a)

	Morning	Evening	Total
Under 30	65	115	180
30 +	36	34	70
Total	101	149	250

b) $\frac{17}{35}$

4)

a)

	Work from home	Work in office	Total
Managers	80	120	200
Employees	150	150	300
Total	230	270	500

b) 92 employees

5)

a)

	Passed	Failed	Total
Consistent Studier	70	20	90
Last-Minute Studier	50	60	110
Total	120	80	200

b) 77.78%

c) 75%

Two-Way Relative Frequency Tables

1) A survey asked students whether they own a pet:

a)

	Has a Pet	No pet	Total
Boys	0.60 (60%)	0.40 (40%)	1.00 (100%)
Girls	0.80 (80%)	0.20 (20%)	1.00 (100%)

b) 60%

2) a): 40% and b): 33.33%
3)
   a) Weekday row relative frequency: Cash: 30% and Card: 70%
   b) Column relative frequency for card: Weekday: 58.33% and Weekend: 41.67%
   c) Cash usage is more likely on weekends than on weekdays.
4)

	Satisfied	Not Satisfied	Total
Fast response	0.8517 (85.71%)	0.1429 (14.29%)	1.000
Slow response	0.3750 (37.5%)	0.6250 (62.5%)	1.000

5)
   a)

	Supports Partly *A*	Support Partly *B*	Neutral	Total
Urban	0.24	0.36	0.10	0.70
Rural	0.16	0.08	0.06	0.30
Total	0.40	0.44	0.16	1.00

   b) $\frac{18}{35} = 0.5143$

Real World Applications

1) Suppose the team has played N games and the current total goal scored is G. So, the mean= $\frac{G}{N}$ and the new mean= $\frac{G+5}{N+1}$. The mean will increase, but the exact change depends on the current total goals and number of games.
2) The mean= 81.625 and the new mean= 84.875. The mean increased by 3.25 points.
3) $Q_1 = 35$, $Q_2 = 43.5$, $Q_3 = 55$ and $IQR = 20$
4) a) Scatter plot:

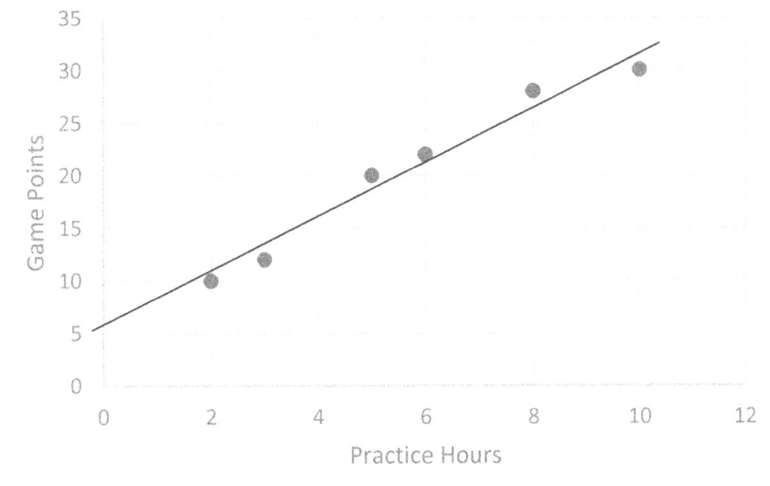

   b) $y \approx 2.53x + 5.07$
   c) Approximately 23 points
5)
   a) Scatter plot.

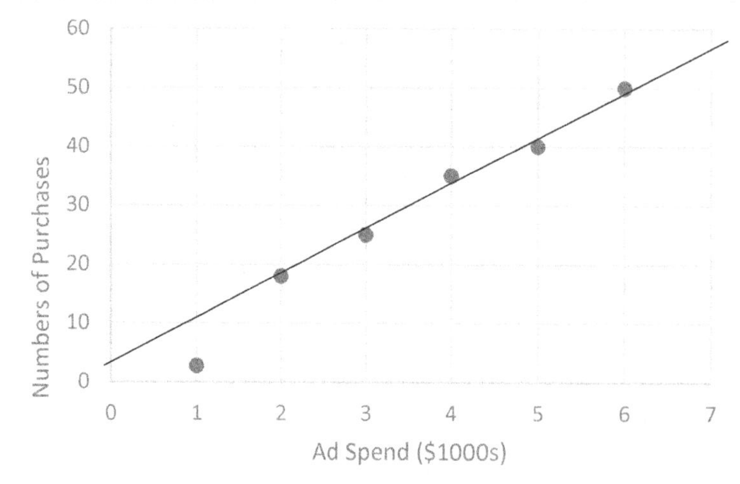

b) $y \approx 8x + 2.14$
c) Approximately 58 purchases

6)

a) 25 students
b) 70% of students
c) $\frac{1}{5}$ of pizza

Chapter 14: Probability

Topics that you'll learn in this chapter:

- ✓ Counting Principle
- ✓ Factorials
- ✓ Permutations and Combinations
- ✓ Probability of Simple events
- ✓ Probability of Opposite, Mutually Exclusive, and Overlapping Events
- ✓ Experimental Probability
- ✓ Make Predictions
- ✓ Probability Using Two-Way Frequency Tables
- ✓ Compound Events: Find the Number of Outcomes
- ✓ Compound Events: Find the Number of Sums
- ✓ Probability of Independent and Dependent Events
- ✓ Real World Applications
- ✓ Worksheets
- ✓ Answer of Worksheets

Counting Principle

The Counting Principle is a fundamental concept in mathematics, used to figure out how many possible outcomes or combinations exist in a situation without having to list all of them.

Concept: If one event has m possible outcomes and another independent event has n possible outcomes, then the total number of outcomes for both events is:

$$m \times n$$

Examples:

1) You are deciding what to eat for lunch. You have 3 sandwich options (chicken, tuna, veggie) and 2 drink options (water, juice). How many possible combinations of sandwiches and drinks can you choose?

   Solution:

   According to the Counting Principle, multiply the number of options for sandwiches by the number of options for drinks: Total combinations $= 3 \times 2 = 6$

   List of combinations:

   - Chicken + Water
   - Chicken + Juice
   - Tuna + Water
   - Tuna + Juice
   - Veggie + Water
   - Veggie + Juice

2) If a password has 3 digits, and each digit can be any number from 0 to 9, what is the total number of possible passwords?

   Solution:

   We have 10 choices in each digit, so the total number of possible passwords is: $10 \times 10 \times 10 = 1000$

Factorials

The factorial of a number is the product of all the whole numbers from that number down to 1. It's written with an exclamation mark (!) after the number. For example, $n!$ means the factorial of n.

Note: By definition, $0! = 1$, even though it doesn't seem like there are any numbers to multiply. This is a special rule in mathematics.

Example: Calculate these factorials: $3!, 5!$ *and* $1!$

Solution:

1. $3! = 3 \times 2 \times 1 = 6$
2. $5! = 5 \times 4 \times 3 \times 2 \times 1 = 120$
3. $1! = 1$

Permutations and Combinations

Permutations (Order matters)

Imagine you're in a race with your friends, and you're trying to figure out who comes in 1st, 2nd, and 3rd place.

All the possible ways they could finish the race are: ABC, ACB, BAC, BCA, CAB and CBA

That's 6 different permutations. In permutations, we care about which order things are in.

☑ If the order matters like positions, rankings, or passwords it's a permutation.

Formula: $Permutation$: $P(n,r) = \frac{n!}{(n-r)!}$; Where:

- $n =$ total number of items
- $r =$ number of items you're choosing

Combinations (Order doesn't matter)

Now let's say you're picking 2 friends to come over and hang out. You don't care about the order just who comes over. If you pick Alice and Ben, it's the same as picking Ben and Alice. So, we don't count both.

Possible combinations of 2 friends from Alice, Ben, and Chloe: AB, AC and BC

☑ If the order doesn't matter like groups, teams, or choosing pizza toppings it's a combination.

Formula: $Combination$: $C(n,r) = \frac{n!}{r! \times (n-r)!}$; Where:

- Still n = total items, and r = how many you choose
- But here, we also divide by $r!$ to remove repeated orders

Examples:

1) Sarah has 5 books and wants to arrange 3 of them on a shelf. In how many different ways can she arrange them?

   Solution:

   The order in which Sarah arranges the 3 books matters. So, we use permutation formula:

   $P(5,3) = \frac{5!}{(5-3)!} = \frac{5\times4\times3\times2\times1}{2\times1} = \frac{120}{2} = 60$

2) If you have 5 friends and you want to pick 3 to invite to your party. In how many different ways can you choose them?

   Solution:

   Order doesn't matter, so this is a combination: $C(5,3) = \frac{5!}{3!\times(5-3)!} = \frac{120}{6\times2} = 10$ ways

Probability of Simple events

As we learned in past years, probability is the chance that something will happen. It tells us how likely an event is. We write probability as a fraction, decimal, or percent.

Simple Event:

A simple event is just one specific outcome. For example, rolling a die and getting a 3, or flipping a coin and getting heads are a simple event.

Formula for the Probability of a Simple Event:

$$Probability = \frac{Number\ of\ favorable\ outcomes}{Total\ number\ of\ possibe\ outcomes}$$

Remember, probability is always between 0 and 1:

- 0 =impossible
- 1 =certain

Example:

A fair 6-sided die is rolled once. What is the probability of each of the following outcomes?

1. Rolling a number greater than 4.
2. Rolling an even number.
3. Rolling a 1 or a 6.

Solution:

1. Rolling a number greater than 4:
   - Number of favorable outcomes= 2 (numbers greater than 4 on a 6-sided die are 5 and 6).
   - Total possible outcomes= 6 (numbers $1, 2, 3, 4, 5, 6$)
   - Probability= $\frac{2}{6} = \frac{1}{3}$
2. Rolling an even number:
   - Number of favorable outcomes= 3 (the even numbers are $2, 4$ and 6).
   - Total possible outcomes= 6 (numbers $1, 2, 3, 4, 5, 6$)
   - Probability= $\frac{3}{6} = \frac{1}{2}$
3. Rolling a 1 or a 6:
   - Number of favorable outcomes= 2 (1 and 6).
   - Total possible outcomes= 6 (numbers $1, 2, 3, 4, 5, 6$)
   - Probability= $\frac{2}{6} = \frac{1}{3}$

Probability of Opposite, Mutually Exclusive, and Overlapping Events

Opposite Events: Opposite events (also called complementary events) are two things that cover all possibilities one must happen if the other doesn't.

Formula: $P(Not\ A) = 1 - P(A)$

Mutually Exclusive Events: Mutually exclusive events are events that cannot happen at the same time. For example, we can't roll a 2 and a 5 at the same time so these events are mutually exclusive.

Formula: $P(A\ or\ B) = P(A) + P(B)$

Overlapping Events: Overlapping events are events that can happen at the same time. For example, in a class, Event A is when a student plays basketball and Event B is when a student plays soccer. Some students might play both sports, so these events overlap.

Formula:

$$P(A\ or\ B) = P(A) + P(B) - P(A\ and\ B)$$

Examples:

1) A school bus arrives on time 4 out of every 5 school day. What is the probability that the bus will not arrive on time on a randomly chosen school day?

   Solution:

   $P(on\ time) = \frac{4}{5} \rightarrow P(not\ on\ time) = 1 - \frac{4}{5} = \frac{1}{5}$

2) A card is drawn from a standard deck of 52 cards. What is the probability that the card is either a King or a Queen?

   Solution:

   There are 4 Kings and 4 Queens in a deck. You can't draw a King and a Queen at the same time, so these are mutually exclusive events.

$$P(King\ or\ Queen) = P(King) + P(Queen) = \frac{4}{52} + \frac{4}{52} = \frac{8}{52} = \frac{2}{13}$$

3) In a class of 30 students: 12 students like math, 15 students like science and 6 students like both math and science. What is the probability that a randomly chosen student likes math or science?

   Solution: This is an overlapping event because some students like both subjects:

$$P(Math\ or\ Science) = P(Math) + P(Science) - P(Both) = \frac{12}{30} + \frac{15}{30} - \frac{6}{30} = \frac{21}{30} = \frac{7}{30}$$

Experimental Probability

Experimental probability is a way to calculate the likelihood of an event based on actual experiments or trials, rather than theoretical calculations.

Definition: Experimental probability is found by conducting an experiment or observing an event multiple time. You use the results of these trials to estimate the probability.

Formula: $Experimental\ Probability = \frac{Number\ of\ times\ the\ event\ happened}{Total\ number\ of\ trials}$

Example: If we flip a coin 20 times and it lands on heads 12 times. What is the experimental probability of getting heads?

Solution: Using formula: $\frac{12}{20} = \frac{3}{5}$; So, based on the experiment, there's a 3 out of 5 chances of getting heads.

Make Predictions

Making predictions means using what you already know (like patterns, data, or probability) to guess what will probably happen next. In probability we can make predictions using experimental or theoretical probability to estimate how often something will happen in the future.

Steps to Make Predictions Using Experimental Probability:

1. **Do an Experiment and Record Results:** Try the event several times and keep track of how often a certain outcome happens.
2. **Find the Experimental Probability:** Use the formula of experimental probability that mentioned above.
3. **Decide How Many Times You Will Repeat the Event:** This is how many times you plan to do the event in the future.
4. **Make the Prediction:** Multiply the experimental probability by the number of future trials:

$$Preduction = Experimental\ Probability \times Number\ of\ Future\ Trials$$

Example:

Liam tosses a coin 40 times. It lands on heads 26 times. If he plans to toss the coin 100 more times, how many times can he expect it to land on heads based on the experimental probability?

Solution:

- He got 26 heads out of 40 tosses, so: Experimental probability of heads= $\frac{26}{40} = 0.65$
- He's going to toss the coin 100 more times, so prediction= $0.65 \times 100 = 65$

Liam can expect the coin to land on heads about 65 times out of the next 100 tosses, based on his experiment.

Probability Using Two-Way Frequency Tables

As we learnt in previous chapter, A two-way frequency table shows data that compares two different categories. It helps organize and analyze data by showing how often combinations of categories happen.

Example Table:

Here's a two-way table showing students and whether they like math or science.

	Like Math	Don't Like Math	Total
Like Science	12	3	15
Don't Like Science	4	1	5
Total	16	4	20

How to Find Probability Using the Table:

Once you have the table, you can use it to calculate probabilities by dividing the relevant frequency by the total number of observations.

$$Probability = \frac{Number\ of\ favorable\ outcomes}{Total\ number\ of\ outcomes}$$

We will do some examples using the table above:

Examples:

1) What is the probability that a randomly selected student likes both math and science?

   Solution:

   Look at where "Like Science" and "Like Math" intersect → 12 students.

$$P(\text{Like Math and Like Science}) = \frac{12}{20} = 0.6$$

2) What is the probability that a student doesn't like math?

   Solution:

   Total who doesn't like math = 3 (like science) + 1 (don't like science) = 4

$$P(\text{Don't Like Math}) = \frac{4}{20} = 0.2$$

3) What is the probability that a student likes science?

   Solution:

   Total who likes science = 12 + 3 = 15

$$P(\text{Like Science}) = \frac{15}{20} = 0.75$$

Compound Events: Find the Number of Outcomes

Compound events in probability refer to situations where you consider the outcomes of two or more events together. To find the number of outcomes for compound events, you use the counting principle or other methods depending on whether the events are dependent or independent.

Steps to Find the Number of Outcomes:

1. **Identify the events:**
   - For example, flipping a coin and rolling a die are two separate events.
2. **Determine the number of outcomes for each event:**
   - Flipping a coin has 2 outcomes (heads or tails).
   - Rolling a 6-sided die has 6 outcomes $(1, 2, 3, 4, 5, \text{or } 6)$.
3. **Multiply the number of outcomes:**
   - If the events are independent (one doesn't affect the other), you can use the Counting Principle: Multiply the outcomes of each event to find the total number of outcomes.
   - For example: Total outcomes = 2 (coin) $\times$ 6 (die) $=$ 12 outcomes

Examples:

1) How many outcomes are there if you flip a coin twice?

   Solution:

   Each coin flip has 2 outcomes (heads or tails). For two flips: Total outcomes $= 2 \times 2 = 4$ outcomes.

2) A bag contains 4 red marbles, 3 blue marble, and 2 green marble. A student randomly draws one marble, records its color, puts it back in the bag, and then draws a second marble. How many possible outcomes are there for the sequence of two draws (e.g., red then blue, blue then green, etc.)?

   Solution:

   Bag has:4 red marbles, 3 blue marbles and 2 green marbles. Total marbles= $4 + 3 + 2 = 9$

   But we're asked only about **colors**, not which exact marble, so each draw has 3 possible outcomes (red, blue, green), so for two draws:

$$\text{Total outcomes} = 3 \times 3 = 9$$

Compound Events: Find the Number of Sums

Imagine you have two events where numbers are involved (like rolling dice or picking numbers). "Finding the number of sums" means listing or calculating all the possible total values (or sums) that can result from combining the outcomes of these events.

Example with Two Dice:

Suppose you roll two 6-sided dice. Each die has numbers from 1 to 6. The possible sums are created by adding the numbers rolled on the two dice.

1. The smallest sum is $1 + 1 = 2$.
2. The largest sum is $6 + 6 = 12$.

If you list all the possible sums (e.g., $2, 3, 4 \ldots up\ to\ 12$), you can count how many ways each sum can occur:

All the Possible Sums:

SUM	WAYS TO GET IT	TOTAL WAYS
2	$(1,1)$	1 Ways
3	$(1,2), (2,1)$	2 Ways
4	$(1,3), (2,2), (3,1)$	3 Ways
5	$(1,4), (2,3), (3,2), (4,1)$	4 Ways
6	$(1,5), (2,4), (3,3), (4,2), (5,1)$	5 Ways
7	$(1,6), (2,5), (3,4), (4,3), (5,2), (6,1)$	6 Ways
8	$(2,6), (3,5), (4,4), (5,3), (6,2)$	5 Ways
9	$(3,6), (4,5), (5,4), (6,3)$	4 Ways
10	$(4,6), (5,5), (6,4)$	3 Ways
11	$(5,6), (6,5)$	2 Ways
12	$(6,6)$	1 Ways

Example: What is the probability of rolling a sum of 7 with two dice?

Solution:

There are 6 ways to get a sum of 7: $(1,6), (2,5), (3,4), (4,3), (5,2), (6,1)$

Each dice has 6 outcomes so total outcomes: $6 \times 6 = 36$

$$P(sum\ of\ 7) = \frac{6}{36} = \frac{1}{6}$$

Probability of Independent and Dependent Events

Independent Events

- Events are independent if the outcome of one event does not affect the outcome of the other.
- **Probability Rule**: If A and B are independent events, the probability of both occurring is:

$$P(A \text{ and } B) = P(A) \times P(B)$$

Example: Flipping a coin and rolling a die. The coin flip doesn't influence the die roll.

Dependent Events

- Events are dependent if the outcome of one event affects the probability of the other.
- **Probability Rule**: If A and B are dependent events, you calculate the probability of both occurring by taking the dependency into account:

$$P(A \text{ and } B) = P(A) \times P(B \mid A)$$

Here, $P(B \mid A)$ is the conditional probability of B, given that A has already occurred.

Example: Drawing two cards from a deck without replacement. The first drawing affects the probability for the second draw.

Examples:

1) Two dice are rolled. What is the probability of rolling a 4 on the first die and a 6 on the second die?

Solution:

Rolling one die has 6 possible outcomes, and the probability of rolling a specific number is:

$P(4) = \frac{1}{6}, P(6) = \frac{1}{6}$ Since these events are independent, multiply the probability:

$P(4 \; and \; 6) = P(4) \times P(6) = \frac{1}{6} \times \frac{1}{6} = \frac{1}{36}$

2) A box contains 5 blue marbles and 3 red marbles. Two marbles are drawn one after another without replacement. What is the probability of drawing a blue marble first and a red marble second?

Solution:

- The probability of drawing a blue marble first is: $P(blue\ first) = \frac{5}{8}$
- After drawing a blue marble, one blue marble is removed, leaving 7 marbles total (4 blue, 3 red). The probability of drawing red marble next is:

  $P(red\ second \mid blue\ first) = \frac{3}{7}$
- Multiply the probabilities:

$P(blue\ first\ and\ red\ second) = P(blue\ first) \cdot P(red\ second \mid blue\ first) = \frac{5}{8} \times \frac{3}{7} = \frac{15}{56}$

Real World Applications

Probability plays a crucial role in countless real-world scenarios. Here are some diverse applications:

1. **Weather Forecasting:** Meteorologists use probability to predict the likelihood of rain, storms, or other weather conditions based on historical data and atmospheric models.
2. **Medicine and Healthcare:**
   - Probability is used to determine the success rates of treatments and surgeries.
   - In epidemiology, it helps predict the spread of diseases and assess risk factors for populations.
3. **Finance and Insurance:**
   - Financial analysts use probability to evaluate the risks and returns of investments.
   - Insurance companies calculate premiums based on probabilities of accidents, health events, or natural disasters.
4. **Sports:** Probability is utilized to assess team performance, predict match outcomes, and even design fair game strategies.
5. **Gaming and Casinos:** Probability ensures fairness in games like poker, roulette, and lotteries, as well as providing strategies for maximizing wins.
6. **Artificial Intelligence and Machine Learning:** AI algorithms often rely on probability, especially in areas like predictive modeling, natural language processing, and image recognition.
7. **Everyday Decision-Making:** Probability helps us make informed choices, like evaluating the risks of buying lottery tickets or planning trips based on weather predictions.

Example:

At a school fair, Maria buys 1 of 100 raffle tickets. Later, she plays a game where she rolls a fair six-sided die. What is the probability that Maria wins the raffle AND rolls a 6?

Solution:

1. Identify the events:
   - Event A: Maria wins the raffle → probability $= \frac{1}{100}$
   - Event B: She rolls a 6 on a fair die → probability $= \frac{1}{6}$

   These are independent events because one does not affect the other. Winning the raffle doesn't change her die roll, and vice versa.
2. Multiply the probabilities: $P(\text{Win raffle and roll a 6}) = P(Win) \times P(Roll\ a\ 6) = \frac{1}{100} \times \frac{1}{6} = \frac{1}{600}$

Worksheets

Counting Principle

Find the answer to each question.

1) A password is made of 1 number from $(1-5)$ and 1 letter from $(A, B, or\ C)$. How many different passwords can be made?
2) A student has a 2-digit locker code. Each digit can be $0-9$. How many different codes are possible?
3) How many 4-digit numbers can be made using the digits $0, 4, 9, 7$, and 6?
4) How many 5-digit even numbers exist without repeating digits?
5) You have 4 shirts and 3 pants. If you can't wear a red shirt with black pants, and there's 1 red shirt and 1 black pants, how many outfit combinations are possible?

Factorials

Calculate the expressions.

1) $6! - 5!$
2) $3! + 4! + 2!$
3) $\frac{7! \times 3!}{6! \times 4!}$
4) $\frac{8!}{6! \times 2!}$

Do the following problems.

5) There are 6 students in line. In how many different ways can they stand in line?
6) 8 friends are sitting around a circular table. How many different seating arrangements are there?
7) How many digits does 10! have?
8) Find the remainder when 9! is divided by 70.
9) Find the sum of the digits of 6!.
10) A group of 7 people want to arrange themselves in a line for a photograph. However, 2 of them refuse to stand next to each other. How many different arrangements are possible?

Permutations and Combinations

Answer the problems.

1) You need to choose 2 friends from a group of 4. How many different pairs can you choose?
2) How many 3-letter passwords can you make from the letters A, B, C, D, E if no letter repeats?
3) There are 7 people, and a committee of 3 people must be selected. How many different committees can be formed?
4) How many ways can the letters in the word "$LEVEL$" be arranged?
5) A Class has 12 students. A group of 4 will be selected for a math competition.
   a) How many different groups of 4 can be made?
   b) If 1 of the 4 must be selected as a team leader, how many different arrangements are possible?

Probability of Simple events

Find the probability.

1) A spinner is divided into 4 equal sections: red, blue, green, and yellow. What is the probability of spinning blue?

2) One letter is chosen at random from the word "*PROBABILITY*." What is the probability that the letter is a vowel?
3) A number card is picked at random from 1 to 20. What is the probability that the number is a multiple of 3?
4) From a standard deck of 52 playing cards, what is the probability of drawing a heart?
5) What is the probability of a four-digit number being even, created using the digits $9, 5, 4$, and 2 without repeating any digit?

Probability of Opposite, Mutually Exclusive, and Overlapping Events

Do the following problems.

1) If the probability of it raining tomorrow is 0.7, what is the probability that it will NOT rain tomorrow?
2) You roll a standard 6-sided die. What is the probability of rolling either a 3 or a 5?
3) In a class of 30 students, 18 like basketball, 12 like soccer, and 8 like both basketball and soccer. What is the probability that a student chosen at random likes either basketball or soccer?
4) A spinner is divided into 6 sections: 2 red, 1 blue, 1 green, and 2 yellow. What is the probability of spinning either red or not red?
5) A card is drawn from a standard deck of 52 cards. What is the probability of drawing a queen or a red card?

Experimental Probability

Find the answers.

1) A card is drawn from a standard deck of 52 cards 100 times, and a face card appears 25 times. What is the experimental probability of drawing a face card?
2) Two dice are rolled 60 times, and doubles (e.g., $1-1, 2-2$) appear 15 times. What is the experimental probability of rolling doubles?
3) A soccer team plays 50 games. They win 30 games, lose 15, and tie 5. What is the experimental probability of a win, loss, and tie? What is the total probability of these outcomes?
4) Over the course of 100 days, it rains 28 days. What is the experimental probability of rain on any given day? Use this to predict how many days of rain might occur in a 365-day year.
5) You roll a fair 6-sided die 50 times and the following outcomes are recorded: 8 ones, 9 twos, 6 threes, 10 fours, 7 fives, 10 sixes. What is the experimental probability of rolling a number less than 5?

Make Predictions

Do the problems.

1) During 50 basketball shots, a player makes 35 successful shots. Predict how many shots they will make in a game with 120 attempts.
2) In a factory quality check, 10 out of 200 products were found to be defective. Predict how many defective products would be found in a batch of 5,000 products.
3) A student flips a coin 200 times and records the results: 120 heads and 80 tails. Based on these results, predict the number of heads the student will get if the coin is flipped 500 times.

4) Over a week of bird observation, 80 birds were sighted, and 32 of them were sparrows. Predict how many sparrows will be sighted in a month if 320 birds are observed.
5) From 100 draws of a shuffled deck, a king appears 7 times. Predict how many kings would appear in 1,000 card draws.

Probability Using Two-Way Frequency Tables

The following table shows the number of students who prefer different types of technology and their grade level. Answer the following questions according to the table.

Type of Technology	6th Grade	7th Grade	8th Grade	Total
Tablet	8	12	10	30
Laptop	10	15	20	45
Smartphone	12	8	5	25
Total	30	35	35	100

1) What is the probability that a student in the 7th grade prefers a laptop?
2) What is the probability that a randomly selected student is either in 7th grade or prefers a tablet?
3) What is the probability that a randomly selected student prefers a smartphone or is in 6th grade?

The following table shows the number of students who prefer different sports and have varying levels of interest:

Sport	High Internet	Moderate Internet	Low Internet	Total
Soccer	10	8	7	25
Basketball	5	15	5	25
Baseball	8	6	6	20
Total	23	29	18	70

4) What is the probability that a randomly selected student has high interest in soccer?
5) What is the probability that a randomly selected student has moderate interest in either soccer or basketball?

Compound Events: Find the Number of Outcomes

Do the following problems.

1) A pizza shop offers 2 crust types (thin, thick) and 3 toppings (pepperoni, mushroom, and cheese). How many different pizzas can be made by selecting one crust and one topping?
2) You roll three dice. How many possible outcomes are there for this compound event?
3) A basketball team has 12 players. The coach must choose 2 players to be team captains. How many different ways can the coach choose the 2 captains?

4) A patient must take 1 type of pill (out of 5 options) and can choose to take it either once or twice a day. There are also 3 different meal timing options (before meals, after meals, or with meals). How many different medication plans are possible for this patient?
5) A baby can be born as either a boy or a girl and can be born in any of the 12 months of the year. How many different gender + birth month combinations are possible for one baby?

Compound Events: Find the Number of Sums

Answer the question below.

1) You roll two 6-sided dice. How many outcomes give a sum greater than 10?
2) You roll two dice. How many outcomes give an even sum?
3) You flip 3 coins and roll a 6-sided die. How many total outcomes give a sum of 4, where heads = 1, tails = 0, and the die add to the coin total?
4) You draw 2 cards numbered 1 to 5 (no repeats) from a deck. How many different unordered combinations give a sum of 6?
5) From the numbers 1 to 10, two numbers are randomly selected. In how many cases will the sum of the two selected numbers be even?

Probability of Independent and Dependent Events

Find the answers.

1) A family has 4 children. What is the probability that all four children are girls?
2) Two cards are drawn from a deck without replacement. What is the probability that both cards are kings?
3) You flip a coin 3 times. What is the probability of getting exactly 2 heads?
4) A school is planning a field trip, and students are randomly divided into groups for a treasure hunt. Here's the scenario:
   There are 30 students, 12 of whom bring their lunches, and 18 who do not. The organizers decide to randomly select two students (one after another) to check if they brought lunch. What is the probability that:
   a) The first selected student brought their lunch.
   b) Both selected students brought their lunches?
   c) The second selected student brought their lunch, given that the first selected student did not.
5) A jar contains 5 red, 4 blue, and 3 green balls. Two balls are drawn without replacement. What is the probability that the first is red and the second is blue?

Real World Applications

Do the problems.

1) In a car with 5 passengers, where all 5 can drive, how many different ways can they be seated in the car?
2) Alice wants to create a 9 character password using the digits of her birth year (2013) and the letters of her name. How many ways can she create a password where the first 4 characters are digits? (No digit or letter may be repeated in the password)
3) In a factory, 3 types of shoes are produced: casual, formal, and athletic. Each type of shoe is made in 4 different sizes. Casual and athletic shoes are produced in 3 colors, while formal shoes are produced in 2 different colors.
   a) How many different shoes are produced in total in this factory?
   b) What is the probability that a person buys an athletic shoe?

4) In a city with a population of 250,000, we want to determine how many people own a vehicle, how many do not, and the type of vehicle they own. To do this, a sample of 200 people was surveyed, and the following data was obtained: 75 people own a car, 60 people own a motorcycle, 20 people own both a motorcycle and a car, and the rest do not own any vehicle.
   a) Approximately how many people in this city do not own a vehicle?
   b) Approximately how many people own only a car?
   c) Approximately how many people in this city own a car?
5) Each student is either sick or healthy. Suppose that if a student is healthy today, the probability of them being healthy tomorrow is 95%. If a student is sick today, the probability of them being sick tomorrow is 55%. If 20% of the students are sick today, what percentage of students will be healthy tomorrow?
6) A lottery ticket requires selecting 6 numbers out of 49.
   a) How many different lottery tickets can be created?
   b) What is the probability of winning if you select one ticket randomly?
7) The following table shows the number of students who participate in different after-school activities and their favorite subjects:

Activity	Math	Science	Total
Debate Club	12	8	20
Band	10	15	25
Robotics	8	7	15
Total	30	30	60

What is the probability that a student will participate in Band and prefers Science?

8) A shopper has a 1 in 5 chance of winning a discount coupon at the checkout. On a single shopping trip, she checks out at 3 different stores, each with the same 1-in-5 chance, and all outcomes are independent.
   a) What is the probability that she will win at least one coupon?
   b) What is the probability that she will win a coupon at exactly two stores?
9) A company is interviewing 5 candidates for 3 available time slots. Each candidate can only be interviewed once, and no time slot can have more than one candidate.
   a) How many different ways can the company schedule 3 out of 5 candidates in order?
   b) If the first slot is given to a female candidate (and there are 2 females among the 5), what is the probability that the rest of the schedule includes only male candidates?
10) A family of 6 people (2 parents, 4 children) is attending a theater. They want to sit in one row, but the parents must sit next to each other. How many different seating arrangements are possible?

Answer of Worksheets

Counting Principle

1) 30
2) 100
3) 96
4) 13,776
5) 11

Factorials

1) 600
2) 32
3) $\frac{7}{4}$
4) 28
5) 720
6) 5,040
7) 7 digits
8) 0
9) 9
10) 3,600

Permutations and Combinations

1) 6
2) 60
3) 35
4) 30
5) a) 495, b) 1980

Probability of Simple events

1) $\frac{1}{4}$
2) $\frac{4}{11}$
3) $\frac{3}{10}$
4) $\frac{1}{4}$
5) $\frac{1}{2}$

Probability of Opposite, Mutually Exclusive, and Overlapping Events

1) 0.3
2) $\frac{1}{3}$
3) $\frac{11}{15}$
4) 1
5) $\frac{7}{13}$

Experimental Probability

1) 0.25
2) 0.25
3) Win: 0.6, Loss: 0.3, Tie: 0.1 and Total probability:1
4) Experimental probability:0.28 and about 102 days of rain might occur in a 365-day
5) 0.66

Make Predictions

1) 84 successful shots
2) 250 defective products
3) 300 heads
4) 128 spsrrows
5) 70 kings

Probability Using Two-Way Frequency Tables

2) $\frac{3}{7}$
2) $\frac{53}{100}$
3) $\frac{43}{100}$
4) $\frac{1}{7}$
5) $\frac{23}{70}$

Compound Events: Find the Number of Outcomes

1) 6
2) 216
3) 66
4) 30
5) 24

Compound Events: Find the Number of Sums

1) 3
2) 18
3) 8
4) 2
5) 20

Probability of Independent and Dependent Events

1) $\frac{1}{16}$
2) $\frac{1}{221}$
3) $\frac{3}{8}$
4) a) $\frac{2}{5}$, b) $\frac{22}{145}$, c) $\frac{12}{29}$
5) $\frac{5}{33}$

Real World Applications

1) 120
2) 2880 possible passwords
3) a) 32, b) $\frac{3}{8}$
4) a) 106,250 people, b) 68,750 people, c) 93,750 people
5) 85% of students
6) a) 13,983,816, b) $\frac{1}{13{,}983{,}816}$
7) $\frac{1}{4}$
8) a) $\frac{61}{125}$, b) $\frac{12}{125}$
9) a) 60, b) $\frac{1}{2}$
10) 240

Chapter 15: Practice Test

Your Future Starts Now; Let's Get Ready!

This is your chance to practice like it's the *real* AK STAR 8th Grade Math Test. Think of it as a warm-up for the big day, one step closer to reaching your goals, building confidence, and proving what you've learned. You've worked hard to get here, and this practice test is your way to show yourself just how far you've come.

Before You Begin:

- Find a quiet spot, grab your pencil, and have scratch paper ready.
- Don't worry about timing—go at your own pace.
- It's okay to guess if you're stuck. No points are taken off for wrong answers.
- **For 8th grade**, calculators *might* be allowed depending on the section, checking the instructions or asking your teacher.

Tips for Taking the Test:

Here are a few smart moves that can really help:

- **Read every question carefully.** Don't rush; understanding the question is half the battle.
- Use your tools! You might have access to a **ruler, protractor, reference sheet**, and **calculator**. These tools are there to help you—*use them!*
- Stay calm and focused. Take one question at a time. You've got this.

After the Test:

Once you're done, check your answers using the key. Don't skip the review part, this is where real growth happens! If you get something wrong, take a minute to understand why. Every mistake is a chance to get better before the real thing.

Your Success Starts Here

This isn't just a test, it's a step toward high school, college, and your future goals. Every question you answer is part of the plan you're building for yourself. So, take this seriously, give it your best effort, and be proud of how far you've come.

You're not just preparing for a test; you're preparing for your future.

Formula Sheet

AK STAR Mathematics Formula Sheet Grade 8

Below are formulas you may find useful as you work on the problems. However, some of the formulas may not be used. You may refer to this page as you take the test.

Formulas

Formula	
Area of a Triangle	$A = \frac{1}{2}bh$
Area of a Parallelogram	$A = bh$
Area of a Circle	$A = \pi r^2$
Circumference of a Circle	$C = \pi d$, or $C = 2\pi r$

Volume of a Prism	$V = Bh$
Volume of a Cylinder	$V = \pi r^2 h$
Volume of a Sphere	$V = \frac{4}{3}\pi r^3$
Volume of a Cone	$V = \frac{1}{3}\pi r^2 h$
Pythagorean Theorem	$a^2 + b^2 = c^2$

Unit Conversions

1 mile = 5,280 feet
1 mile = 1,760 yards
1 pound = 16 ounces
1 ton = 2,000 pounds

1 cup = 8 fluid ounces
1 pint = 2 cups
1 quart = 2 pints
1 gallon = 4 quarts
1 liter = 1,000cubic centimeters

AK STAR Practice Test

1. Evaluate the expression: $-3^2 + 6 \div (-2) \times 4 - |-5 + 2|$

   A. -19

   B. -24

   C. -17

   D. -25

2. Which of the following fractions is equal to the repeating decimal $0.\overline{36}$?

   A. $\frac{4}{11}$

   B. $\frac{2}{5}$

   C. $\frac{12}{37}$

   D. $\frac{18}{49}$

3. Which of the following is true?

   A. $\sqrt{50} > 7.5$

   B. $\sqrt{2} < 1.3$

   C. $\frac{22}{7} < \pi$

   D. $\sqrt{10} > 2.9$

4. Evaluate the expression: $\frac{-\frac{2}{3}+\frac{1}{4}\div 2}{3\frac{3}{5}\times\frac{10}{21}}$

   A. $-\frac{35}{48}$

   B. $\frac{34}{21}$

   C. $-\frac{91}{288}$

   D. $-\frac{35}{288}$

5. Evaluate the expression: $\frac{1}{\left(-\frac{1}{2}\right)^{-3}}$

   A. $\frac{1}{8}$

   B. $-\frac{1}{8}$

   C. 8

   D. -8

6. Write the following expression in scientific notation: $4200000 - 1.7 \times 10^5$
   A. 2.5×10^6
   B. 4.03×10^5
   C. 2.5×10^5
   D. 4.03×10^6
7. Simplify $\sqrt{50} + 2\sqrt{18} - \sqrt{8}$
   A. $9\sqrt{2}$
   B. $7\sqrt{2}$
   C. $11\sqrt{2}$
   D. $13\sqrt{2}$
8. Simplify $\sqrt[3]{54x^9y^6}$
   A. $3x^3y^2\sqrt[3]{3}$
   B. $3x^3y^2\sqrt[3]{6}$
   C. $3x^3y^2\sqrt[3]{2}$
   D. $2x^2y^3\sqrt[3]{27}$
9. Mia earns \$6.50 per hour and works 30 hours. Her sister earns \$7.75 per hour. How many hours must her sister work to match Mia's earnings?
   A. 25.2 hours
   B. 28.5 hours
   C. 30 hours
   D. 22.6 hours
10. A store increased the price of a \$40 jacket by 25%. A week later, they discounted it by 25%. What is the final price?
    A. \$30.00
    B. \$37.50
    C. \$40.00
    D. \$42.50
11. Factor completely: $6x^2 - 17x + 12$
    A. $(6x - 8)(x - 1.5)$
    B. $(3x - 4)(2x - 3)$
    C. $(3x - 3)(2x - 4)$
    D. $(6x - 9)(x - 1.3)$

12. A map scale is 3 cm to 2 km. If a park has an area of $54cm^2$ on the map, what is its actual area?

    A. $24\ km^2$

    B. $36\ km^2$

    C. $48\ km^2$

    D. $72\ km^2$

13. Solve for x: $-2(3x-1) = -x+5$

    A. $-\frac{5}{6}$

    B. -1

    C. 1

    D. $-\frac{3}{5}$

14. Which graph represents a proportional relationship with a constant of proportionality of 0.8?

A.

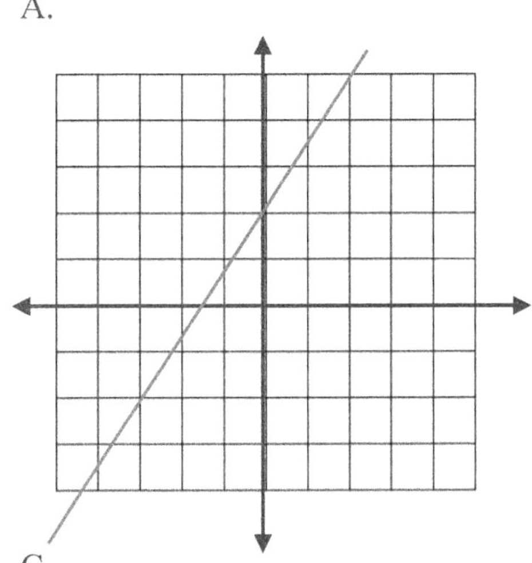

B.

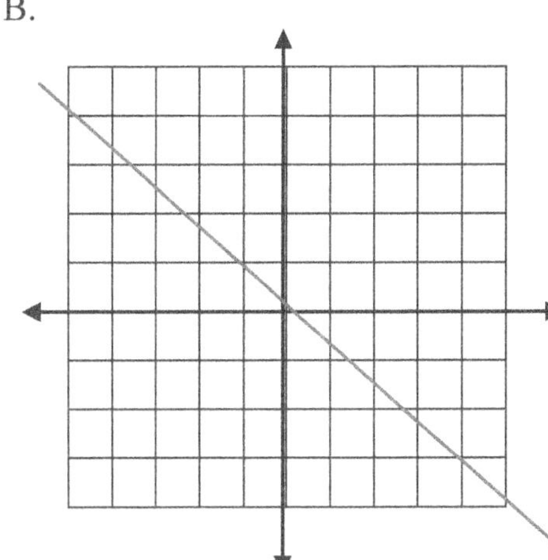

C.

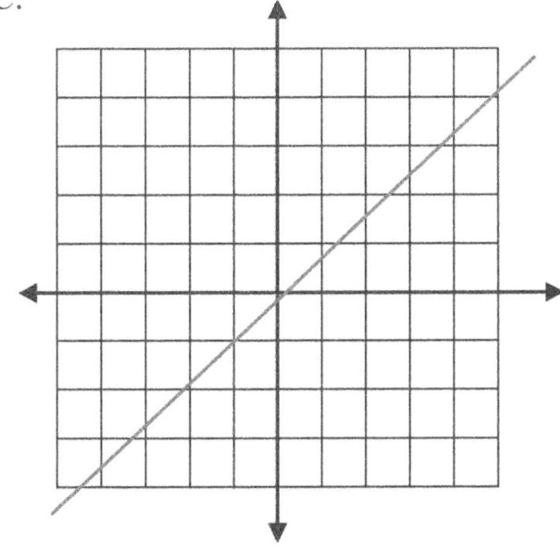

D.

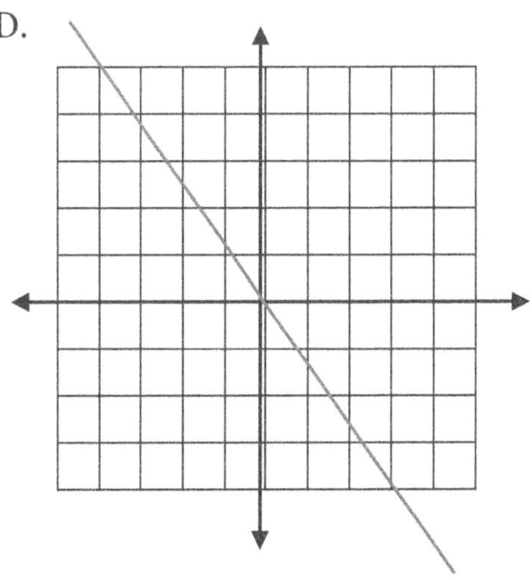

15. Which equation represents the statement: "A number divided by 5, increased by 12, equals the number minus 3"?

A. $\frac{x}{5} + 12 = x - 3$

B. $\frac{x}{5} - 12 = x + 3$

C. $5x + 12 = x - 3$

D. $\frac{x}{5} + 3 = x - 12$

16. A deposit of $2,500 earns 3.5% simple interest annually. How much interest is earned after 3 years?

A. $250

B. $225

C. $275

D. $262.5

17. If $x = 3$ and $y = -2$, what is the value of $\frac{x^2 - 2xy}{y^2 - x}$?

A. $\frac{1}{21}$

B. $\frac{8}{3}$

C. 21

D. $\frac{16}{5}$

18. Which equation has infinitely many solutions?

A. $6x = 5x + 1$

B. $2(x + 3) = 2x + 6$

C. $4x - 8 = 4(x - 2) + 1$

D. $\frac{x}{2} = 10$

19. Which equation represents the line with slope -2 and y-intercept 4 in slope-intercept form?

A. $2x + y = 4$

B. $y = -2x + 4$

C. $y = 4x - 2$

D. $y = -2x - 4$

20. Solve for m: $2m + 5 < 4m - 3$

A. $m > 4$

B. $m < 4$

C. $m > 1$

D. $m < 1$

21. Solve the system: $\begin{cases} y - 2x = -1 \\ 2y + 3x = 12 \end{cases}$

A. (1,1)

B. (3,5)

C. (2,3)

D. (4,7)

22. a is inversely proportional to $(2b - 5)$. If a = 8 and b = 9, express a in terms of b.

A. $a = 2b - 18$

B. $a = 108(2b - 5)$

C. $a = \frac{104}{2b-5}$

D. $a = \frac{2b-5}{104}$

23. Find the equation of the line through points $(2, 3)$ and $(4, 7)$.

A. $y = 2x - 1$

B. $y = 2x + 1$

C. $y = x + 1$

D. $y = x - 1$

24. Which of the following equations represents the given line?

A. $y = \frac{2}{3}x - 3$

B. $y = -\frac{3}{2}x - 2$

C. $y = \frac{3}{2}x + 3$

D. $y = -\frac{3}{2}x - 3$

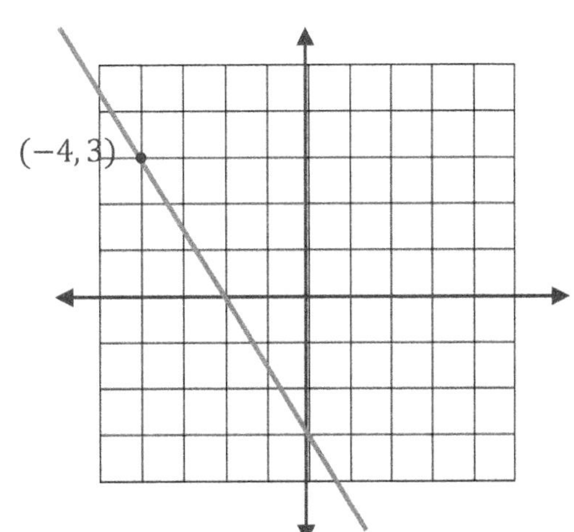

25. Emma and Noah buy 20 toys. Emma buys 6 more than Noah. How many does Emma buy?

A. 7

B. 10

C. 13

D. 16

26. Simplify: $-3x(2x^2 + 5x - 7) - 2(x^2 - x + 1)$

A. $-6x^3 - 17x^2 + 23x - 2$

B. $-3x^3 - 17x^2 + 23x - 23$

C. $-6x^3 + 13x^2 - 7x - 2$

D. $-3x^3 - 7x^2 - 23x - 6$

27. Find the quotient of division: $(x^4 - 2x^3 + 3x^2) \div x - 1$

A. $x^3 - 2x^2 - x + 2$

B. $x^3 - 2x^2 + x + 1$

C. $x^3 + x^2 - 2x - 2$

D. $x^3 - x^2 + 2x + 2$

28. If $f(x) = 3x + 6$ and $5f(2a) = 150$, what is a?

A. 3

B. 4

C. 5

D. 6

29. If $f(x) = 2x - 1$ and $g(x) = x^2 + 3$, find $f(g(2))$.

A. 9

B. 13

C. 19

D. 21

30. Solve using the quadratic formula: $2x^2 - 9x + 4$

A. $x = -4$ and $\frac{1}{2}$

B. $x = -\frac{1}{2}$ and 4

C. $x = \frac{1}{4}$ and 2

D. $x = \frac{1}{2}$ and 4

31. What is the domain of $f(x) = \sqrt{2x - 6}$

A. $x \geq 3$

B. $x \geq 6$

C. $x \leq 3$

D. All real numbers

32. Which graph represent the parabola $x^2 + 2x - 3$

A.

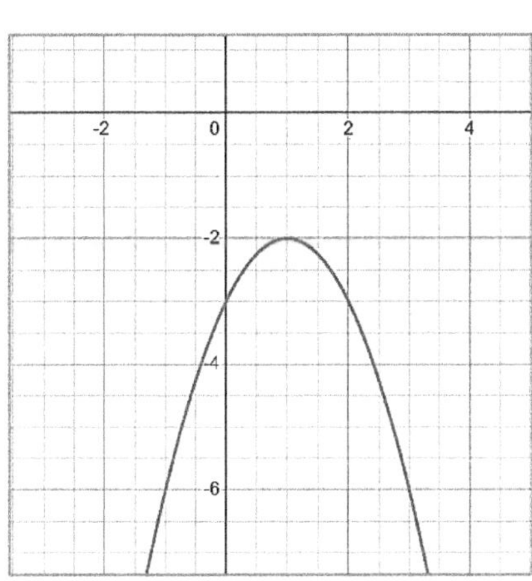

B.

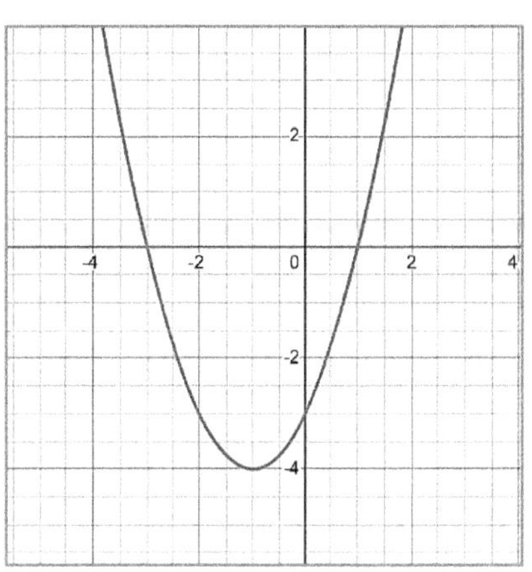

C.

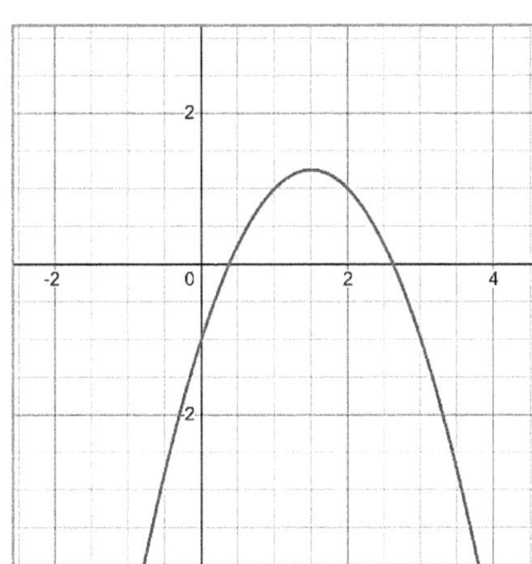

D.

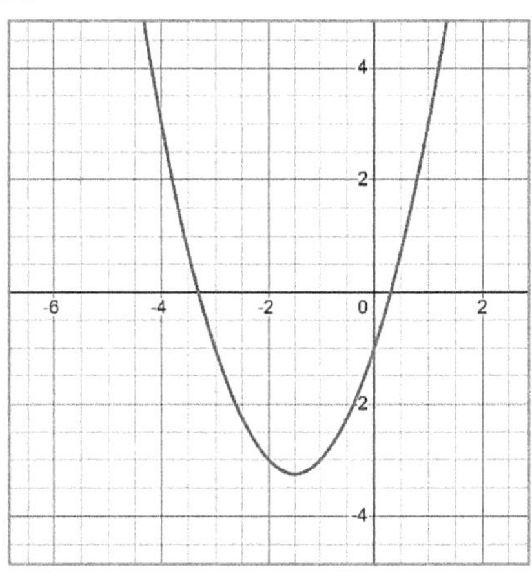

33. Find the general formula and the sum of the first 20 terms of 4,9,14,19, ...

A. $a_n = 5n - 1$ and Sum= 1030

B. $a_n = 5n$ and Sum= 1130

C. $a_n = 5n - 1$ and Sum= 1130

D. $a_n = 4n - 5$ and Sum= 1030

34. Lily is 12 years old, and her twin siblings are 9 years old each. What is the mean age of the three children?

    A. 13

    B. 12

    C. 11

    D. 10

35. Point $B(2,5)$ is rotated 90° counterclockwise about the origin. What are its new coordinates?

    A. $(-5, 2)$

    B. $(5, -2)$

    C. $(-2, -5)$

    D. $(5, 2)$

36. A triangle has vertices at $(2,1)$, $(4,1)$, and $(3,4)$. If dilated by a scale factor of 3 about the origin, what is the new coordinate of $(3,4)$?

    A. $(6, 12)$

    B. $(9, 12)$

    C. $(9, 4)$

    D. $(3, 12)$

37. A rectangle has an area of 24 square units. After a dilation, its area is 96 square units. What is the scale factor?

    A. $\frac{1}{2}$

    B. 4

    C. $\sqrt{2}$

    D. 2

38. Find the value of x in figure below:

    A. 19.5

    B. 9.5

    C. 5.9

    D. 15.1

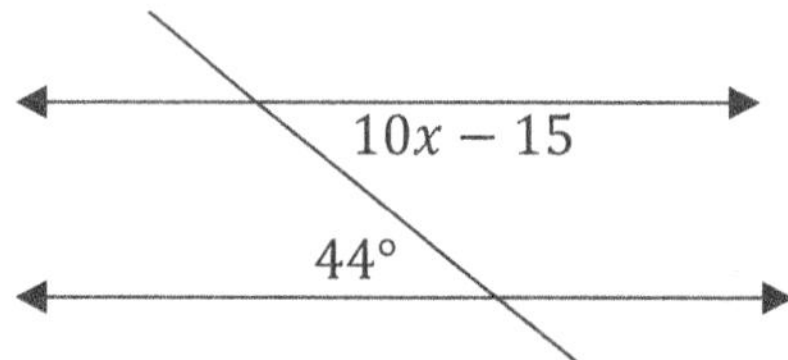

39. A cylinder has radius 5 cm and height 10 cm. What is its total surface area?

A. 100π

B. 150π

C. 200π

D. 250π

40. A Sphere has a volume of $36\pi\ cm^3$. What is its diameter?

A. $6\ cm$

B. $3\ cm$

C. $9\ cm$

D. $12\ cm$

41. The $3rd$ of a geometric sequence is 18, and the common ratio is 2. What is the $6th$ term?

A. 576

B. 288

C. 144

D. 1152

42. A survey was conducted among 200 high school students to find out their preferences for types of movies and whether they prefer to watch movies at home or in theaters. The results are shown in the two-way frequency table below:

	Home	**Theater**	**Total**
Action	48	32	80
Comedy	40	20	60
Drama	30	30	60
Total	118	82	200

Based on the table, what is the probability that a randomly selected student either prefers comedy movies or prefers to watch movies at home?

A. 0.69

B. 0.65

C. 0.70

D. 0.79

43. Which correlation coefficient (r) best describes a strong negative linear relationship?

A. $r = 0.9$

B. $r = -0.2$

C. $r = -0.8$

D. $r = 0.5$

44. How many ways can 4 books be arranged on a shelf?

A. 12

B. 16

C. 24

D. 256

45. The bag has 3 red, 5 blue, and 2 green marbles. What is the probability of drawing red or green?

A. $\frac{1}{2}$

B. $\frac{1}{3}$

C. $\frac{1}{4}$

D. $\frac{1}{5}$

46. Which function machine maps:

Input: $\{-3, -1, 0, 2, 4\}$

Output: $\{9, 1, 0, 4, 8\}$

Rule: If $x < 0$, output is x^2. If $x \geq 0$, output is $2x$. Is this a function?

A. No, because different inputs give same outputs

B. Yes, each input has exactly one output

C. No, because it uses a piecewise rule

D. Yes, but only for $x \geq 0$

47. Which Without algebra, determine the intersection of $y = 0.5x + 1.5$ and $2x - 3y = 6$:

A. $(3, 3)$

B. $(21, 12.5)$

C. $(5, 4)$

D. $(9, 6)$

48. A 13-foot ladder is leaning against a wall. The base of the ladder is 5 feet away from the wall. How high up the wall does the ladder reach?

A. 12 ft

B. 10 ft

C. 11 ft

D. $\sqrt{(169 - 25)}$ ft

49. Which table shows a nonlinear relationship between x and y?

Table A:

x	1	2	3	4	5
y	3	6	9	12	15

Table B:

x	1	2	3	4	5
y	2	5	10	17	26

A. Table A

B. Table B

C. Both

D. Neither

50. In a triangle ABC, angle $A = 40°$, angle $B = 65°$. What is the measure of the exterior angle at vertex C?

A. 115°

B. 105°

C. 130°

D. 145°

Answer Key

Now it's time to check your answers! Review your results to understand any mistakes and find out which areas you can improve on.

Practice Test					
1	B	18	B	35	A
2	D	19	B	34	B
3	D	20	A	37	D
4	B	21	C	38	C
5	B	22	C	39	B
6	D	23	A	40	A
7	A	24	D	41	C
8	C	25	C	42	A
9	A	26	A	43	C
10	B	27	D	44	C
11	B	28	B	45	A
12	A	29	B	46	B
13	D	30	D	47	B
14	C	31	A	48	A
15	A	32	B	49	B
16	D	33	A	50	B
17	C	34	D		

Answers and Explanations

1) **Order of Operations: Answer: B**

According to acronym **PEMDAS** we have:

$-3^2 + 6 \div (-2) \times 4 - |-5 + 2| = -9 + (-3) \times 4 - 3 = -9 - 12 - 3 = -24$

If we did $-2\times4=-8$ first, then $6\div-8=-0.756$,

we would **violate PEMDAS**, because we skipped the **left-to-right rule** for MD.

2) **Convert Repeating Decimals: Answer:** A

Let $x = 0.\overline{36} \Rightarrow 100x = 36.\overline{36} \Rightarrow$ subtract: $100x - x = 36.\overline{36} - 0.\overline{36} \Rightarrow 99x = 36 \Rightarrow$ $x = \frac{36}{99} = \frac{4}{11}$

3) **Compare Rational and Irrational Numbers: Answer:** D

$\sqrt{9} < \sqrt{10} < \sqrt{16} \Rightarrow 3 < \sqrt{10} < 4 \Rightarrow \sqrt{10}$ is greater than 3, so $\sqrt{10} > 2.9$ is true and other choices are wrong.

4) **Mixed Operations on Rational Numbers: Answer:** D

Numerator: $-\frac{2}{3} + \frac{1}{4} \div 2 = -\frac{2}{3} + \left(\frac{1}{4} \div 2\right) = -\frac{2}{3} + \frac{1}{8} = -\frac{16}{24} + \frac{3}{24} = -\frac{13}{24}.$

Denominator: $3\frac{3}{5} \times \frac{10}{21} = \frac{18}{5} \times \frac{10}{21} = \frac{6}{1} \times \frac{2}{7} = \frac{12}{7}$

Final division: $-\frac{13}{24} \div \frac{12}{7} = -\frac{13}{24} \times \frac{7}{12} = -\frac{91}{288}$

5) **Integer Exponents: Answer:** B

$$\frac{1}{\left(-\frac{1}{2}\right)^{-3}} = \frac{1}{\left(-\frac{2}{1}\right)^{3}} = \frac{1}{(-2)^3} = \frac{1}{-8}$$

6) **Scientific Notation: Answer:** D

$4200000 = 4.2 \times 10^6$. Rewrite both numbers with the same exponent: $1.7 \times 10^5 = 0.17 \times 10^6$.Then: $4.2 \times 10^6 - 0.17 \times 10^6 = (4.2 - 0.17) \times 10^6 = 4.03 \times 10^6$

7) **Radical Expressions: Answer:** A

Simplify each term: $\sqrt{50} = 5\sqrt{2}, 2\sqrt{18} = 6\sqrt{2}$ and $\sqrt{8} = 2\sqrt{2}$. Combine: $5\sqrt{2} + 6\sqrt{2} - 2\sqrt{2} = 9\sqrt{2}$

8) **Cube Root: Answer:** C

$$\sqrt[3]{54x^9y^6} = \sqrt[3]{27 \times 2 \times (x^3)^3 \times (y^2)^3} = 3x^3y^2\sqrt[3]{2}$$

9) **Proportional Relationship: Answer:** A

Mia's earnings: $30 \times 6.50 = 195$. Sister's hours: $\frac{195}{7.75} \approx 25.16 \approx 25.2$ hours.

10) **Percent Increase: Answer:** B

Increase: $40 + (0.25 \times 40) = \$50$. Discount: $50 - (0.25 \times 50) = \$37.50$

11) **Factoring Quadratics: Answer:** B

Find two numbers that multiply to $6 \times 12 = 72$ and add to $-17 \rightarrow -8$ and -9.

Rewrite middle term and factor by grouping: $6x^2 - 8x - 9x + 12 = 2x(3x - 4) - 3(3x - 4) = (3x - 4)(2x - 3)$

12) **Rations and Proportions: Answer:** A

Scale: $3\ cm = 2\ km$. Since area scales by the square of the linear scale factor, we square the scale factor: $\left(\frac{2}{3}\right)^2 = \frac{4}{9}$ ⇒Actual area= $54 \times \frac{4}{9} = 6 \times 4 = 24\ km^2$

13) **Equations: Answer:** D

$$-2(3x-1) = -x+5 \rightarrow -6x+2 = -x+5 \rightarrow -6x+x = 5-2 \rightarrow -5x = 3 \rightarrow x = -\frac{3}{5}$$

14) **Proportional Relationship on Graph: Answer:** C

For proportionality:

1) Must pass through $(0,0)$

2) Slope $(k) = \frac{y}{x}$

A: Dose not pass through (0, 0). B: Slope = $-\frac{4}{5}$. C: Slope = $\frac{4}{5} = 0.8\ (true). D$: Slope = $-\frac{5}{4}$

15) **Translating Expressions: Answer:** A

Let x be the number. The phrase "a number divided by 5" is $\frac{x}{5}$ "increased by 12" means $+12$, and "equals the number minus 3" is $=\ x\ -\ 3$. Thus, the equation is $\frac{x}{5} + 12 = x - 3$. To verify, solve: Multiply by 5 to clear the fraction: $x + 60 = 5x - 15$. Rearrange: $60 + 15 = 5x - x$, so $75 = 4x$ and $x = \frac{75}{4}$ this satisfies the equation.

16) **Simple Interest: Answer:** D

Simple Interest (I) = $P \times r \times t. P$ =principle= $\$2,500, r$ =annual interest rate= $3.5\% = 0.035$ *and* t =time in years= $3 \Rightarrow I = 2500 \times 0.035 \times 3 = 262.50$

17) **Evaluating Rational Expressions: Answer:** C

Numerator: $3^2 - 2(3)(-2) = 9 + 12 = 21$. Denominator: $(-2)^2 - 3 = 4 - 3 = 1$. Final answer: $\frac{21}{1} = 21$

18) **Equation with Infinitely Many Solutions: Answer:** B

Option B simplifies to $2x + 6 = 2x + 6$, which is always true.

19) **Forms of Linear Equations: Answer:** B

Slope-intercept form is $y = mx + b$. Given $m = -2$ and $b = 4$: $y = -2x + 4$

20) **Inequalities: Answer:** A

$$2m + 5 < 4m - 3 \rightarrow 2m - 4m < -3 - 5 \rightarrow -2m < -8 \rightarrow m > -\frac{8}{-4} \rightarrow m > 4$$

21) **System of Equations: Answer:** C

Substitute $y = 2x - 1$ into the second equation: $3x + 2(2x - 1) = 12$.

Simplify: $7x - 2 = 12 \rightarrow x = 2$. Find y: $y = 2(2) - 1 = 3$

22) **Expressions and Equations: Answer: C**

Inversely proportional (means recognizing that as one quantity increases, the other decreases proportionally, following the formula $a = \frac{k}{x}$.): $a = \frac{k}{2b-5}$ (k is a constant of proportionality) substitute a and b: $8 = \frac{k}{2(9)-5} \rightarrow k = 8(13) = 104; a = \frac{104}{2b-5}$

23) **Linear Equations: Answer:** A

Slope $= \frac{7-3}{4-2} = 2$. Using point-slope form at: (2,3): $y - 3 = 2(x - 2)$, so $y = 2x - 4 + 3 = 2x - 1 \rightarrow y = 2x - 1$.

24) **Graphing a Line: Answer:** D

Using two points $(-4, 3)$ and $(0, -3)$ from graph and point-slope form we have: Slope $= \frac{3-(-3)}{-4-0} = -\frac{3}{2} \rightarrow y - (-3) = -\frac{3}{2}(x - 0)$, so $y = -\frac{3}{2}x - 3$

25) **System of Equations: Answer:** C

Let Noah buy x toys. Emma buys $x + 6$. Total: $x + (x + 6) = 20$. Solve: $2x + 6 = 20 \rightarrow 2x = 14 \rightarrow x = 7$. Emma: $7 + 6 = 13$.

26) **Simplifying Polynomials: Answer:** A

Distribute $-3x$ and -2 and then combine like terms $-3x(2x^2 + 5x - 7) - 2(x^2 - x + 1) = -6x^3 - 15x^2 + 21x - 2x^2 + 2x - 2 = -6x^3 - 17x^2 + 23x - 2$.

27) **Division of Polynomials: Answer:** D

Use polynomial long division or synthetic division.

28) **Function Evaluation: Answer:** B

Compute $f(2a) = 3(2a) + 6 = 6a + 6$. Then, $5f(2a) = 5(6a + 6) = 30a + 30 = 150$. Solve: $30a = 120 \rightarrow a = 4$.

29) **Composition of Functions: Answer:** B

Find $g(2) = 2^2 + 3 = 7$. Compute $f(7) = 2(7) - 1 = 13$.

30) **Quadratic Formula: Answer:** D

Apply the quadratic formula $x = \frac{-b\pm\sqrt{b^2-4ac}}{2a}$: $a = 2, b = -9$ and $c = 4 \Rightarrow x = \frac{-(-9)\pm\sqrt{(-9)^2-4(2)(4)}}{2(2)} = \frac{9\pm\sqrt{49}}{4} = \frac{9\pm7}{4} \rightarrow x = 4$ or $x = \frac{1}{2}$

31) **Domain from an Equation: Answer:** A

The expression under the square root must be non-negative: $2x - 6 \geq 0 \rightarrow x \geq 3$.

32) **Graphing Quadratic Function: Answer:** B

$a = 1 > 0 \rightarrow$The parabola is upward. Axis of symmetry: $x = -\frac{b}{2a} = -\frac{2}{2(1)} = -1$. Vertex: substitute $x = -1$ into equation: $y = (-1)^2 + 2(-1) - 3 = -4$ so, the vertex is $(-1, -4)$.
X-intercepts: solve $x^2 + 2x - 3$ by factoring: $(x + 3)(x - 1) = 0 \Rightarrow x = -3, x = 1$.
So, graph B is the answer.

33) **Arithmetic Sequence: Answer:** A

Common difference $d = 9 - 4 = 5$. General formula: $a_n = a_1 + (n - 1)d \rightarrow a_n = 4 + (n - 1)5 = 5n - 1$. Sum formula$= \frac{n}{2}(2a_1 + (n - 1)d) = \frac{20}{2}(2(4) + (20 - 1)5) = 10(8 + 95) = 1030$.

34) **Mean: Answer:** D

Mean $\frac{12+9+9}{3} = \frac{30}{3} = 10$

35) **Transformation: Answer:** A

Rotation rule for 90°: $(x, y) \rightarrow (-y, x) \Rightarrow (2,5) \rightarrow (-5, 2)$

36) **Dilations & Scale Factor: Answer:** B

Dilation rule: $(x, y) \to (kx, ky)$, where $k = 3 \Rightarrow (3,4) \to (9,12)$

37) **Scale Factor: Answer:** D

Area scales by $k^2. k^2 = \frac{96}{24} = 4 \to k - 2$

38) **Transversals & Parallel Lines: Answer:** C

Using Alternate interior angles rule: $10x - 15 = 44 \to 10x = 44 + 15 = 5 \to x = \frac{59}{10} = 5.9$

39) **Surface Area of a Cylinder: Answer:** B

Total SA = $2\pi r^2 + 2\pi rh = 2\pi(25) + 2\pi(50) = 150\pi\ cm^2$.

40) **Volume of Sphere: Answer:** A

Volume = $\frac{4}{3}\pi r^3 = 36\pi \Rightarrow \frac{4}{3}r^3 = 36 \Rightarrow r^3 = 27 \Rightarrow r = 3$ and diameter is $2r = 2(3) = 6$

41) **Geometric Sequence: Answer:** C

Use $a_n = a_1 \times r^{n-1}$:Find a_1: $18 = a_1 \times 2^2 \to a_1 = \frac{18}{4} = 4.5$. Compute: $a_6 = 4.5 \times 2^5 = 4.5 \times 32 = 144$

42) **Two-way Frequency Table: Answer:** A

P(Comedy or Home)=P(Comedy)+P(Home) −P(Comedy and Home).

From Table: P(Comedy) $= \frac{60}{200}$, P(Home) $= \frac{118}{200}$, and P(Comedy and Home)$= \frac{40}{200}$.

Now substitute: P(Comedy or Home)$= \frac{60}{200} + \frac{118}{200} - \frac{40}{200} = \frac{138}{200} = 0.69$

43) **Correlation Coefficient: Answer:** C

Strong negative correlation: r close to -1

44) **Permutation: Answer:** C

Number of permutations= $4! = 24$

45) **Probability: Answer:** A

Mutually exclusive events: P (red or green)$= \frac{3}{10} + \frac{2}{10} = \frac{5}{10} = \frac{1}{2}$

46) **Function or Not? Answer: B**

Each input maps to only one output, so it is a function.

47) **Graphing Systems of Equations:** Answer: B

Substitute y: $2x - 3(0.5x + 1.5) = 6 \to 0.5x = 10.5 \to x = 21$; then $y = 12.5$.

48) **Pythagorean Theorem: Answer: A**

Use $a^2 + b^2 = c^2$ with c = 13 and b = 5. So, $a^2 = 169 - 25 = 144 \Rightarrow a = \sqrt{144} = 12$ ft.

49) **Compare Linear vs. Nonlinear Functions:** Answer: B

Table A is linear (constant change). Table B has increasing second differences (quadratic).

50) **Angle Relationships – Exterior Triangle: Answer: B**

Exterior angle = sum of opposite interior angles = $40° + 65° = 105°$.

"END"

www.ingramcontent.com/pod-product-compliance
Lightning Source LLC
Chambersburg PA
CBHW081804200326
41597CB00023B/4142
9781636203256